This volume records the lectures and symposia of the 12th International Conference on General Relativity and Gravitation. Plenary lecturers reviewed the major advances since the previous conference in 1986. The reviews cover classical and quantum theory of gravity, colliding gravitational waves, gravitational lensing, relativistic effects on pulsars, tests of the inverse square law, numerical relativity, cosmic microwave background radiation, experimental tests of gravity theory, gravitational wave detectors, and cosmology. These provide a useful summary of research areas in which there is intense current activity.

The plenary lectures are complemented by summaries of symposia, provided by the chairmen. Almost 700 contributed papers were presented at these and they cover an even wider range of topics than the plenary talks.

The book provides a comprehensive guide to research activity in both experimental and theoretical gravitation and its applications in astrophysics and cosmology. It will be essential reading for research workers in these fields, as well as theoretical and experimental physicists, astronomers, and mathematicians who wish to be acquainted with modern developments in gravitational theory and general relativity.

GENERAL RELATIVITY AND GRAVITATION, 1989

General Relativity and Gravitation, 1989

Proceedings of the 12th International Conference on
General Relativity and Gravitation
University of Colorado at Boulder, July 2–8, 1989

Edited by

Neil Ashby, David F Bartlett and
Walter Wyss

Department of Physics, University of Colorado at
Boulder

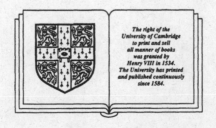

The right of the
University of Cambridge
to print and sell
all manner of books
was granted by
Henry VIII in 1534.
The University has printed
and published continuously
since 1584.

CAMBRIDGE UNIVERSITY PRESS

Cambridge
New York   Port Chester   Melbourne   Sydney

CAMBRIDGE UNIVERSITY PRESS
Cambridge, New York, Melbourne, Madrid, Cape Town, Singapore, São Paulo

Cambridge University Press
The Edinburgh Building, Cambridge CB2 2RU, UK

Published in the United States of America by Cambridge University Press, New York

www.cambridge.org
Information on this title: www.cambridge.org/9780521384285

First published 1990
This digitally printed first paperback version 2005

A catalogue record for this publication is available from the British Library

ISBN-13 978-0-521-38428-5 hardback
ISBN-10 0-521-38428-1 hardback

ISBN-13 978-0-521-02079-4 paperback
ISBN-10 0-521-02079-4 paperback

Contents

WORKSHOPS

Part B. Relativistic astrophysics, early universe, and classical cosmology 131

INVITED TALKS

WORKSHOPS

Cosmic strings
Microwave background radiation

INVITED TALKS

   E. Adelberger

   Introduction
   Searches for violations of the universality of free fall
   Searches for violations of the inverse square law
   Conclusions

   G. Pizzella

   Interaction of a gravitational wave with a resonant antenna
   The electromechanical transducer
   The noise
   Optimum filtering and effective temperature
   Present and future sensitivity
   Brief review of the bar experiments
   Ultimate bar sensitivity

   I. I. Shapiro

   Introduction
   Past light cone
   Future light cone
   Epilogue

   R. Weiss

Contents

WORKSHOPS

> Introduction
> Laser interferometer projects
> Technology development for the long baseline detectors
> Finding the signals in the noise
> Space experiments
> Prototype detectors
> Conclusion

Part D. Quantum gravity, superstrings, quantum cosmology

INVITED TALKS

> Introduction
> Hamiltonian framework
> Quantum theory
> Discussions

The aims of quantum cosmology
The quantum mechanics of cosmology
Proposals for a law of initial conditions
Predictions
The place of quantum cosmology in general relativity

Introduction
Connection between string theory and general relativity
Quantum perturbation expansion
Classical string theory
Speculations
Notes

WORKSHOPS

Introduction
Developments in the classical theory
The role of classical general relativity in the quantum theory
Developments in quantum general relativity
A wild speculation

Coarse-grained effective action
Quantum violations of energy conditions
Renormalized stress tensors and black holes

Preface

We note with sadness that GR-12 was the last important conference for two of the significant figures in physics in the last half of the twentieth century: William M. Fairbank and Eduardo Amaldi. Ironically, they were raised in traditions far removed from general relativity but both had made important experimental contributions to the field during the past twenty years. Fairbank, with Schiff, Cannon, and Everitt, started the investigation that we now call the Stanford Relativity Gyroscope experiment and helped bring it to the point that it will be put into orbit as NASA's Gravity Probe B. He then, with Hamilton, started the cryogenic gravity wave detection project at Stanford to further advance the pioneering experiments of Weber. Amaldi joined with Pizzella to build tuned gravity wave detectors in Italy. When the Italian bureaucracy became too difficult he helped move the experimental laboratory to CERN where it has become the world's strongest program.

In the evening of the next to last day of GR-12 both Fairbank and Amaldi attended a small informal meeting of experimentalists representing all of the major tuned bar groups. The focus of the meeting was to establish times when all of the experiments would be operated in coincidence and to establish protocols for exchanging the data that would be generated by these coincidence experiments. They both expressed great confidence that such a coordinated effort would lead us to the discovery of gravitational waves and the development of gravitational wave astronomy. And they both felt that when gravity waves were discovered, the discovery would be by their group and that the other groups would confirm the discovery.

Their enthusiasm and instinct for physics will be sorely missed.

We are grateful to William O. Hamilton for stepping in to write the report on the C3 Workshop, on resonant bar and microwave gravitational wave experiments, after the death of Bill Fairbank, the Chairman of the workshop. We also thank him for assisting with the above biographical notes about Amaldi and Fairbank.

The speakers and workshop chairs have done outstanding work with the articles presented here, and the editors are particularly grateful to them for their contributions. The fortunate readers of these Proceedings will find the articles to be comprehensive, excellent reviews and reports on the current status of the various subfields. We regret only that an article on Cosmic Strings was not received and could not be included in these Proceedings.

We also express our appreciation to the GR-12 Conference's Scientific Program Committee and the Local Organizing Committee, whose members are listed on the following page, for the meticulous planning and careful selection which made the conference such a success.

Wanda Derushia provided able assistance during the conference and early part of the preparation of these Proceedings. We also acknowledge the help of Sandy Rush with preparation of some of the manuscripts.

Financial assistance, without which the conference could not have taken place, was received from:

The National Science Foundation
The National Aeronautics and Space Administration
The International Society on General Relativity and Gravitation
The International Union of Pure and Applied Physics
The National Institute of Standards and Technology
The Gravity Research Foundation
The University of Colorado at Boulder

Boulder, Colorado, May 1990

Neil Ashby
David F. Bartlett
Walter Wyss

Scientific Program Committee

G. F. R. Ellis, *Chairman*

A. J. Anderson

N. Ashby

A. Ashtekar

V. Belinsky

P. L. Bender

B. Bertotti

V. B. Braginsky

M. A. Castagnino

Y. Choquet-Bruhat

J. Ehlers

J. Faller

L. Z. Fang

H. Fleming

J. B. Hartle

A. Held (Secretary,
  GRG Society)

G. Horowitz

M. A. H. MacCallum

T. Nakamura

E. T. Newman (President,
  GRG Society)

M. Novello

R. B. Partridge

D. W. Sciama

I. D. Soares

R. M. Wald

J. Wilson

Local Organizing Committee

N. Ashby, *Chairman*

D. F. Bartlett

P. L. Bender

J. Faller

D. Wineland

W. Wyss

Part A.

Classical relativity and gravitation theory

1

Colliding waves
in general relativity

Valeria Ferrari
International Center for Relativistic Astrophysics - ICRA,
Dipartimento di Fisica "G. Marconi,"
Universitá di Roma, Rome, Italy

1. Introduction

The gravitational interaction between waves is a phenomenon in which the richness and the originality of the theory of general relativity are explicitly manifested. It became apparent in 1970-71 when Khan, Penrose[1] and Szekeres[2] found the first exact solutions describing the collision of pure gravitational waves: it was shown that when two plane gravitational waves with collinear polarization, and with a step or an impulsive profile collide, their subsequent interaction culminates in the creation of a curvature singularity, an event unpredicted by any linearized version of the theory of gravity. As we shall see, this is only a particular result, although probably the most remarkable, of the interaction of gravitational waves. Similar behaviors are also manifested when waves of a different nature collide. This is due to the fact that any kind of energy generates a gravitational field. As a consequence, when two arbitrary waves collide, a gravitational interaction will accompany, as a side effect, the interaction which is peculiar to the particular fields considered. These gravitational effects, though negligible to some extent, are nevertheless relevant from a theoretical point of view. In this lecture we shall investigate the main features of the scattering of plane waves in terms of exact solutions of Einstein's equations. Therefore, let us start by explaining what gravitational plane waves are and how to find exact solutions of Einstein's equations describing their interaction. The methods I shall

3

describe can be generalized when one is dealing with other kinds of null fields, as for example, electromagnetic or massless scalar fields.

2.   Colliding wave solutions

A gravitational plane wave is a region of spacetime confined between two parallel planes, in which the curvature is different from zero and which propagates through the spacetime, in the direction normal to the planes, at the speed of light. This is usually referred to as a "sandwich wave." When only one wave is present, the spacetime is flat before and after the passage of the wave and curved inside the sandwich. Since we are considering vacuum solutions, the Ricci tensor is zero everywhere. More rigorously, a plane wave is a non-flat solution of Einstein's equations in vacuum, which admits a five parameter group of motions,[3] namely the same symmetries of an electromagnetic wave.

When the thickness of the sandwich tends to zero, the Riemann tensor remains finite on the hypersurface perpendicular to the direction of propagation and the Ricci tensor remains zero everywhere, the sandwich wave becomes an impulsive wave, and the corresponding Riemann tensor becomes proportional to a δ-function:

$$R^{\alpha}_{\beta\delta\gamma} \sim \delta(x - t),$$

where x is the direction of propagation. This is the most simple model of a gravitational plane wave. It is apparent that these waves are idealized models. Firstly they are plane, thus they represent to some approximation, the field far from radiating sources. Secondly, they have an infinite wavefront. This assumption certainly imposes severe restrictions on the global behavior of the solutions which describe their collision. However, these solutions provide interesting information on the role played by the nonlinearity of Einstein's equations in these scattering processes, and they should act as a guide for the investigation of more realistic situations.

As previously mentioned, the first exact solutions describing the interaction of gravitational waves date back to 1970 and 1971. In 1972 Szekeres[4] showed how to state the problem of colliding waves as an initial data problem, in the case of collinear polarization. In 1977 Nutku and Halil[5] generalized the Khan-Penrose solution to the case when the impulsive waves have non-collinear polarization. In 1984 much attention has again been focused on these problems, due to an alternative approach suggested by Chandrasekhar and the author.[6] We showed that the mathematical theory of colliding waves can be constructed in a way similar to the mathematical theory of black holes, due to certain reciprocal relations existing between stationary axisymmetric spacetimes and spacetimes with two spacelike Killing vectors. The application of this theory, whose main features I shall briefly outline, allows us to find exact solutions describing the region where two plane waves interact, and it has been applied successfully during the past five years in obtaining a variety of new solutions.

Let us assume that two plane waves travel along the same direction x, one against the other. Due to the symmetry of the problem, the metric is expected to depend on $t \pm x$ only, and to be independent of y and z, which are assumed to be the coordinates on the wavefront. In other words, we require that the solution possess two spacelike Killing vectors, $\frac{\partial}{\partial y}$, $\frac{\partial}{\partial z}$, which span the wavefront. With these assumptions, a suitable choice of the gauge allows the metric to be cast in the following form:

$$
ds^2 = f\left[\frac{dt^2}{1-t^2} - \frac{dx^2}{1-x^2}\right] - \sqrt{(1-x^2)(1-t^2)}\left(\left[\frac{1-|E|^2}{|1-E|^2}\right]dy^2 \right.
$$
$$
\left. + \frac{|1-E|^2}{1-|E|^2}\left[dz + i\frac{(E-E^*)}{|1-E|^2}dy\right]^2\right),
\tag{1}
$$

where $f = f(t \pm x)$ is real and $E = E(t \pm x)$ is a complex function. Thus, by the exclusive use of the symmetries, the number of unknown components of the metric tensor is reduced to 3. E and f must be found by solving Einstein's field equations in vacuum. Since we are considering waves travelling along the same direction, namely the "head-on" collision between plane waves, one might question whether or not this assumption is too restrictive. The answer is no, since if the two waves propagate along arbitrary directions, it is always possible to make a transformation to a frame of reference in which they appear to approach each other from exact opposite spatial directions. Therefore, the case of the head-on collision we are considering is not restrictive.

It is possible to demonstrate that the entire set of Einstein's equations splits into two blocks. E satisfies a **non-linear** equation:

$$
(1 - EE^*)\left\{\left[(1-t^2)E_{,t}\right]_{,t} - \left[(1-x^2)E_{,x}\right]_{,x}\right\} =
$$
$$
- 2E^*\left\{(1-t^2)(E_{,t})^2 - (1-x^2)(E_{,x})^2\right\},
\tag{2}
$$

and the function f satisfies a set of **linear**, first order, partial differential equations, whose driving terms are given by the derivatives of the function E:

$$\frac{x}{1-x^2}F_{,t} + \frac{t}{1-t^2}F_{,x} = -\frac{2}{(1-|E|^2)^2}(E_{,t}E_{,x}^* + E_{,x}E_{,t}^*),$$

$$2tF_{,t} + 2xF_{,x} = \frac{3}{1-t^2} + \frac{1}{1-x^2} \qquad (3)$$

$$-\frac{4}{(1-|E|^2)^2}\left[(1-t^2)|E_{,t}|^2 + (1-x^2)|E_{,x}|^2\right],$$

where

$$F = log\frac{f}{\sqrt{1-t^2}}.$$

Once we know a solution for the function E, the linear system for f can be solved by quadrature. Thus the heart of the problem is the solution of the equation for E. The function E is complex, therefore the physical problem we are solving possesses two degrees of freedom corresponding to the two states of polarization of the colliding waves.

A point should be stressed. The equation satisfied by E is already known in a different context as the Ernst equation. In fact, in 1968 Ernst[7,8] showed that if one is looking for stationary axisymmetric solutions of Einstein's equations, the fundamental components of the metric can be combined into a single complex function which satisfies the Ernst equation, indeed. In that case, of course, the function E will depend on the radial distance from the center, and on the polar angle θ. However the equation is, formally, the exact same equation (2) derived in the context of colliding waves. The remaining equations for the component of the metric equivalent to our function f, are similar in structure to the set of equations (3). Thus, there exists a formal analogy between the mathematical theory of colliding waves and the mathematical theory of black holes, since both stationary axisymmetric spacetimes and spacetimes with two spacelike Killing vectors admit the same Ernst equation. Now, the study of the gravitational field of massive bodies is the natural context in which the validity of a theory of gravity must be tested. Therefore, since the general theory of relativity was formulated in 1916, stationary axisymmetric solutions have been extensively studied, and an enormous amount of work has been done in elaborating techniques and methods for solving these problems. Due to the analogy described above, it is clear that we can use all of the methods developed for stationary axisymmetric solutions over the past sixty years, to find solutions for the collision of gravitational waves. I will say nothing more about these techniques (for an extensive review see, for example, Ref. 9.) I shall now describe the general features of the wave-wave interaction process, as they emerge from the variety of solutions that have been found. Before starting, we need to remark upon the following fact. Let us introduce a couple of null coordinates $u = t - x$ and $v = t + x$. By the use of the aforementioned techniques, we are able to find the solution in the region where two waves interact, corresponding to region I in the two-dimensional diagram represented in Fig. 1.

Then, by using an algorithm introduced by Penrose, we extend the solution in the precollision regions II, III, and IV. The algorithm consists in assuming that in the regions before the collision, where respectively only an outgoing wave (II) or an ingoing wave (III) are present, the metric depends on u or on v only; that region IV, representing the spacetime between the two incoming waves, is flat, and that

$$g_{\mu\nu}^{II}(u) = g_{\mu\nu}^{I}(u, v = 0),$$
$$g_{\mu\nu}^{III}(v) = g_{\mu\nu}^{I}(u = 0, v), \tag{4}$$
$$g_{\mu\nu}^{IV} = g_{\mu\nu}^{I}(u = 0, v = 0).$$

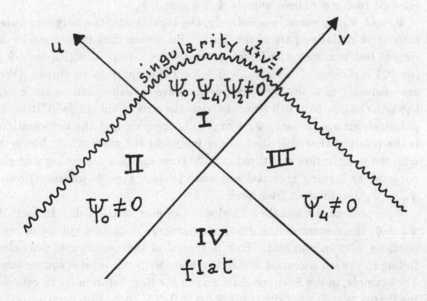

Fig. 1 The spacetime resulting from the collision of plane gravitational waves.

The requirements (4) are accomplished by the formal substitution

$$u \to uH(H), \qquad v \to vH(v)$$

(where H is the Heaviside step function) in the expression of the metric valid in the interaction region.

The procedure we use for finding the complete solution is certainly unusual. We proceed, in some sense, as shrimps: find the solution in the region of interaction, then extend it back to the past to know what is the profile of the waves whose collision produced that interaction. However, even if we use an approach which could appear as exotic, the problem is well posed in the sense that the initial data on the null boundaries $u = 0$ and $v = 0$ uniquely determine the solution in the region of interaction (see Ref. 4). With the clarification of these points, we are now in a position to describe the common features of these solutions.

We can investigate their behavior by using the Weyl scalars, which provide essential information on the nature of the free part of the gravitational field. They are five complex scalars, constructed by projecting the Weyl tensor onto a suitable chosen null tetrad, and each of them carries particular information. We shall restrict ourselves to those scalars relevant to our problem, namely Ψ_0, Ψ_4, and Ψ_2.

Ψ_0 and Ψ_4 represent, respectively, the ingoing and the outgoing pure transverse radiative part of the field. This means that if we consider a ring of test particles, the forces generated by a field in which only Ψ_0 (or Ψ_4) is different from zero, will deform the ring into an ellipse. (We are assuming that the direction of the wave is orthogonal to the ring; the polarization axis will coincide with the axis of the ellipse.) If both polarizations are present, Ψ_0 (or Ψ_4) is complex and the deformation is the result of that described above including the same effect, however with the polarization axis tilted at 45^0. For a solution describing a single outgoing or ingoing gravitational wave (Petrov type N solutions) only Ψ_0 or Ψ_4 are different from zero.

Ψ_2 represents to so-called Coulomb-like part of the field. In fact, if $\Psi_2 \neq 0$, the corresponding gravitational force distorts a sphere of test particles into an ellipsoid. This is typical of the behavior of particles falling in toward a central attracting body with the inverse square law. For example, in the Schwarzschild and in the Kerr solution (or in general for Petrov type D solutions) only Ψ_2 is $\neq 0$. This beautiful description of the nature of a gravitational field in terms of the Weyl scalars was given by Szekeres in 1965[10] by analysing the equations of geodesic deviation.

We find the following situation when two gravitational waves collide. In the region before the collision respectively, only Ψ_0 and Ψ_4 are different from zero, due to the presence of an ingoing and an outgoing wave (see Fig. 1). In the region of interaction we still expect some mixture of ingoing and outgoing radiation and in fact $\Psi_0 \neq 0$ and $\Psi_4 \neq 0$, but in addition, we find that a Coulomb-like component of the field develops, since Ψ_2 turns out to be different from zero. Thus, the two waves do not pass through one another, namely, they do not superimpose and the non-linearity of the interaction manifests itself in the appearance of the Coulomb-like part of the field. This is a true interaction term. We can also say something more by analyzing the structure of the Weyl scalars. Let us do this by considering the simplest case, the Khan-Penrose solution, which describes the collision of two impulsive gravitational waves with collinear polarizations. Although this solution is simple, it exhibits most of the typical features present in more complicated solutions.

Imagine, for example, that the amplitude of both impulsive waves is $A =$ const. Before the collision we have:

$$\Psi_0 = A\delta(u) \qquad \text{in region II} ,$$
$$\Psi_4 = A\delta(v) \qquad \text{in region III} .$$

After the collision, in the region of interaction:[11]

$$\Psi_0 = g(u)\delta(v) + k(u,v) ,$$
$$\Psi_4 = g(v)\delta(u) + k(v,u) ,$$
$$\Psi_2 = s(u,v)\theta(u)\theta(v) .$$

From these expressions we deduce that the two impulsive waves continue after the collision, but with the amplitude scaled by the function g. In addition, they develop a tail given by the function $k(u,v)$, and it is interesting to note that k is a function of u and v in Ψ_0, and of v and u in Ψ_4. The conclusion is that part of the waves are transmitted in the region of interaction, part are reflected by each other and part of the incoming radiation transforms into a Coulomb-like gravitational field.

However, the most remarkable consequence of the interaction is the following: at a finite time from the instant of collision, and at a finite distance from the surface where the collision takes place, a curvature singularity appears, and the Riemann tensor diverges. The singularity is a spacelike singularity and it occurs on the hypersurface

$$u^2 + v^2 = 1 .$$

In addition, it is a global singularity, because any test particle which
is invested by one of the two waves, is forced to enter in the region of
interaction, and to reach the singularity in a finite interval of proper
time: nothing can escape the fate of being terminated onto the spacelike
singularity.

It should also be stressed that the amplitude of the colliding waves in
no way affects the creation of the singularity: even if the amplitude is
very small, the singularity, sooner or later, appears. However the amplitude of the colliding waves, together with the other physical parameter
at our disposal, the angle α between the two directions of polarization,
determine the timescale for the creation of the singularity, according to
the equation:

$$\Delta t = \frac{1}{A^2}\sqrt{1 + \cos^2 \alpha}, \tag{5}$$

where Δt is the time interval between the instant of collision and the
formation of the singularity, and A is the amplitude of the colliding
waves.[12] Thus the time required in creating the singularity is inversely
proportional to the intensity of the wave. A further delay is introduced
if the two waves have non-collinear polarization.

However, the collision of gravitational plane waves does not necessarily
produce a singularity as a final result. In fact there are solutions in
which the physical singularity on the surface $u^2 + v^2 = 1$ is replaced by a
coordinate singularity and precisely, by a Killing-Cauchy horizon, similar
to the horizon appearing in black hole solutions. In the black hole case
the horizon is a compact hypersurface surrounding the singularity, while
in the case of colliding waves it is a non-compact hypersurface. However,
it possesses all the features which characterize a horizon, according to
the standard definition: it is a smooth, null hypersurface on which the
vector, which becomes null, is a Killing vector, and it is a one-way
membrane.

The first solutions which were found to exhibit this non-singular behavior are described in Refs. 13-16. They present an interesting property: the region of interaction is isometric, respectively to a part of the
Kerr, of the Schwarzschild, and of the Taub-Nut solutions inside the
horizon. The isometry is only local since the Killing vectors have open
orbits in our case and closed orbits in the case of black holes. It is
interesting to note that in these solutions, the only nonvanishing Weyl
scalar in the region of interaction is Ψ_2 and this means that the incident
radiation completely transforms into a Coulomb-like field. The second

interesting information is that, since the hypersurface $u^2 + v^2 = 1$ is only a coordinate singularity, these solutions can be extended across the horizon in a way which is similar to the Kruskal extension for the Schwarzschild metric or the equivalent extension for the Kerr metric. When the extension is performed, a singularity always appears. However, it can be a different type of singularity, for example, timelike in the solution found by Chandrasekhar and Xanthopoulos in Ref. 13. Finally, it should be noted that due to the presence of impulsive waves, these solutions always present a singularity at the points $u = 1$, $v = 0$, and $u = 0$, $v = 1$.[17] Other horizon-like solutions have been found, or analyzed, in Refs. 18 and 19. The last remark I would like to make on these solutions concerns their stability. Let us consider for example the solutions which are isometric to the Schwarzschild and the Taub-Nut metrics. They belong to a large class of soliton solutions identified by a set of parameters, and they correspond to a particular choice of those parameters. If one of the parameters is slightly changed, even by an infinitesimal amount, the solution immediately becomes singular. The choice of a particular set of parameters corresponds to a particular choice of the initial data on the null boundaries. Therefore, a perturbation of the initial data (or equivalently, a plane symmetric perturbation of these solutions) transforms the horizon into the usual spacelike singularity. In addition, Chandrasekhar and Xanthopoulos[20] have shown that the presence of an arbitrary small amount of null dust in the region of interaction would immediately change the horizon into a curvature singularity. These are clear indications of instability of the Killing-Cauchy horizons. (The problem of the stability of horizon-like solutions against small perturbations has also been analyzed in Refs. 21 and 22.)

3. Creation of curvature singularities

Apart from some exceptions, we have seen that the final result of the collision of gravitational plane waves is the creation of a curvature singularity. Therefore, the next question to answer is: why does the collision of gravitational plane waves generate such an infinite gravitational field? This occurrence is certainly related to a process of mutual focusing of the two colliding waves. This can be understood, for example, by studying the behaviour of null geodesics in the field of a single wave. If we consider a tube of neighboring null geodesics and take and infinitesimal ring orthogonal to them, compute how this ring expands or contracts, how it is rotated or sheared while moving in the gravitational field (for example, by computing the optical scalars in the Newman-Penrose formalism) we will see that as the null congruence passes through a plane gravitational wave, it starts to contract and to shear. The shear axis is determined by the polarization of the wave. Thus, as Penrose first suggested in 1966,[23] there is an anastigmatic focusing of the null congruence induced by the gravitational wave. It would be interesting to understand whether the extreme efficiency of the process of focusing may be due to the global plane symmetry of these waves. One possible way to get an indication on this problem, is to study the behavior of finite fronted gravitational waves.

Thus, firstly, let us see how to construct a model of a finite fronted gravitational wave.

In 1969 Bonnor[24] found an exact solution describing the plane- fronted impulsive wave generated by a massless particle moving in flat spacetime at the speed of light. The solution was rederived in 1971 by Aichelburg and Sexl[25] using the following interesting approach. They subjected the Schwarzschild metric for a spherical massive particle to a Lorentz transformation along a definite direction. The gravitational field of a moving particle has the same characteristic behavior as the electromagnetic field of a charged boosted particle: it is compressed in the direction orthogonal to the line of motion. Then they considered the limit when the velocity of the boost tends to the velocity of light, allowing the mass to become null, but maintaining finite the energy of the particle, according to their relativistic expressions:

$$ m = \frac{p}{\gamma V}, \qquad\qquad \gamma = \frac{1}{\sqrt{1 - V^2/c^2}}. $$

Thus, when the velocity of the boost, V, tends to c, then $m \to 0$, but p, the momentum of the particle, remains finite. The lime of course, must be performed very carefully, since the Lorentz transformation becomes singular in that extreme regime. The final result is the occurrence of a sort of sonic boom: the gravitational field gets squashed on the surface orthogonal to the line of motion, which becomes the wavefront, and we are left with an impulsive plane wave generated by a massless particle. This solution has also been rederived and studied by Dray and 't Hooft.[26] The final form of the metric is:

$$ ds^2 = -dudv - 4\pi G \log(r)\delta(u)du^2 + dr^2 + r^2 d\theta^2 \,, \qquad (6) $$

where r and θ are polar coordinates on the wavefront, and the only nonvanishing component of the corresponding energy momentum tensor is:

$$T_{uu} = 8\pi p\delta(u)\delta(r)\,, \tag{7}$$

where $u = 0$ is the equation of the wavefront, and $r = 0$ is the location of the null particle. This solution can easily now be generalized to represent the gravitational wave associated with a distribution of null particles, all flying parallel along the same direction. We can write the metric as:

$$ds^2 = -dudv + f(r,\theta)\delta(u)du^2 + dr^2 + r^2d\theta^2\,. \tag{8}$$

The corresponding energy momentum tensor is:

$$T_{uu} \sim \rho(r,\theta)\delta(u)\,, \tag{9}$$

and ρ is connected to the function f by the Einstein equations:

$$\rho(r,\theta) = -\frac{1}{16\pi G}\Delta f(r,\theta)\,, \tag{10}$$

where Δ is the Laplace operator in the (r,θ) plane. From (9) it is clear that ρ is the energy density of the distribution of null particles on the wavefront. Thus, if we choose a particular profile for the function f, Eq. (10) gives the corresponding energy distribution of null particles.

Following this procedure, one can compute, for example, the metric of the gravitational wave associated with a homogeneous beam of null particles, assuming $\rho = const$ within a certain radius $0 \leq r \leq R$ and zero outside. The corresponding f, inside the beam, is

$$f = -4\pi G\rho r^2\,.$$

This is what we call a beamlike gravitational wave. It is apparent that the gravitational waves associated with finite distributions of null particles are finite fronted waves.

Now our wish was to find an exact solution describing the collision of two such beams, but we haven't been able to find that solution so far. As a preliminary to the study of the full collision problem, we have considered the behavior of null geodesics in the gravitational field of a single beamlike gravitational wave[27] and some interesting properties of these solutions have emerged.

If we assume that the beam is homogeneous, the result is the following: the wave acts on the null incident geodesics as a perfect converging lens, for which we can define a focus, which depends exclusively on the energy density of the null particles on the wavefront:

$$F = \frac{1}{8\pi\rho},$$ (11)

and a law of conjugate points

$$\frac{1}{u_1} + \frac{1}{u_2} = \frac{1}{F}.$$ (12)

This means that incident null rays, travelling in the direction opposite to the wave and parallel to the axis of the beam, after the impact with the wave, will be focused in F; and that all null rays emitted by a source located in u_1, after the impact, will be focused in u_2 according to the law of conjugate points.

These results seem to suggest that an infinite energy density is expected to form also in the collision of finite fronted waves, provided, at least, that one of them is homogeneous.

Thus, the appearance of a singularity might be connected to the constancy of the energy density on the wavefront, rather than to the infinite extent of the colliding waves. There is another intriguing indication in support of this conjecture: the location of the focus of a beamlike wave coincides with the location of the singularity. In other words, in the solutions for infinite fronted gravitational waves, the singularity forms at a certain distance from the plane where the collision takes place, and after some time; this distance and this time coincide with the focal distance and the time needed to focus null geodesics after the impact with a beamlike gravitational wave. Of course this is true if the energy density of the beam is equal to the energy density of the infinite fronted wave. This can be seen from Eq. (5) in the case $\alpha = 0$, by replacing the amplitude, A, by the corresponding energy density on the wavefront (see also Ref. 4). And finally, we can understand why, when two infinite fronted waves collide, the singularity appears on a surface. In the case of a beam, the null incident congruence focuses to a point on the axis of the beam, since there is an axis of symmetry and in some sens, the geodesics "know" where to focus. But in the case of infinite fronted waves, there is no such preferred axis, thus for the incident congruence, all points at a distance $u = F$ from the plane of collision are equivalent; all these points are focal points and this is the reason why the singularity, instead of being pointlike, is a surface singularity.

But now one could raise the following objection: the conjectures made on the creation of the singularity have been inferred from the study of the behavior of null geodesics in the field of a finite fronted gravitational wave; but how faithfully does geodesics' behavior simulate the real collision of two finite fronted waves? In fact, geodesics, by definition, do not interact with each other. In the actual collision problem, however, the null particles do interact after being deflected, and the interaction becomes stronger and stronger as far as they approach each other. One might argue that this interaction should accelerate the process of focusing, possibly ending in the formation of a black hole (independently of the amplitude of the colliding waves?) This is of course a fascinating conjecture, but our analysis in terms of null geodesics doesn't arrive so far. An answer to these questions can be given only in terms of exact solutions.

4. Exact solutions including "null dust"

The last part of this talk will be dedicated to the description of what is usually called the "null dust."

The null dust is a gas of pressureless, non-interacting massless particles, characterized by the following energy momentum tensor:

$$T_{\mu\nu} = \rho(u)K_\mu K_\nu , \qquad (13)$$

where ρ is a positive scalar function of the null coordinate u, representing the energy density of the null particles, and K_μ is a null vector. In this respect, it is clear that the finite fronted waves previously described are some kind of null dust. But, as we shall see, the notion of null dust includes a larger class of fields. Exact solutions are known which describe the interaction of two infinite fronted clouds of null dust. Due to the particular behavior they exhibit, their interpretation has been a matter of an intense discussion especially during the last year. Let us consider the following three examples.

1) Chandrasekhar-Xanthopoulos (1986):[28] Two shells of null dust, following the leading edges of two impulsive gravitational waves, collide. The resulting interaction region is filled with a mixture of null dust moving in opposite directions.

2) Chandrasekhar-Xanthopoulos (1985):[29] The same waves considered in the previous solution collide and produce an interaction region which is filled with a perfect fluid with an extreme equation of state: the energy density of the fluid, ϵ, is equal to the pressure p and the velocity of sound is equal to the velocity of light. The matter produced is therefore extremely rigid and for this reason it is called stiff matter.

3) Ferrari-Ibañez (1989):[30-31] In this class of solutions, again two shells of null dust collide and produce a "non-perfect fluid" with an anisotropic distribution of the pressure, the components orthogonal to the direction of propagation being different from the component parallel to the line of motion. These solutions are free from impulsive components.

All solutions describing the collision of dust end in a physical singularity. (Solutions exhibiting similar behavior have been found or analyzed also in Refs. 32-35.)

The situation is as follows: in the first two solutions the same initial data seem to produce a different source in the region of interaction. In the third case, the initial data are different but similar in their nature, in the sense that we are considering null dust, as well, and the source produced in the region of interaction is different, it is a non-perfect fluid or, at least, something which looks like a non-perfect fluid. In order to clarify the origin of this ambiguity, we need to understand the mathematical reason and the physical reason for this behavior. Regarding the mathematical reason, we should remember the procedure we used for finding solutions: first we find a solution in the region of interaction, $g_{\mu\nu}(u, v)$, then extend the solution back to the past (see Eqs. (4)). For example, on the null boundary $v = 0$, the metric becomes $g_{\mu\nu}(u, 0)$: we assume that this will be the solution in the entire region II. In this way the metric in region II is a function of $u = x - t$ only and will represent some kind of progressive wave. Similarly, we proceed for the region III. Now, when one is searching for vacuum solutions, the initial data on the boundaries $u = 0$ and $v = 0$ are sufficient to uniquely determine the solution in the region of interaction. But in solving Einstein's equation in the presence of a source, as we are doing in these cases, the mathematical structure of Einstein's equations is such that the solution is not uniquely determined by the initial data, unless we supplement our equations with some additional conditions. These arguments have been recently pointed out by Taub.[36,37] But what are those conditions? To answer this question we need to investigate in some more detail the physics of the problem. Thus, the appropriate question is: what is the null dust? We have defined the null dust by giving its energy momentum tensor (Eq. (13)). but this definition presents an intrinsic ambiguity. In fact, this form of the energy momentum tensor can represent any kind of massless particles propagating as plane waves along a definite direction. These can be photons, scalar waves, electromagnetic waves, neutrinos, or the null particles as described earlier, which we obtained by boosting massive particles at the speed of light. Thus, unless we do not specify the physical nature of the null dust by assigning the specific equations satisfied by the field, the ambiguity cannot be removed. For example, if the null colliding dusts are composed by photons, the Maxwell equations must be satisfied not only before but also after the collision.

Thus, in the first of the three solutions described above the two streams of null particles pass through each other maintaining their individuality. The energy momentum tensor in the interaction region is the superposition of the two tensors corresponding to each stream. Therefore, it is reasonable to say that they are two different kinds of fields, for example, photons and scalar waves. In this case we can say little more.

The second and the third cases are very intriguing and can be discussed together since they are both examples of what has been called a "gravitationally induced transformation of null dust into a fluid." This definition is in some sense appropriate, because in the initial "spray" of null particles, the streamlines are null trajectories, while in the source produced after the interaction, the streamlines are timelike trajectories. Then, from an hydrodynamical point of view, the source generated in the region of interaction possesses properties different from those of the two colliding fields: the properties of a matter fluid. However, if we analyze the solution (3) in more detail, we find that the source produced in the region of interaction is identified by a scalar function ϕ, which satisfies the Klein-Gordon equation for a massless scalar field. This function extends continuously with its first derivatives in the precollision regions where the Klein-Gordon equation is identically satisfied. With this information we are able to identify the nature of the null colliding fields and to remove the original ambiguity: our class of solutions represents the gravitational interaction of two scalar waves. After the collision, a source is produced which, at a macroscopical level, behaves as a fluid, in the sense that the hydrodynamical equations are satisfied, that we can define, and measure, the energy density and the pressure and that the streamlines of the fluid are timelike trajectories. However this source is still composed at a microscopical level by massless scalar particles, as the Klein-Gordon equation explicitly reveals.

The fact that a scalar field mimics the behavior of a fluid is fascinating but not entirely surprising if one remembers that different types of matter can be represented by the same energy momentum tensor; and we have used the energy momentum tensor to deduce the hydrodynamical behavior of the source.

This occurrence was investigated by Tabenski and Taub,[38] and by Belinski and Khalatnikov.[39] They showed that the energy momentum tensor of a massless scalar field

$$T_{ab} = \phi_{,a}\phi_{,b}\,, \qquad \phi^{;a}_{;b} = 0\,,$$

can be written in the form of the energy momentum tensor of an ideal fluid with the equation of state $\epsilon = p$, and with the velocity field and the pressure given by:

$$u_i = \frac{\phi_{;i}}{\sqrt{-\phi_{;k}\phi^{;k}}}\,,$$

and

$$p = -\frac{1}{2}\phi_{;k}\phi^{;k}$$

$\epsilon = p$ is the equation of state satisfied in the region of interaction by the solution (2) found by Chandrasekhar and Xanthopoulos, and this is the reason why their solution also allows an interpretation in terms of colliding scalar waves.

As a perspective, it would be interesting to investigate whether the solutions describing the interaction of scalar waves can have some relevance to the physics of the early universe. In fact, it is believed that at the time when the various interactions were unified, the universe contained predominantly massless particles. At that time, phase transitions are suggested to have occurred which generated bubbles of the new broken-symmetry phase. These bubbles, accelerated by the energy released in the transition, possibly collided, and generated pairs of scalar waves whose following gravitational interaction is usually neglected in these models.[40] The results presented above seem to suggest that a deeper investigation of the problem would enlarge our understanding of the past history of our universe.

And finally, I would like to make a general remark. The theory of general relativity was formulated more than seventy years ago. The first exact solution, the Schwarzschild solution, was found in 1916, and its charged generalization by Reissner and Nordstrom, two years later. Then one has to wait until 1963 for the Kerr solution and two years more for its charged generalization, the Kerr-Newman solution. Since then, an enormous amount of work has been done to understand the meaning of these solutions, and one might be tempted to think that the original content of general relativity is manifested and exhausted by the black hole solutions. However, if one resists such a temptation and is not afraid to confront the mathematical complexity of the theory, the new and fascinating predictions and implications that are disclosed show clearly that the full content of the theory is far from having been exhaustively explored.

References

1. K. A. Khan and R. Penrose (1971), *Nature* **229**, 185.
2. P. Szekeres (1970), *Nature* **228**, 1183.
3. H. Bondi, F. A. E. Pirani and I. Robinson (1959), *Proc. Roy. Soc. Lond.* **A251**, 519.
4. P. Szekeres (1972), *J. Math. Phys.* **13(3)**, 286.
5. Y. Nutku and M. Halil (1977), *Phys. Rev. Lett.* **39**, 1379.
6. S. Chandrasekhar and V. Ferrari (1984), *Proc. Roy. Soc. Lond.* **A396**, 55.
7. F. J. Ernst (1968), *Phys. Rev.* **167**, 1175.
8. F. J. Ernst (1968), *Phys. Rev.* **168**, 1415.
9. D. Kramer, H. S. Stephani, M. MacCallum and E. Herlt (1980), in *Exact solutions of Einstein's field equations*, ed. E. Schmutzer (VEB Verlag der Wissenschafte, Berlin).
10. P. Szekeres (1965), *J. Math. Phys.* **6**, 1387-1391.
11. J. B. Griffiths (1990). *Colliding waves in general relativity* (Oxford University Press, to be published, 1990).
12. V. Ferrari (1988), *Phys. Rev.* **D37**, 3061.
13. S. Chandrasekhar and B. C. Xanthopoulos (1986), *Proc. Roy. Soc. Lond.* **A408**, 175.
14. V. Ferrari and J. Ibañez (1987), *Gen. Relat. Grav.* **19(4)**, 405.
15. V. Ferrari, J. Ibañez and M. Bruni (1987), *Phys. Rev* **D36(4)**, 1053.
16. V. Ferrari and J. Ibañez (1988), *Proc. Roy. Soc. Lond.* **A417**, 417.
17. This kind of singularity also appears in the Bell-Szekeres solution, in which two electromagnetic waves with a step wavefunction collide and interact; P. Bell and P. Szekeres (1974), *Gen. Relat. Grav.* **5**, 275.
18. S. Chandrasekhar and B. C. Xanthopoulos (1987), *Proc. Roy. Soc. Lond.* **A411**, 311.
19. A. Feinstein, J. Ibañez (1989), *Phys. Rev.* **D39**, 470.
20. S. Chandrasekhar and B. C. Xanthopoulos (1987), *Proc. Roy. Soc. Lond.* **A414**, 1.
21. S. Chandrasekhar and B. C. Xanthopoulos (1988), *Proc. Roy. Soc. Lond.* **A420**, 93.
22. U. Yurtsever (1987), *Phys. Rev.* **D36**, 1662.
23. R. Penrose (1966), , in *Perspectives in geometry and relativity*, ed. B. Hoffman (Indiana University Press, Bloomington, IN).
24. W. B. Bonnor (1981), *Commun. Math. Phys.* **13**, 29.
25. P. C. Aichelburg and R. U. Sexl (1971), *Gen. Relat. Grav.* **2**, 303.
26. T. Dray and G. 't Hooft (1985), *Nuc. Phys.* **B253**, 173.
27. V. Ferrari, P. Pendenze, G. Veneziano (1988), *Gen. Relat. Grav.* **20(11)**, 1185-1191.
28. S. Chandrasekhar and B. C. Xanthopoulos (1986), *Proc. Roy. Soc. Lond.* **A403**, 189.

29. S. Chandrasekhar and B. C. Xanthopoulos (1985), *Proc. Roy. Soc. Lond.* **A402**, 37.
30. V. Ferrari and J. Ibañez (1990), Gravitational interaction of massless particles, *Phys. Lett A.*, to be published.
31. V. Ferrari and J. Ibañez (1990), On the gravitational interaction of null fields, *Class. Quant. Grav.*, to be published.
32. S. Chandrasekhar and B. C. Xanthopoulos (1985), *Proc. Roy. Soc. Lond.* **A402**, 205.
33. T. Dray and G. 't Hooft (1986), *Class. Quant. Grav.* **3**, 825.
34. D. Babala (1987), *Class. Quant. Grav.* **4**, 189.
35. A. Feinstein, M. A. H. MacCallum and J. M. M. Senovilla (1989), On the ambiguous evolution and production of matter in space-times with colliding waves, preprint.
36. A. H. Taub (1988), *J. Math. Phys.* **29**, 690.
37. A. H. Taub (1988), *J. Math. Phys.* **29**, 2622.
38. R. R. Tabenski and A. H. Taub (1973), *Commun. Math. Phys.* **29**, 61.
39. V. A. Belinskii and I. M. Khalatnikov (1973), *Sov. Phys. JETP* **36(4)**, 591.
40. S. W. Hawking, I. G. Moss and J. M. Stewart (1982), *Phys. Rev.* **D26(10)**, 2681.

2

How fast can pulsars spin?

John L. Friedman
Department of Physics,
University of Wisconsin,
Milwaukee, Wisconsin 53201 USA

The upper limit set by gravity on the rotation of neutron stars is sensitive to the equation of state of matter at high density. No uniformly rotating equilibrium can have angular velocity greater than that of a particle in circular orbit at its equator, and, for a given baryon mass, the configuration with maximum angular velocity rotates at this Keplerian frequency. The limiting frequency decreases with increasing stiffness in the equation of state, because (for a given mass) models of neutron stars constructed from equations of state that are stiff above nuclear density have substantially larger radii and moments of inertia than models based on the softer equations of state. While for cold neutron stars the Keplerian frequency is the gravitational limit on angular velocity, for hotter stars ($T > 10$ K), viscosity is apparently low enough that gravitational instability to nonaxisymmetric perturbations sets in slightly earlier. The corresponding constraint on the equation of state is more stringent; and if the 1968 Hz frequency seen in optical emission from SN 1987A is the angular velocity of a newly formed pulsar, neutron star matter must be unexpectedly soft above nuclear density. Too soft an equation of state, however, cannot support a spherical neutron star with mass as large as the observed 1.44 solar mass member of the binary pulsar 1913+16. A rather narrow range of equations of state survives the two observational constraints.

Quark stars and stars with pion or kaon condensates are possible alternatives. It appears, however, that quark matter cannot be stable at zero pressure, without there being enough quark matter (strangelets) in the interstellar medium that all neutron stars would be born as quark stars.

A limiting frequency as large as 1968 Hz has other implications. The mass and baryon mass for all models rotating at 1968 Hz exceed 1.5 and 1.7 solar masses. Of a broad set of equations of state, including all those in the Arnett-Bowers collection, the equations of state that allow 1968 Hz models have, for spherical stars, a stringent upper mass limit below 1.7 solar masses. Any observation of a spherical neutron star with mass much above 1.7 solar masses would therefore be difficult to reconcile with the existence of a half-millisecond pulsar. Finally, each model at 1968 Hz has mass that exceeds the spherical upper mass limit for its EOS, implying collapse upon spin down.

1. Introduction

Until 1982, the fastest known pulsar was the Crab, whose spin of frequency 190 s^{-1} (period 33 ms) was too slow to compete with pressure in supporting the star. Few serious efforts had been made to observe more rapidly rotating neutron stars, either as pulsars or X-ray sources; perhaps the fact the Crab was slowly rotating at birth made it seem improbable that supernovae could produce rapidly rotating offspring, and that conclusion was consistent with the statistics of hundreds of other pulsars. In November of 1982, however, the observation of a pulsar PSR 1937+21, with frequency 4033 s^{-1} (period 1.56 ms) precipitated a search for millisecond pulsars. The spin down rate of PSR 1937+21 is quite slow,[2] $\dot{P} = 1.05 \times 10^{-19}$, implying a weak magnetic field, and strengthening the hope for a class of pulsars whose rotation is limited by gravity rather than by their magnetic fields. It seemed unlikely that 4000 s^{-1} would be the limiting frequency, because most equations of state proposed for neutron star matter yield stars that are compact enough (at $M > 1.4 M_\odot$) to spin at 6000 s^{-1} or faster. But the uncertainty in the equation of state was too large to decide the question. In the next six years, several additional pulsars with periods shorter than 10 milliseconds were found; but none were faster than half the frequency of the first "fast" pulsar, and Imamura, Middleditch, and Stieman-Cameron[3] suggested that "If an excess of pulsars is found with periods on the order of 1.6 ms, then a strong inference could be made that the neutron stars are hovering near GRR-driven instability points" - that they rotate at the gravitational limit. Then, in March, 1988, a nearly identical pulsar

(PSR 1957+20) was in fact discovered,[4] with a period 1.607 ms (frequency 3900 s^{-1}) within 3% of the fastest. Because the magnetic fields of the two pulsars differ by a factor of 3, the 3% agreement would have to be largely accidental if the frequencies are magnetically determined. It looked as if 4000 s^{-1} might really be the gravitational limit on neutron star rotation.[5] As we will see, that would imply that the equation of state of neutron star matter is rather stiff.

In January of 1989, however, a group of observers[6] who, for 18 months, had been seeking a newborn pulsar in the remnant SN 1987A (of the supernova in the Large Magellanic Cloud), saw for nine hours what appeared to be a pulsar with frequency 12,370 s^{-1} (period 0.508 ms). If this turns out to be the frequency of a rotating neutron star, the conclusion is exactly reversed. The equation of state of neutron star matter must be surprisingly soft, for density in the range 2×10^{14} - 10^{15} g/cm^3 , in order to accommodate a spin this fast.

Approximating a neutron star by a uniformly rotating perfect fluid

Despite its l km thick crust, and a superfluid interior that confines vorticity to microscopic tubes, for lengths > l cm a neutron star is accurately coarse-grained as a perfect fluid. Using a perfect fluid energy momentum tensor,

$$T^{ab} = \epsilon u^a u^b + p q^{ab} \qquad (1.1)$$

where q^{ab} is the projection orthogonal to the fluid's 4-velocity u^a,

$$q^{ab} = g^{ab} + u^a u^b , \qquad (1.2)$$

is equivalent to assuming that all stresses are isotropic in the rest frame of the fluid. One can estimate the error due to anisotropy in the crust by observations of glitches: as a neutron star spins down, its crust initially remains frozen in the shape of a fluid equilibrium corresponding to the earlier, more rapid, rotation. The anisotropic stress needed to support a departure from equilibrium grows with spin-down until it reaches the maximum value that the crust can support. At that point the crust breaks and abruptly assumes a shape appropriate to the smaller angular momentum. Because this starquake brings a sudden decrease in the star's moment of inertia, one observes a sudden increase in its angular velocity. The maximum fractional strain is approximated by the fractional displacement of the crust: if ℓ is the crustal thickness,

$$\text{fractional anisotropic strain} \sim \frac{\delta R}{\ell} = \frac{R}{\ell}\frac{\delta R}{R} \sim \frac{R}{\ell}\frac{\delta\Omega}{\Omega} \lesssim 10^{-5}. \qquad (1.3)$$

Corroborating this **observational** estimate of the maximum **strain** is a **theoretical** estimate of the maximum **stress**.[6] The crust is assumed to be an ordinary solid, whose anisotropic stresses are electromagnetic. The lattice energy density is of order

$$\epsilon_{\text{lat}} \sim \frac{(Ze)^2}{a} n \sim 10^{28} \text{erg/cm}^3.$$

Then an average crustal pressure, $p \gtrsim 10^{33} \text{erg/cm}^3$, implies a fractional anisotropic stress of order

$$\epsilon_{\text{lat}}/p \lesssim 10^{-5}. \qquad (1.4)$$

The second error one makes in approximating a neutron star as a perfect fluid is to coarse-grain the fluid's velocity field as uniform rotation. In the star's interior neutrons and protons are expected to form superfluids, and in a superfluid, a velocity field is curl-free except in vortex lines. For a uniformly rotating fluid with angular velocity Ω, one has

$$\nabla \times \mathbf{v} = 2\Omega\hat{z}. \qquad (1.5)$$

The star's interior approximates this field by a set of vortices, each of which carries vorticity quantized in units

$$\frac{\pi\hbar}{m_n}. \qquad (1.6)$$

The number of vortex lines per unit area is then

$$\frac{2\Omega}{\left(\frac{\pi\hbar}{m_n}\right)} = 1.0 \times 10^3 \text{ cm}^{-2} \frac{\Omega}{1\,\text{rad/s}}. \qquad (1.7)$$

Although the approximation of uniform rotation is consequently invalid on scales shorter than 1 cm, the error in computing the structure of the star on larger scales is negligible. That is, with T^{ab} approximated by a value, $< T^{ab} >$ averaged over several cm, the error in computing the metric is of order

$$\delta g_{ab} \sim \left(\frac{1\text{cm}}{R}\right)^2 \sim 10^{-11}. \qquad (1.8)$$

(For a time independent geometry, the field equations, $G_{ab} = 8\pi T_{ab}$, can be written as coupled elliptic equations for g_{ab} with source T_{ab}. The error δg_{ab} satisfies a second order elliptic equation whose source, $\delta T_{ab} =$

$T_{ab}- < T_{ab} >$, is rapidly varying. Then, writing $\partial^2 g \approx R^{-2}g, \partial^2 \delta g \approx \lambda^{-2}\delta g$, where $\lambda = 1$ cm is the characteristic wavelength of δT, we recover Eq. (1.8).)

Spacetime of a rotating star

The metric g_{ab} of a stationary axisymmetric rotating fluid has two commuting Killing vectors, ϕ^a and t^a, generating rotations and asymptotic time-translations. The fluid's 4-velocity has the form,

$$u^a = N(t^a + \Omega\phi^2),\qquad (1.9)$$

implying[8] the existence of a family of 2-surfaces orthogonal to the Killing vectors. The metric, g_{ab}, can be written in terms of dot products of the Killing vectors,

$$t \cdot t, \qquad t \cdot \phi, \qquad \phi \cdot \phi,$$

and a conformal factor, $e^{2\mu}$, that characterizes the geometry of the orthogonal 2-surfaces:

$$ds^2 = -e^{2\nu}dt^2 + e^{2\psi}(d\phi - \omega dt)^2 + e^{2\mu}(dr^2 + r^2 d\theta^2),\qquad (1.10)$$

where,

$$e^{2\psi} = \phi_a \cdot \phi^a, \qquad \omega = \frac{-t^a \phi_a}{\phi^c \phi_c}, \qquad e^{2\nu} = -t_a t^a + \frac{(t^b \phi_b)^2}{\phi^c \phi_c}.\qquad (1.11)$$

Field equations

Within days after formation, neutrino emission cools neutron stars to 10^{10} K $\simeq 1$ MeV. This is much smaller than the Fermi energy of the interior, in which a density greater than nuclear (.18 fm^{-3}) implies a Fermi energy greater than ϵ_F (.18 fm^{-3}) ≈ 60 MeV. Because the star is in this sense cold, and because nuclear reaction times are shorter than the cooling time, one can use a 1-parameter, zero-temperature equation of state (EOS) to describe the matter:

$$\epsilon = \epsilon(p), \quad \text{or equivalently}, \quad \epsilon = \epsilon(n), \quad p = p(n).$$

Then the equations of motion,

$$\frac{\nabla_b T_a^b}{\epsilon + p} = \frac{\nabla_a p}{\epsilon + p} + u \cdot \nabla u = 0,\qquad (1.12)$$

have a first integral (a relativistic version of the Bernoulli equation),

$$\int_0^p \frac{dp}{\epsilon + p} - \log(u \cdot \nabla t) = \text{constant} = \nu\big|_{pole} . \qquad (1.13)$$

For a given EOS, there is a 2-parameter family of equilibria, with each model specified by, for example, Ω and $\nu\big|_{pole}$ or Ω and ϵ_c (central density). Having specified a model by the values of the two parameters, one solves (1.13) and the remaining 4 independent components of the equation, $G_{ab} = 8\pi T_{ab}$, for p and the metric potentials ν, ω, ψ, and μ.

2. Constructing models of rapidly rotating relativistic stars

Models described here were computed using a code developed by Butterworth and Ipser[9] (1975) and slightly modified to handle tabulated EOS's by Ipser, and by Friedman and Parker.[10] Earlier codes for various idealized EOS's are described by Bardeen and Wagoner,[11] Bonazzola and Schneider,[12] and Wilson;[13] and a recent code for polytropes was obtained by Komatsu, Eriguchi and Hachisu.[14] A slow-rotation approximation was developed by Hartle and used by Hartle and Thorne to model stars based on a few early EOS's.[15] In last several years Ray, Datta, and Kapoor have used the formalism to carry out an extensive investigation of models based on a number of more recently proposed EOS's.[16]

In the Butterworth-Ipser code, models are computed iteratively by the Newton-Raphson method. Recall that to find a solution to the nonlinear equation

$$f_i(\overrightarrow{x}) = 0 , \qquad (2.1)$$

one guesses a solution $x^{(n)}$, expands $f(x^{(n)} + \delta x)$ to second order in a Taylor series about $x^{(n)}$, and solves for δx. That is, given $x^{(n)}$, one obtains $x^{(n+1)} = x^{(n)} + \delta x^{(n+1)}$ by writing,

$$f(x^{(n)}) + f'(x^{(n)}) \cdot \delta x^{(n+1)} = 0 \Rightarrow \delta x^{(n+1)} = -(f'(x^{(n)})^{-1} \cdot f(x^{(n)}) , \qquad (2.2)$$

where f' is the Jacobian matrix of the function f, $f' \equiv \partial_j f_i$. For the system of equations (2.1), (2.2), describing a rotating star, with p, g_{ab} specified on a finite grid, $\{(i,j)\}$, we have

$$\overrightarrow{x} = \{p(i,j), g_{ab}(i,j)\} \qquad (2.3)$$

and

$$f(x^{(n)}) = \begin{cases} \int \dfrac{dp}{\epsilon + p} + \log u^t - \nu\big|_{pole} & (p^{(n)}, g^{(n)}) \\ G_{ab} - 8\pi T_{ab} & (p^{(n)}, g^{(n)}) \end{cases} . \qquad (2.4)$$

Then δf is the result of linearizing f about $(p^{(n)}, g^{(n)})$:

$$\delta f = f'(x^{(n)}) \cdot \delta x = \begin{cases} \delta\left[\int \dfrac{dp}{\epsilon + p} + \log u^t \right] \\ \delta\left[G_{ab} - 8\pi T_{ab} \right] \end{cases} . \qquad (2.5)$$

with $\nu\big|_{pole}$ and Ω fixed. One inverts the matrix f' to find δp and δg, and continues the iteration, typically until convergence to 1 part in 10^6 is obtained:

$$p^{(n+1)} = p^{(n)} + \delta p, \quad g^{(n+1)} = g^{(n)} + \delta g . \qquad (2.6)$$

The code is quite efficient, with convergence to 1 part in 10^6 ordinarily requiring between 5 and 25 iterations, depending on the rotation and central density of the star.

Comparison with spherical and slowly rotating models and several internal checks give consistent measures of error. For integrated quantities, we have

$$\frac{\delta g}{g}, \frac{\delta p}{p}, \frac{\delta M}{M}, \frac{\delta J}{J} \sim 1\%; \qquad (2.7)$$

the typical grid we used had about 25 radial grid points inside the star, with corresponding errors in the radius R and redshift Z given by

$$\frac{\delta R}{R}, \frac{\delta Z}{Z} \leq 5\%. \qquad (2.8)$$

3. Ω_{max} for stars stable against collapse

Gravity sets a hard upper limit on the rotation of a neutron star, the Kepler frequency Ω_K. No equilibrium configuration can have an angular velocity greater than that of a particle in circular orbit at its equator, and, for a given mass, M, the configuration with maximum angular velocity rotates at the angular velocity for which

$$\Omega_{\text{satellite}} = \Omega_{\text{star}} \equiv \Omega_K(M). \qquad (3.1)$$

Because of the increasing oblateness that accompanies rotation, $\Omega_K(M)$ is typically only about half the frequency of a particle in circular orbit about the nonrotating model of the same mass.

Stability against collapse

A simple, turning point criterion governs secular stability against axisymmetric perturbations:[17] along a sequence of rotating neutron stars with constant angular momentum, J, and increasing central density, the configuration with maximum mass marks the onset of secular instability to collapse. The growth time of the secular instability is the time required to redistribute the stars' angular momentum, which one presumably observes as the relaxation time following a glitch. Within the accuracy of our models, the configuration with maximum mass, among the 2-parameter space of equilibria is also an extremum of mass along a curve of constant J. If this is exact, then the configuration with maximum mass has the largest value of Ω_K, among all models stable against collapse (but ignoring for the moment the question of stability against nonaxisymmetric perturbations):

$$\Omega_{\max} = \Omega_K(M_{\max}), \qquad (3.2)$$

where $M_{\max}$ is the maximum mass among all rotating models, 13-20% greater than the maximum mass for a spherical star with the same EOS.

Nonaxisymmetric stability

For hot ($T > 10^7$ K) neutron stars, nonaxisymmetric instability sets an upper limit on rotation more stringent than Ω_K. The instability is driven by gravitational waves, and arises in the following way.[18] For slowly rotating stars, gravitational waves carry positive J from a forward mode and negative J from a backward mode, thus damping all nonaxisymmetric perturbations.

But when $m\Omega \sim \sigma$, the mode that moves backward relative to the star is dragged forward relative to an inertial observer. Gravitational waves now carry negative J from a mode that already has $J < 0$ (the perturbed star has lower J because the perturbation moves backward relative to the star) so gravitational radiation drives the mode. The Dedekind bar instability found by Chandrasekhar[19] is the $m = 2$ case of this mechanism, but higher modes are unstable first, and neutron stars appear to reach the Keplerian frequency before the $m = 2$ mode is unstable.

Viscosity damps out the instability when the viscous damping time is shorter than the radiation reaction time, $\tau < \tau_{GRR}$.[20,21] For neutron stars, shear viscosity is proportional to T^{-2} and rises as the star cools. At the same time, the bulk viscosity, arising from neutrino emission from hyperon-producing nuclear reactions in the core is proportional to T^6, and falls as the star cools.[22] As a result, one expects instability only for temperatures in the range $10^7 \, \text{K} \lesssim T \lesssim 2 \times 10^9 \, \text{K}$. In particular for the pulsar SN 1987A, standard (URCA) cooling[23] would imply $T \sim 10^9$ K and the pulsar in SN 1987A may well be unstable, spinning down in a time given by[10,24]

$$\tau \sim 10^6 \, \text{s} \left(\frac{M}{1.4 M_\odot} \right)^4 \left(\frac{R}{10 \, \text{km}} \right)^5 \left(\frac{\Omega - \Omega_c}{0.1 \, \Omega_c} \right)^{-(2\ell+1)} . \tag{3.3}$$

If, however, the star has a quark core or a pion condensate, the cooling rate is likely to be much more rapid.[25] In that case the viscosity would quickly rise to damp out the instability, and the Kepler frequency accurately measures the limiting rotation.

Equations of state considered

There is a dramatic difference between model neutron stars built from the softer of the proposed EOS's and those built from stiff EOS's. Figure 1 shows, for stars of about $1.4 M_\odot$, the much greater central condensation of models built from a soft EOS (G in the list below) compared to almost uniform density models constructed from stiffer EOS's. We constructed several hundred models based on the 17 proposed EOS's listed below,[28] obtaining for 14 of these explicit models with maximum mass and angular velocity consistent with stability against collapse. While the nuclear physics used to compute many of these EOS's has been subsequently revised, the collection can be regarded as sampling the space of possible EOS's above nuclear density. The letters A-O are labels used by Arnett and Bowers[29] in their study of spherical models, and abbreviations used in the reviews by Baym and Pethick[30] and Shapiro and Teukolsky[31] are given in parentheses.

| A(R) | Reid soft core, Pandharipande (1971) |
| B | Reid soft core with hyperons (1971) |

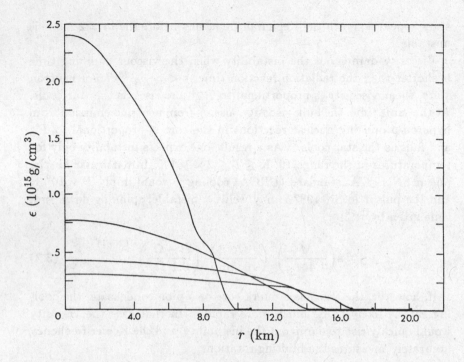

Fig. 1. Energy density in the equatorial plane vs. the radial coordinate r for four models rotating with maximum angular velocity, $\Omega \approx \Omega_K$, and having rest mass $M_0 \approx 1.4 M_\odot$.

C(BJI)	Bethe-Johnson I	
D(BJV)	Bethe-Johnson V	} (1974)
F	Arponen (1972)	
G	Canuto-Chitre (1974)	
L(MF)	Mean field, Pandharipande-Smith (1975)	
M(TI)	Tensor interaction, Pandharipande-Smith (1975)	
N*(RMF)	Relativistic mean field, Serot (1979)	
O	Bowers, Gleeson, Pedigo (1975)	
FP(TNI)	Three-nucleon interaction,	
	Friedman-Pandharipande (1981)	
PAL1	Prakash-Ainsworth-Lattimer Parameterized EOS,	
PAL3	$K = 120$ MeV (1988)	
GL	Glendenning Parameterized EOS,	
	$K = 210$ MeV (1986)	
WFF	Wiringa-Fiks-Fabrocini Modification of F-P (1988)	

Q   Quark EOS, Glendenning (1989)
π   Weise-Brown pion-condensate (1974)

Four EOS's with varying stiffness are depicted in Figure 2. Because the models based on the soft EOS, B (like those based on EOS G in Fig. 1), are much more centrally condensed than those based on, say, EOS L, the softer EOS allows a much greater angular velocity for the softer EOS. As shown in Figure 3, the maximum rotation therefore rises as the EOS gets softer, despite the fact that stiffer EOS's can support more mass. Making the EOS stiff at the highest neutron-star densities and soft at low and medium density may maximize rotation, by providing models in which a stiff core is able to support a massive, compact star.

Curves of $\Omega_K(M)$, the maximum rotation for each mass M, are shown in Figure 4 for the larger set of EOS's that we considered. For each of these 14 EOS's, the maximum rotation can be estimated from the

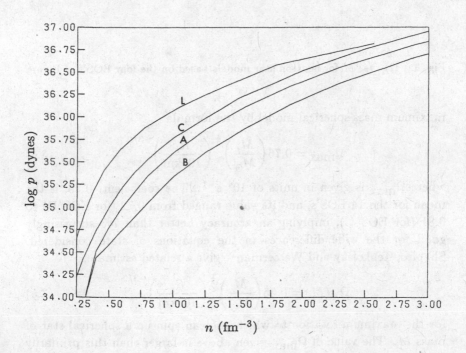

Fig. 2. Log(pressure) vs. number density of baryons for four EOS's.

John L. Friedman

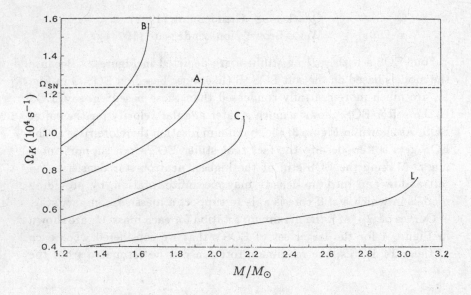

Fig. 3. Ω_K vs. M for neutron star models based on the four EOS's of Figure 2.

maximum mass spherical model by the formula

$$\Omega_{\text{max}} = 0.76 \left(\frac{M_s}{M_\odot} \right)^{\frac{1}{2}} \left(\frac{R_s}{10 \text{ km}} \right)^{-3/2} \tag{3.4}$$

where Ω_{max} is given in units of 10^4 s^{-1}. The coefficient, 0.76, is the mean for the 14 EOS's, and its value ranged from 0.72 (for EOS D) to 0.81 (for EOS C), implying an accuracy better than 7%–surprisingly good for the wide differences in the equations of state considered. Shapiro, Teukolsky and Wasserman[26] give a related estimate,

$$\Omega_K(M) = 0.65 \left(\frac{M}{M_\odot} \right)^{\frac{1}{2}} \left(\frac{R_s}{10 \text{ km}} \right)^{-3/2} \tag{3.5}$$

for the maximum rotation to which one can spin up a spherical star of mass M. The value of Ω_{max} given above is larger than this primarily because the maximum mass for rotating models exceeds that for the spherical models by 13-20%. (Eq. (3.4), with slightly different values of the coefficient was given independently by Haensel and Zdunik[27] and by Friedman, Ipser and Parker.[10])

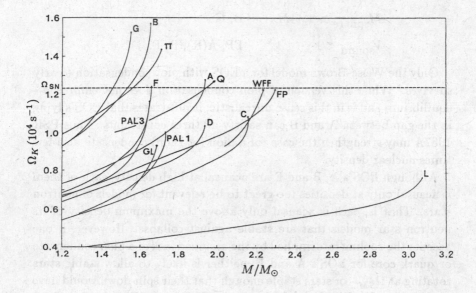

Fig. 4. Ω_K vs. M for 14 EOS's.

4.   Conclusions

Constraints on the EOS of matter above nuclear density.

As a number of authors have observed, the existence of a half-millisecond pulsar stringently restricts the EOS governing neutron star matter.[32,10,26,27] If PSR 1987A is rotating with angular velocity $\Omega_{SN} = 1.237 \times 10^4 \, \text{s}^{-1}$, the EOS must satisfy at least the following three constraints: It must allow spherical stars with mass $M > 1.44 M_\odot$, because one component of the binary pulsar PSR 1913+16 has an observed mass $1.441 \pm .003 M_\odot$. It must allow rotating stars with angular velocity $\Omega \geq \Omega_{SN}$. It must satisfy $V_{\text{sound}} < 1$. These constraints are sufficient to rule out nearly all the proposed EOS's:

Constraint	EOS's Ruled Out
$\Omega > \Omega_{SN}$	C(BJI), D(BJV) L(MF), M(TI), N*(RMF), O, PAL1, PAL3, GL, WFF;

$$M_{\text{sphere}} > 1.44 M_\odot \qquad \text{B, G;}$$

$$V_{\text{sound}} < 1 \qquad \text{FP, A(R), B, F.}$$

Only the Weise-Brown model for a EOS with pion condensation clearly survives. While this version of pion condensation does not enforce β - equilibrium and is in this sense unrealistic, it illustrates that EOS's lying in the gap between A and B can satisfy all three constraints. Thus PSR 1987A may strengthen the case for a pion (or kaon) condensate at a few times nuclear density.

Although EOS's A, B and F are acausal at high density, each of them is acausal only at densities too great to be relevant for models of neutron stars: That is, each is acausal only above the maximum density of its neutron star models that are stable against collapse. However, if one accepts the cooling rate implied by the assumed lack of a pion condensate or quark core for EOS's A and F, neither is likely to allow stable stars rotating at Ω_{SM} - or stars stable enough that their spin-down would have eluded detection in the nine-hour observation time of PSR 1987A. (EOS B is already eliminated by the $M > 1.44 M_\odot$ constraint for spherical models.)

Finally, as will be discussed shortly, it does not quite seem to be possible to rule out quark stars based on the three constraints listed above.

A further constraint on the EOS is provided by measurements of the spin-down of the Crab. By assuming the pulsar's rotational energy is the sole source of energy for replenishing short-lived electrons in the nebula, Trimble[33] obtains a lower limit on the moment of inertia of the Crab pulsar of $7 \pm 1 \times 10^{44}$ gcm^2. However, this constraint is less secure than the $1.44 M_\odot$ requirement, and it appears to be slightly weaker: EOS's that are too soft to allow $I > 6 \times 10^{44}$ gcm^2 are too soft to allow $M > 1.44 M_\odot$.

Comments on quark matter

At sufficiently high density, one expects nucleons to coalesce, forming a fluid of approximately free quarks. It is plausible that such a phase transition to quark matter occurs at several times nuclear density and a number of authors considered the properties of neutron stars with quark cores and/or the possibility of a sequence of stable quark stars.[34] More recently, the suggestion that even at zero pressure quark matter might be the lowest energy state of baryons[35] led to increased interest in pure quark stars.[36] (These are commonly referred to as "strange stars" because the number of strange quarks could be comparable to the number of down quarks.)

Before discussing the question of whether the existence of quark stars or stars with quark cores might make it easier to satisfy the constraints discussed above, I would like to argue that quark matter <u>cannot</u> be the ground state at zero pressure. The reason is that if a small fraction f_Q of neutron stars spontaneously convert to quark stars, enough quark matter would now contaminate the interstellar medium that all observed "neutron stars" would have to be quark stars. (Neutron stars with quark seeds are expected to rapidly convert to quark stars.) Because the expected crust of normal matter surrounding a quark star is too thin to account for the energy of glitches, observations of these starquakes appear to rule out the possibility that all stars at or above nuclear density are quark stars.[37] The contamination of the interstellar medium by quark matter arises from coalescence of binary systems in which one member is a quark star and the other a neutron star or black hole (Witten, Ref. 35b). Clark, van den Heuvel and Sutanyou[38] estimate the rate of formation of close neutron star binaries in the galaxy at $3 \pm 1.6 \times 10^{-4}$ y^{-1} . While the uncertainty may be greater than this, we can conservatively estimate the rate of coalescence of binaries with one quark star and one neutron star at more than $f_Q \cdot 10^{-4}$ y^{-1}. Then an expected dispersion to infinity of more than .01 $M_\odot$ in each coalescence[39] implies a present interstellar mass of quark matter

$$M \geq (f_Q \times 10^{-4} \, y^{-1})(.01 M_\odot)(10^{10} \, y) = f_Q \times 10^4 M_\odot \, .$$

The fractional density of quark matter is then $\rho_{\text{QUARK}}/\rho > f_Q \times 10^{-7}$, and each newly formed neutron star will contain at least a mass $m_Q \equiv f_Q 10^{-7} \cdot M_\odot = f_Q \cdot 2 \times 10^{26}$ g unless all the material ejected in coalescence is in clumps larger than m_Q (or smaller than the minimum size needed for stability–a few hundreds or thousands of nucleons, if quark matter is stable at zero pressure). It thus appears that if PSR 1987A is a quark star, f_Q must be more than large enough to imply that all neutron stars are quark stars.

Even if quark matter is not stable at zero pressure, it may be stable at low enough pressure that PSR 1987A consists largely of quarks. One could either suppose that in most neutron stars the density is too low to overcome the barrier against a phase transition, but that the star's nucleon interiors are only metastable; or that all neutron stars have large quark interiors (allowing a large enough normal crust to accommodate glitches). The first alternative allows the greatest freedom to avoid constraints: one could have an EOS for (metastable) nuclear matter that allowed $1.44 M_\odot$ spherical stars and had, say, a limiting rotational frequency near 0.4×10^4 s^{-1} (like that of PSR 1937+21); and an EOS for (stable) quark matter with a limiting frequency of $\Omega_{SN} = 1.24 \times 10^4$ s^{-1}. It may, however, be that a bag constant large enough to permit $\Omega_K \geq \Omega_{SN}$ is so large that quark matter will not be stable over most of the star. Some effort has been made to decide whether quark stars are plausible candidates for SN 1987A,[40] but it is not yet clear that the physically allowed ranges of bag constant, strange quark mass and coupling constant are consistent with the rotation speed Ω_{SN}. The particular quark EOS plotted in Fig. 4, for which Ω_K barely exceeds Ω_{SN}, succeeds because its bag constant, $B = (170 \,\mathrm{MeV})^4$, is somewhat larger than the value of $(145 \,\mathrm{MeV})^4$ that would be required to make quark matter stable out to the edge of the star according to the standard criterion (matching the Gibbs free energy of the competing phases). Glendenning[40] argues that the inaccuracy of the criterion due to the finite size of a nucleon–as well as the uncertainty in the EOS of strange quark matter–is sufficient to allow all or nearly all of the star to be quark matter with this value of B.

Additional EOS's have been examined in a recent preprint by Lattimer et al.[41] (received after this conference), reporting results from an independently constructed version of the Butterworth-Ipser[9] code. Their results appear to be consistent with the discussion here and with the estimate (3.4).

A trap

The difficulty in finding an EOS that can satisfy the three constraints discussed above has led to speculation that the observed emission at frequency Ω_{SN} might result from oscillation rather than rotation.[42] Although the 0.5 ms period is within the expected range[43] for oscillation of a $1.4 M_\odot$ star, the reconstructed pulse shape, its harmonic content and high luminosity, and the apparent existence of at least one very close companion are all more naturally explained by rotation.[44]

If we assume that PSR 1987A has a 0.5 ms period of rotation, then it is caught in a trap. For every EOS we have examined, the mass of any neutron star rotating with $\Omega \geq \Omega_{SN}$ exceeds the minimum nonrotating mass. (The excess mass ranges from 3% for EOS π, to 17% for EOS A.) If spun down sufficiently far, say by a pulsar mechanism, such an object will collapse (Refs. 10, 26, and Sato and Suzuki, and Goldman in Ref. 32). Hence one is led to speculate on the existence of a class of black holes formed in this way with masses $\sim 1.5 M_\odot$ to $2 M_\odot$. A magnetic field of 10^9 G (corresponding to the observed luminosity of the remnant SN 1987A) could be expected to spin such a pulsar down in $10^7 - 10^8$ yr, while a field of 4×10^7 G would be sufficient for spin down within the age of the Universe (10^{10} yr). If a similarly rapid pulsar is the typical outcome of the supernova of a similar progenitor star (initially a main sequence star of mass $< 20 M_\odot$), one expects at least 10^6 black holes to have formed in this way in the lifetime of the galaxy.

Time spent in preparing this review was supported in part by the National Science Foundation.

References

1. D. C. Backer, S. R. Kulkarni, C. Heiles, M. M. Davis and W. M. Gross (1982), *Nature* **300**, 615-618.
2. J. H. Taylor and D. R. 10b.A. Stinebring (1986), *Rev. Astr. Astrophys.* **24**, 285-327.
3. J. N. Imamura, J. Middleditch and T. Y. Steiman-Cameron (1987), *Astrophys. J.* **314**, L11-L13.
4. A. S. Fruchter, D. R. Stinebring and J. H. Taylor (1988), *Nature* **333**, 237-238.
5. J. L. Friedman, J. N. Imamura, R. H. Durisen and L. Parker (1988), *Nature* **336**, 560.
6. J. Kristian, C. R. Pennypacker, J. Middleditch, M. A. Hamuy, J. N. Imamura, W. E. Kunkel, R. Lucinio, D. E. Morris, R. A. Muller, S. Perlmutter, S. J. Rawlings, T. P. Sasseen, I. K. Shelton, T. Y. Steiman-Cameron and I. R. Tuohy (1989), *Nature* **338**, 234.
7. G. Baym and D. Pines (1971), *Ann. Phys.* **66**, 816.
8. B. Carter (1969), *J. Math. Phys.* **10**, 70.
9. E. M. Butterworth and J. R. Ipser (1976), *Ap. J.* **204**, 200.
10a. J. L. Friedman, J. R. Ipser and L. Parker (1986), *Ap. J.* **304**, 115.
10b. J. L. Friedman, J. R. Ipser and L. Parker (1989), *Phys. Rev. Lett.* **62**, 3015.
11. J. M. Bardeen and R. V. Wagoner (1974), *Ap. J.* **167**, 359.
12. S. Bonnazola and J. Schneider (1974), *Ap. J.* **191**, 273.
13. J. R. Wilson (1972), *Ap. J.* **176**, 195.
14. H. Komatsu, Y. Eriguchi, and I. Hachisu, *Astron. Ap.*, in press.
15. J. B. Hartle (1967), *Ap. J.* **150**, 1005.
16. J. B. Hartle and K. J. Thorne (1968), *Ap. J.* **153**, 807.
17. J. L. Friedman, J. R. Ipser and R. Sorkin (1988), *Ap. J.* **325**, 722.
18a. J. L. Friedman and B. F. Schutz (1978), *Ap. J.* **222**, 281.
18b. J. L. Friedman (1978), *Comm. Math. Phys.* **62**, 247.
19. S. Chandrasekhar (1970), *Phys. Rev. Lett.* **24**, 611.

20. The first result of this nature is due to L. Lindblom and S. Detweiler, (1977), *Ap. J.* **211**, 565; a recent analysis is given in C. Cutler and L. Lindblom, (1987), *Ap. J.* **314**, 234. Estimates of the instability points can be found in R. A. Managan (1985), *Ap. J.* **294**, 463; J. N. Imamura, R. H. Durisen and J. L. Friedman (1985), *Ap. J.* **294**, 474; and Ref. 21.

21. J. R. Ipser and L. Lindblom (1989), *Phys. Rev. Lett.* **62**, 2777.

22. R. F. Sawyer, preprints UCSBTH-89-09, UCSBTH-89-33 (University of California at Santa Barbara).

23. K. Nomoto and S. Tsuruta (1987), *Ap. J.* **312**, 711.

24. W. Kluzniak, L. Lindblom, P. Michelson and R. V. Wagoner (1989), *Nature* **339**, 19.

25a. R. C. Duncan, S. L. Shapiro and T. Wasserman (1983), *Ap. J.* **267**, 358.

25b. N. Iwamoto (1980), *Phys. Rev. Lett.* **44**, 1637.

25c. A. Burrows (1980), *Phys. Rev. Lett.* **44**, 1640.

26a. S. Shapiro, S. A. Teukolsky and I. Wasserman (1989), *Nature* **340**, 451.

26b. S. Shapiro, S. A. Teukolsky and I. Wasserman (1983), *Ap. J.* **272**, 702.

27. P. Haensel and J. L. Zdunik (1989), *Nature* **340**, 617.

28. References to the EOS's are, in the order listed: V. R. Pandharipande, (1971), *Nuc. Phys.* **A207**, 298, and (1971), **A174**, 641; H. A. Bethe and M. Johnson (1974), *Nuc. Phys.* **A230**, 1; J. Arponen (1972), *Nuc. Phys.* **A191**, 257; V. Canuto and S. M. Chitre (1974), *Phys. Rev.* **D9**, 1587; V. R. Pandharipande and R. A. Smith (1975), *Nuc. Phys.* **A175**, 225; B. D. Serot (1979), *Phys. Lett.* **86B**, 146, and (1979), **87B**, 403; R. L. Bowers, A. M. Gleeson and R. Pedigo (1975), *Phys. Rev.* **D12**, 3043; B. Friedman and V. R. Pandharipande (1981), *Nuc. Phys.* **A361**, 502; M. Prakash, T. L. Ainsworth and J. M. Lattimer (1988), *Phys. Rev. Lett.* **61**, 2518; N. K. Glendenning (1986), *Phys. Rev. Lett.* **57**, 1120; R. B. Wiringa, V. Fiks and A. Fabrocini (1988), *Phys. Rev.* **C38**, 1010; N. K. Glendenning, preprint, LBL-27446; W. Weise and G. E. Brown (1975), *Phys. Lett.* **55B**, 300.

29. J. H. Taylor and J. M. Weisberg (1989), *Ap. J.* **253**, 908.

30. G. Baym and C. Pethick (1979), *Ann. Rev. Astron. Ap.* **17**, 415.

31. S. L. Shapiro and S. A. Teukolsky (1983), *Black Holes, White Dwarfs, and Neutron Stars*, (Wiley-Interscience, New York).

32. K. Sato and H. Suzuki (1989), *Prog. Theor. Phys.*, in press; D. Bhattacharya and G. Srinivasan (1989), *Current Science* **58**, 280; J. I. Katz (1989), *Internal Structure of PSR0535-69 (SN1987A)*, Washington University, preprint; I. Goldman (1989), *SN1987A Pulsar: Magnetic Field, Mass, and Equation of State of the Neutron Star*, Tel Aviv University, preprint; K. Chen (1989), *How Fast Can a Neutron Star Rotate?*, preprint; S. Bonazzola and J. A. Marck (1989), *Abstracts of Contributed Papers, Twelfth International Conference On General Relativity and Gravitation*, Boulder, CO, Vol. II.

33. V. Trimble (1987), in *High Energy Phenomena Around Collapsed Stars*, ed. F. Pacini (Reidel, Amsterdam), p. 105.

34. N. Itoh (1980), *Prog. Theor. Phys.* **44**, 291; J. R. Ipser, M. B. Kislinger and P. D. Morley (1975), *Quark-Bag Equation of State and a Possible Third Regime of Stable Cold Stars*, Enrico Fermi Institute Report EFI, 75-38; J. C. Collins and M. J. Perry (1973), *Phys. Rev. Lett.* **34**, 1353; G. Chapline and M. Nauenberg (1976), *Nature* **264**, 235; G. Baym and S. Chiu (1976), *Phys. Lett.* **62B**, 241; K. Brecher and G. Caparaso (1976), *Nature* **259**, 377; W. B. Fechner and P. C. Joss (1978), *Nature* **274**, 347; B. Freedman and D.

McLerran (1978), *Phys. Rev.* **D17**, 1109; B. D. Serot and H. Uechi (1987), *Ann. Phys.* **179**, 272.

35a. A. R. Bodmer (1971), *Phys. Rev.* **D4**, 1601.

35b. E. Witten (1984), *Phys. Rev.* **D30**, 272.

36. P. Bhattacharjee (1986), *Fortschr. Phys.* **34**, 817; C. Alcock, E. Farhi and A. Olinto (1986), *Ap. J.* **310**, 261; P. Haensel, J. L. Zdunik and R. Schaeffer (1986), *Astron. Ap.,* **160**, 121; P. Haensel (1987), *Acta. Phys. Pol.* **B18**, 739.

37. M. A. Alpar (1987), *Phys. Rev. Lett.* **58**, 2151.

38. J. P. A. Clark, E. P. J. van den Heuvel, and W. Sutantyo (1979), *Astron. and Ap.* **72**, 120.

39a. D. Eichler, M. Livio, T. Piran, and D. N. Schramm, *Nature*, in press.

39b. J. P. A. Clark and D. M. Eardley (1977), *Ap. J.* **215**, 311.

40. J. A. Friedman and A. V. Olinto (1989), *Nature* **341**, 635; N. K. Glendenning (1989), *J. Phys.* **G15**, L225, and *Fast pulsar in SN1987A: Candidate for strange quark matter*, LBL-27446; F. Grassi (1989), *The pulsar in SN1987A and the equation of state of dense matter*, University of Illinois preprint P/89/10/140.

41. J. M. Lattimer, M. Prakash, D. Masak, and A. Yahil (1989), *Rapidly rotating pulsars and the equation of state*, preprint.

42a. Q. Wang, K. Chen, T. T. Hamilton, M. Ruderman and J. Shaham (1989), *Nature* **338**, 319.

42b. W. Kluzniak, L. Lindblom, P. Michelson, and R. V. Wagoner (1989), Columbia University preprint **CAL 371 G ITP-858.**

42c. R. C. Duncan (1989), preprint.

43. L. Lindblom and E. N. Glass (1983), *Ap. J. Suppl.* **53**, 93.

44. J. Middleditch, C. R. Pennypacker, J. Kristian, J. R. Graham, S. Heathcote, J. N. Imamura, W. E. Kunkel, D. E. Morris, R. A. Muller, S. Perlmutter, S. J. Rawlings, T. P. Sasseen, T. Y. Steiman-Cameron and I. R. Tuohy, *Nature*, in press.

3

Global properties of solutions to Einstein's field equations

Helmut Friedrich

Max-Planck-Institut für Astrophysik,
Karl-Schwarzschild-Str. 1,
8046 Garching, Federal Republic of Germany

1. Introduction

The methods invented so far to develop conjectures on and to work
out the details of the time evolution of gravitational fields fall into two
classes. The first class comprises "physical considerations," techniques
of differential geometry and topology and other ideas which do not take
into account the field equations while the remaining methods like: study
of explicit solutions, of the Cauchy problem local in time (with its important related notions like "domain of dependence," "Cauchy stability,"
etc.), of formal expansion type solutions, of approximation procedures,
etc., have been used to derive information about the evolution more
or less directly from the field equations. The division above is somewhat artificial, as is illustrated e.g. by the use of certain positivity
assumptions together with Raychaudhuri's equations in the proof of the
Hawking-Penrose singularity theorems. However, since the field equations are used in this case in a very weak way one obtains quite general
results about the occurrence of a non-complete geodesic but almost no
information about the expected "singularity."

In spite of all the ingenuity with which the methods indicated above
have been employed, the important open problems of classical general
relativity, e.g. the development and (causal) structure of singularities
and the formation of horizons etc. for space-times arising from regular
data, the asymptotic behavior of gravitational fields and the relation
between the far fields and the structure of the sources etc., remained to

a large extent unsolved. This suggest that the techniques listed above are insufficient to analyze these problems.

Missing are tools which allow us to get a grip on the long time evolution of the fields and to derive rigorous statements of some generality about solutions of the field equations in the large and near singularities.

Various efforts have been made in this direction in recent years (cf. e.g. Christodoulou (1986); Klainerman (1988); and Moncrief (1987)) which in the end should flow together to yield a general picture of the global structure of solutions to the field equations. At present it is too early for any attempt to discuss the various approaches in a systematic way. Therefore I shall concentrate in the following on some lines of research in this area in which I have been interested myself for some time.

2. The asymptotic behavior of gravitational fields

In an attempt to give a precise meaning to notions like "gravitational radiation," "gravitational radiation field," "loss of mass due to gravitational radiation" etc. for gravitational fields which were to represent the far fields of "isolated gravitating systems," various authors (Bondi et al. (1962); Sachs (1962); Newman and Penrose (1962)) analyzed formal expansion type solutions of the field equations. In this analysis they had to assume a certain fall-off behavior of the gravitational field along outgoing null geodesics. Their assumptions were supported by the asymptotic behavior of a few explicit solutions of the field equations and by the fact that they were able to draw under these assumptions a rather consistent picture of the situation they wanted to model, which is particular allowed one to introduce the notions mentioned above in a rigorous way. This development came to a certain end when Penrose (1963) showed that the assumed fall-off in light-like directions could be derived from the requirement that the asymptotic behavior of the fields were such, that the conformal structure of the fields could be extended smoothly through "null infinity" (denoted in the following by $\mathcal{I}$). The set $\mathcal{I}$ is thought of here as a collection of ideal future (past) endpoints of outgoing (incoming, respectively) null geodesics.

In much of the subsequent work on the far fields of isolated gravitational systems Penrose's conditions (to which we will refer in the following as the "$\mathcal{I}$-concept") were assumed from the outset and the physical and geometrical implications of the assumed behavior was analyzed. Moreover, the conformal techniques introduced in the study of null infinity were extended to analyze other asymptotic regions of space-times. This development has been discussed by Ashtekar (1984) at GR-10.

Occasionally doubts were raised as to whether the $\mathcal{I}$-concept was natural for the fields of isolated gravitating systems (cf. Blanchet and Damour (1986)). These doubts arose mainly from the results of certain approximation calculations. In view of the fact that the global conformal structure of exact asymptotically flat solutions to Einstein's field equations may deviate drastically from that of Minkowski space (cf. Penrose (1980)), it appears difficult to come here to a definite conclusion about the appropriateness of the $\mathcal{I}$-concept.

The basic open problem, which will be discussed in this article, is the relation between the geometrically introduced $\mathcal{I}$-concept together with its implications on the conformal structure of space-time and the consequences (of the "conformal structure") of the field equations. An analysis of this relation should give:

– Information about the class of solutions allowing a smooth conformal structure at null infinity;

– A derivation of this asymptotic behavior as a consequence of the evolution process defined by the field equations, where the field develops from a "well-understood" situation, e.g. from standard Cauchy data which are asymptotically flat at space-like infinity and possibly satisfy further conditions;

– Insight into the "conformal structure of the field equations" and its consequences on the conformal (and thence on the causal) structure of their solutions.

It should be clear that such a program does not only require the proof of global existence theorems *but on top of that control of the asymptotic behavior of the fields which is sufficiently sharp to allow one to decide whether they permit a smooth structure at null infinity.*

3.   The conformal boundary

To motivate the latter discussion it is convenient to give a quick discussion of the $\mathcal{I}$-concept. Let $(\tilde{M}, \tilde{g}_{\mu\nu})$ be a 4-dimensional Lorentz-space, the "physical space." Assume this may be obtained in the following way: there is another 4-dimensional Lorentz-space $(M, g^{\mu\nu})$, the "non-physical space," where M is a smooth manifold with boundary $\mathcal{I}$. On M there is a smooth real-valued function Ω, the "conformal factor," such that $\Omega > 0$ on $M \backslash \mathcal{I}$, $\Omega = 0$ and $d\Omega \neq 0$ on $\mathcal{I}$. The manifold $\tilde{M}$ is diffeomorphic to

$M\backslash\mathcal{I}$ and after a diffeomorphic identification such that $\tilde{M} = M\backslash\mathcal{I}$ we have

$$g_{\mu\nu} = \Omega^2 \tilde{g}_{\mu\nu} \quad \text{on} \quad \tilde{M}. \tag{1}$$

If in this discussion all objects can be assumed to be smooth we shall say that $\mathcal{I}$ represents a smooth conformal boundary for $(\tilde{M}, \tilde{g}_{\mu\nu})$. It represents a surface at infinity for the physical space-time, since any physical null-geodesic which approaches $\mathcal{I}$ in the future (say) will be future complete. If $\tilde{g}_{\mu\nu}$ satisfies in addition near $\mathcal{I}$ Einstein's field equations

$$\tilde{R}_{\mu\nu} - \frac{1}{2}\tilde{R}\tilde{g}_{\mu\nu} + \lambda\tilde{g}_{\mu\nu} = 0, \tag{2}$$

with cosmological constant λ, then it follows that the hypersurface $\mathcal{I}$ is space-like, time-like or null if λ is positive, negative or zero respectively (assuming the signature +2).

The simply connected, geodesically complete, conformally flat standard solutions of einstein's field equations which allow the construction of a smooth conformal boundary in the sense outlined above are:

Case $\lambda > 0$: de Sitter space, where the space-like conformal boundary consists of two components $\mathcal{I}^+, \mathcal{I}^-$ (each of which is diffeomorphic to the sphere S^3) which represent the future or past endpoints, respectively, of the time-like and null geodesics.

Case $\lambda = 0$: Minkowski space, where the light-like conformal boundary consists again of two components $\mathcal{I}^+, \mathcal{I}^-$ (each of which is diffeomorphic to $\mathbb{R} \times S^2$) which represent the future or past, respectively, endpoints of the null geodesics. In suitable smooth conformal extensions of this solution appear 3 additional "points at infinity": i^+ and i^-, which represent the future and past, respectively, endpoints of all the time-like geodesics, and i^0 which represents space-like infinity. In such an extension, $\mathcal{I}^+$ is obtained as the past light cone of i^+ and the future light cone of i^0 while $\mathcal{I}^-$ is obtained as the future light cone of i^- and the past light cone of i^0. We shall shall refer in the following to such a situation as the "regularity of future or past, respectively, time-like infinity" and to the "regularity of spacelike infinity." As an illustration of to what extent the metrical relations of Minkowski space are distorted by a conformal rescaling which allows a smooth extension of the rescaled metric through $\mathcal{I}$, may serve the observation, that the unit hyperbolae in Minkowski space may be extended through $\mathcal{I}$ as spacelike hypersurfaces which thus are transverse to $\mathcal{I}$ as well as to the extended null cone through the "origin" of Minkowski space.

Case $\lambda < 0$: Anti-de Sitter space, where the time-like conformal boundary $\mathcal{I}$ (diffeomorphic to $\mathbb{R} \times S^2$) represents the endpoints of the space-like and the future as well as the past endpoints of the null geodesics.

In the following I shall only consider the cases $\lambda \geq 0$, the reason for this being not that the methods which I shall discuss later do not apply to the case $\lambda > 0$ but simply that no effort has been made so far to apply them.

In generalizing the notion of the conformal boundary from the standard cases above, where the boundary may be constructed essentially by group theoretical arguments, to the case of solutions which are not conformally flat it will be necessary to make some "completeness assumptions" in order not to leave out parts of the boundary which may possibly be constructed. In the case $\lambda > 0$, where it is natural to assume that the space-time has a compact Cauchy hypersurface, the obvious assumption is that $\mathcal{I}^+$ or $\mathcal{I}^-$, respectively, is diffeomorphic to this Cauchy surface. In the case $\lambda = 0$ two conditions are considered. The first is the requirement that the null generators of $\mathcal{I}^+$, $\mathcal{I}^-$ are complete in a choice of the conformal factor (which is always possible) where the null congruence on null infinity is expansion free. The second is the requirement, that the space of null generators of $\mathcal{I}^-$ or $\mathcal{I}^+$, respectively, is diffeomorphic to the sphere S^2. If that condition is satisfied, it can be shown (Penrose (1965)) that the Weyl tensor $C^{\mu}_{\nu\lambda\rho}$, which is the same for the physical as well as for the unphysical field, must vanish on null infinity and, provided all fields are smooth, the tensor field $d^{\mu}_{\nu\lambda\rho} = \Omega^{-1} C^{\mu}_{\nu\lambda\rho}$ extends smoothly to null infinity. A certain component of the tensor $d^{\mu}_{\nu\lambda\rho}$ on $\mathcal{I}$ in a frame which is suitably adapted to $\mathcal{I}$ attains then a natural interpretation as the "gravitational radiation field."

Since the $\mathcal{I}$-concept translates fall-off conditions for the physical field into smoothness conditions for the nonphysical field at the conformal boundary, the question which degree of smoothness may be expected at null infinity is of decisive importance for the idea of the conformal boundary. The concept would lose much of its beauty if it turned out too strong a requirement that the radiation field exists and is at least continuous on $\mathcal{I}$.

To come to a decision here one would like to start from suitable Cauchy data, follow the evolution of the field, and analyze the global conformal structure of the resulting space-time. To this purpose I studied first what may be called the "conformal structure of the field equations" and tried to analyze subsequently which conclusions may be drawn from that about the behavior of the solutions in the large.

4. The conformal structure of the field equations

Conformal methods have been used in various contexts to obtain information about the existence of solutions to the field equations and about their behavior in the large. The first application is probably due to Lichnerowicz who investigated the problem of solving the constraint equations which are implied on a space-like hypersurface. This has subsequently been developed into a very elegant formalism by various people (cf. Choquet-Bruhat and York (1980)). Quite a different use of conformal methods has been made in the analysis of the behavior of static and stationary solutions of the Einstein vacuum field equations near space-like infinity (cf. Beig (1981)), where it has been shown under weak assumptions on the smoothness and fall-off behavior of the solutions, that the conformal structure extends analytically through the point which represents space-like infinity. This analysis of 3-dimensional Riemannian geometries should be distinguished from the following discussion of the 4-dimensional Lorentzian case. However it should be interesting to study more systematically the interrelationship between these different investigations. That there is still more to be understood about this relation will be illustrated at the end of this article by some recent results on a curious relationship between static and radiative space-times.

The full 4-dimensional Einstein equations (2) are obviously not invariant under the conformal rescalings (1): they were designed to determine isometry classes of metrics. Formally this follows from the transformation law

$$\tilde{R}_{\mu\nu} - \frac{1}{2}\tilde{g}_{\mu\nu}\tilde{R} + \lambda\tilde{g}_{\mu\nu} = R_{\mu\nu} - \frac{1}{2}g_{\mu\nu}R + 2\Omega^{-1}\nabla_\mu\nabla_\nu\Omega + g_{\mu\nu}\{\Omega^{-2}\lambda$$
$$+ (3\Omega^{-2}\nabla_\lambda\Omega\nabla_\rho\Omega - 2\Omega^{-1}\nabla_\lambda\nabla_\rho\Omega)g^{\lambda\rho}\}.$$
$$(3)$$

However there is a conformally invariant part of the field equations. Using the invariance of the conformal Weyl-tensor

$$\tilde{C}^\mu_{\nu\lambda\rho} = C^\mu_{\nu\lambda\rho}$$

under the rescaling (1) one finds the relation (Penrose (1963))

$$\nabla_\mu(\Omega^{-1}C^\mu_{\nu\lambda\rho}) = \Omega^{-1}\tilde{\nabla}_\mu\tilde{C}^\mu_{\nu\lambda\rho}.$$
$$(4)$$

For our purpose the most remarkable feature of the relation (3) is the occurrence of the factors Ω^{-1}, Ω^{-2} on the right hand side. If we assume that (2) holds, then the vanishing of the right hand side of (3) yields a partial differential equation for the nonphysical metric and the conformal factor (don't worry at the moment where the conformal factor comes from). If we want to solve this equation near and on the conformal boundary, we get difficulties, because Ω vanishes there.

It is quite a surprising property of Einstein's field equations, that this problem can be circumvented. Introducing as unknowns the tensor fields

$$g_{\mu\nu}; \quad \Omega; \quad s \equiv \frac{1}{4}\nabla_\mu\nabla^\mu\Omega;$$
$$s_{\mu\nu} \equiv \frac{1}{2}(R_{\mu\nu} - \frac{1}{4}Rg_{\mu\nu}); \quad d^\mu_{\nu\lambda\rho} \equiv \Omega^{-1}C^\mu_{\nu\lambda\rho}, \tag{5}$$

one finds the

REGULARITY THEOREM (Friedrich (1979, 1981a,b):

Assuming that relation (1) and $\Omega > 0$ hold, then Einstein's field equations (2) for the unknown $\tilde{g}_{\mu\nu}$ is equivalent to the following "regular conformal field equations" for the unknown tensor fields (5):

$$R_{\mu\nu\lambda\rho} = \Omega d_{\mu\nu\lambda\rho} + 2\{g_{\mu[\lambda}s_{\rho]\nu} + s_{\mu[\lambda}g_{\rho]\nu}\} + \frac{1}{6}Rg_{\mu[\lambda}g_{\rho]\nu}, \tag{6}$$

$$\nabla_\mu d^\mu_{\nu\lambda\rho} = 0, \tag{7}$$

$$2\nabla_{[\lambda}s_{\rho]\nu} = \nabla_\mu\Omega d^\mu_{\nu\lambda\rho} + \frac{1}{12}g_{\nu[\lambda}\nabla_{\rho]}R, \tag{8}$$

$$\nabla_\mu\nabla_\nu\Omega = -\Omega s_{\mu\nu} + sg_{\mu\nu}, \tag{9}$$

$$\nabla_\mu s = -s_{\mu\nu}\nabla^\nu\Omega - \frac{1}{12}R\nabla_\mu\Omega - \frac{1}{24}\Omega\nabla_\mu R, \tag{10}$$

together with the "constraint equation"

$$\lambda = 6\Omega s - 3\nabla_\mu\Omega\nabla^\mu\Omega + \frac{1}{4}R\Omega^2. \tag{11}$$

Without going into the details of the derivation of the equations (6-11) from Einstein's field equation's (2) a few comments are appropriate to understand this statement better.

Eq. (11) is called the constraint equation, since it is a consequence of Eqs. (6-10) if it is satisfied at one point. The cosmological constant may thus be considered here as a constant of integration and we need to discuss in the following only Eqs. (6-10).

These equations are regular in the sense that no factor Ω appears in front of the principal parts and no factor Ω^{-1} appears on the right hand sides of the equations. They are equivalent to Einstein's equations (2) for negative as well as for positive values of Ω and represent thus the slightest possible generalization of Einstein's equations which is regular for all values of Ω. The regularity result extends also to Einstein's equations coupled to (conformally invariant) zero rest mass equations like, e.g., the Yang-Mills equations.

By introducing the conformal factor Ω, which up to now was completely arbitrary, a new "gauge freedom"

$$(\Omega, g_{\mu\nu}) \rightarrow (\Theta\Omega, \Theta^2 g_{\mu\nu}) \quad \text{with } \Theta \text{ a positive function}$$

has been introduced. It can be shown that near some initial surface the Ricci-scalar R can be prescribed arbitrarily and may thus be considered as the "gauge source function" for the conformal factor (Friedrich (1985)). With this interpretation Eqs. (6-10) may now be read as field equations for the unknowns (5).

To illustrate how special the conformal structure of Einstein's field equations is, we may compare them with two other types of field equations.

If we consider the conformally invariant space-time field equations where the Lagrangian density is given by $C_{\mu\nu\lambda\rho}C^{\mu\nu\lambda\rho}$ one finds spherically symmetric solutions which look similar to the Schwarzschild solution in standard Schwarzschild coordinates. By a similar procedure as in the Schwarzschild case one can construct a smooth "null infinity" for these solutions. But then one finds, that the Weyl tensor need not vanish on that null infinity and that the field equations are in fact satisfied everywhere through null infinity. This "null infinity" is in this case essentially meaningless.

If one studies on Minkowski space near null infinity solutions of the massive Klein-Gordon equations which extend smoothly to null infinity, one finds that these solutions vanish there to all orders and there is no regular extension of these equations to null infinity.

Einstein's equations show a conformal behavior which lies in a subtle way between these two extremes. If null infinity is smooth and satisfies the completeness condition described above, the conformal Weyl tensor vanishes on null infinity but the radiation field is free to vary (respectively, may be prescribed freely, as will be seen below). Furthermore the regular conformal field equations extend through null infinity. It is likely that any sharp results concerning the asymptotic behavior of solutions to Einstein's equations must take into account, directly or indirectly, this very particular behavior of the equations and their solutions.

To study the consequences implied by the possibility of having a regular extension of Einstein's equations on the structure of their solutions we need to know more about the nature of the regular conformal field equations. It is immediately seen that they are overdetermined and at first sight they do not resemble any of the systems of known type. It turns out that one can show the following theorem:

HYPERBOLICITY THEOREM (Friedrich (1979, 1981a,b, 1985, 1988b)):

The regular conformal field equations imply hyperbolic evolution equations which propagate the constraints.

To illustrate this statement I shall briefly sketch the procedure one has to go through (in principle, it is in fact sufficient to do it once and this has been done in the articles quoted above) if one wants to solve an initial value problem for the regular conformal field equations with data given on a space-like hypersurface S:

  – Choose a normalized spin frame $[\delta_a]_{a=0,1}$, express all tensors and equations with respect to this spin frame. Then the vector field τ with components

$$\tau^{aa'} = \epsilon_0^a \epsilon_{0'}^{a'} + \epsilon_1^a \epsilon_{1'}^{a'},$$

is timelike and we may assume that the spinframe has been chosen such, that τ is orthogonal to S. With the spinframe is associated a pseudo-orthonormal frame $e_{aa'} = \delta_a \overline{\delta}_a$, which has in a suitably chosen coordinate system x^μ the components $e_{aa'}^\mu = e_{aa'}(x^\mu)$. The introduction of the vectorfield τ allows us to perform the splitting

$$2\tau_b^{a'} \nabla_{aa'} = \epsilon_{ab} P + 2D_{ab}$$

of the directional covariant derivative operators $\nabla_{aa'}$ into the differential operators P, which acts in the direction of τ, and $D_{ab} = \tau_{(b}^{a'} \nabla_{a)a'}$, which act in directions orthogonal to τ.

  – We may now decompose the field equations algebraically with respect to τ. For example, the equation (7) which translates into the equation

$$\nabla_{a'}^f \phi_{abef} = 0$$

for the symmetric rescaled Weyl spinor ϕ_{abef} is then found to be algebraically equivalent to the "constraint equations"

$$D^{hf}\phi_{abhf} = 0 \tag{12}$$

and the "propagation equations"

$$\binom{4}{a+b+e+f}\{P\phi_{abef} - 2D^h_{(f}\phi_{abe)h}\} = 0, \quad a+b+e+f = 0, \ldots 4. \tag{13}$$

All the other equations may be split in a similar way into constraint and propagation equations. The binomial coefficients have been introduced in Eq. (13) because then the equations are symmetric hyperbolic (they are in fact also strictly hyperbolic). The complete system of propagation equations derived from the regular conformal field equation by this procedure is then symmetric hyperbolic.

– In the course of this splitting procedure there occur the "gauge source functions"

$$R, \qquad F^\mu = \nabla^{bb'}e^\mu_{bb'}, \qquad F_{ac} = \nabla^{bb'}\epsilon(\nabla_{bb'}\delta_a\delta_c),$$

in the equations, which may be prescribed arbitrarily.

– For (sufficiently smooth) data on S which solve the "conformal constraints" implied on S by the regular conformal field equations (these constraints are equivalent to the constraints implied by the Einstein equations) we may now assume that there exists a solution to the symmetric hyperbolic evolution equations. It can then be shown that this solution satisfies in fact the complete system (6-11).

I want to finish this discussion with a few comments. The spin frame formalism is not necessary but extremely convenient for the analysis of the equations and the detailed discussion of the solutions. I have described the splitting procedure in some detail, because it allows one to analyze the propagation properties of spinor equations in a simple and natural way and should be useful in many space-time contexts. In the spin frame formalism it is also easy to see, that one can derive from Eqs. (6-10) evolution equations of wave equation type. These equations have been considered, without the use of spinors, by Choquet-Bruhat and Novello (1987). There is no need to be particularly careful with the choice of the gauge source functions in a local problem but in a global problem their choice may be decisive. The symmetric hyperbolic form of the equations may also be achieved by fixing the gauge dependent quantities by other conditions, e.g. by choosing Gauss coordinates etc. (Friedrich (1983)). Furthermore characteristic initial value problems for the regular conformal field equations, where data are given on two transverse null hypersurfaces, can be reduced to initial value problems for symmetric hyperbolic systems in a similar way. Finally it may be pointed out, that the hyperbolicity result extends also to the conformal Einstein equations, coupled to the Maxwell-Yang-Mills equations and probably also to other zero rest mass equations (Cutler and Wald (1989); Friedrich (1988b)). One of the reasons which makes it so difficult to obtain global existence results in the case of Einstein's field equations is the fact that there is no conserved energy or pseudoenergy available. It may be worthwhile however to analyze in more detail the fact, that the integrand in the energy estimate obtained for the symmetric hyperbolic system (13) is (in the case where this equation is derived from the vacuum Bianchi identities) given by a certain component of the Bel-Robinson tensor.

5. The global conformal structure of solutions: Case $\lambda > 0$

In the following I want to discuss some of the results on the asymptotic behavior and on the global or semi-global existence, respectively, of solutions to Einstein's field equations which have been derived as a consequence of the properties of the equations presented above.

As we have seen above the conformal boundary is space-like in the case of solutions to Einstein's equations with positive cosmological constant. This suggests the use of past null and time-like infinity as a Cauchy surface for the regular conformal field equations. This leads to the following result.

SEMI-GLOBAL EXISTENCE (CASE $\lambda > 0$) (Friedrich (1986b, 1988b):

Given $\lambda > 0$, a 3-dimensional orientable, compact Riemannian space with manifold S and metric h_{ab} (and associated Levi-Cevita derivative operator D), and a symmetric h-trace free tensor d_{ab} such that $D^a d_{ab} = 0$ on S, then, provided all fields are sufficiently smooth, there exists a unique solution to Einstein's equations (2) on a manifold diffeomorphic to $S \times \mathbb{R}$ which has a smooth conformal boundary $\mathcal{I}^-$ diffeomorphic to S which represents past null and time-like infinity for that solution. The solution implies certain fields on $\mathcal{I}^-$ which after suitable diffeomorphic identification of $\mathcal{I}^-$ with S and possibly a suitable conformal rescaling

coincide with the given data h_{ab}, d_{ab} on S. A similar result is obtained
for the Einstein-Maxwell-Yang-Mills equations.

The fact that all null geodesics in these solutions are complete in
past directions justifies calling the solutions "semi-global." By simple
function-counting one might get the impression that the freedom to give
data is too large. It turns out however, that initial data sets as described
above which are conformally related in a certain sense lead to isomorphic
solutions. This is related to the fact, that one need not solve an analogue
of the usual Hamiltonian constraint if one prescribes asymptotic data
on $\mathcal{I}^-$. Apart from this difference the freedom to give data here is
essentially the same as in the standard Cauchy problem.

Though I want to concentrate in this article on the existence problem
I should like to point out a method which makes it extremely easy to
obtain more detailed information on the solutions near past conformal
infinity. By using "conformal geodesics" one can construct certain systems of physical Gauss coordinates which are adapted to $\mathcal{I}^-$ in a way
which considerably simplifies the translation of calculations made in the
non-physical space into the physical space. The physical space-time may
then be described as follows (Friedrich and Schmidt (1987)).

Suppose e_a, $a = 1, 2, 3$, is an h-orthonormal frame on S which has
components $e_a^\alpha = e_a(x^\alpha)$ with respect to a local coordinate system x^α,
$\alpha = 1, 2, 3$ on S. We express all tensors which may be calculated on
S from our data with respect to the frame e_a. Let R_{ab} be the h-Ricci
tensor and let Σ be positive such that $3\Sigma^2 = \lambda$. Then there is a system
of Gauss coordinates (t, z^α) for the physical metric, with t taking values
$< t_0$ for some $t_0 \in \mathbb{R}$, such that the slices $S_c = \{t = c = const\}$ approach
$\mathcal{I}^-$ in a uniform way if $t \to -\infty$. Furthermore there is a frame $\tilde{e}_a$ for the
tangent vectors on the surfaces S_t which together with $\tilde{e}_0 = \partial_t$ forms
an orthonormal frame for the physical metric such that the components
$\tilde{e}_a^\alpha = \tilde{e}_a(z^\alpha)$ of the vectors $\tilde{e}_a$ are given by

$$\tilde{e}_a^\alpha = \Sigma e_b^\alpha \{ h_a^b e^{\Sigma t} + \frac{1}{2}(R_a^b - \frac{1}{4} R h_a^b) e^{3\Sigma t} - \frac{1}{3} \Sigma e_a^b d^{4\Sigma t} \} + O(e^{5\Sigma t}) \quad (14)$$

for $t \to -\infty$ after setting $x^\alpha = z^\alpha$.

In the coordinate system used here it is quite simple to calculate
further coefficients in the expansion of $\tilde{e}_a^\alpha$ from the given data h and d.
It may be emphasized here that by the existence result above, (14) is
not a formal expansion but represents a Taylor expansion of a smooth
solution to Einstein's equations (2) if the data h and d are smooth and

in fact gives the first terms in an expansion of an analytic solution if the data are analytic.

If we try to extend the solutions considered above further into the future, we may expect singularities to develop. To show the existence of solutions which allow a smooth conformal boundary in the future as well as in the past, it appears therefore advisable to consider solutions which are "not too far away" from known solutions with smooth conformal boundary.

GLOBAL EXISTENCE, STABILITY OF $\mathcal{I}$ (Friedrich (1986c), (1988b)):

If an asymptotic initial data set (S, h, d) of the type considered in the previous theorem determines a solution of Einstein's equations (2) which admits a smooth conformal boundary in the future as well as in the past, then the same will be true for asymptotic initial data (S, h', d') which are "sufficiently close" to the data (S, h, d). A similar result holds for the Einstein-Maxwell-Yang-Mills equations.

A corresponding result is obtained by working with standard Cauchy data. Since the de Sitter solution has the described smooth conformal boundary, this result implies the existence of non-conformally flat solutions with smooth conformal boundary. These are truly global solutions since all null geodesics are complete. The "closeness" condition may be made precise in terms of Sobolev norms on the data.

The basic idea of the proof of this result is to apply standard stability results for solutions of symmetric hyperbolic systems to suitably formulated initial value problems for the conformal field equations. By using the semi-global existence results and the smoothness assumptions on the conformal boundary of the solution determined by the data (S, h, d), one may conclude that there exists an extension of that solution *as a solution of the regular conformal field equations* up to and beyond future conformal infinity. I shall call the resultant nonphysical space-time the "reference solution" and denote the components of its conformal boundary by $\mathcal{I}_r^+$ and $\mathcal{I}_r^-$. By the semi-global result the asymptotic initial data set (S, h', d') determines a solution of the regular conformal field equation which has a conformal boundary $\mathcal{I}^-$ in the past which together with certain fields implied on it by this solution may be diffeomorphically identified with (S, h', d'). when formulating the initial value problem for the regular conformal field equations to obtain this solution we may use the freedom to fix the coordinate gauge source function to construct a harmonic map from this solution onto the comparison solution, which maps $\mathcal{I}^-$ onto $\mathcal{I}^+_r$ and defines at least near $\mathcal{I}^-$ a diffeomorphism. By this map we may diffeomorphically carry all fields we want to construct onto the reference solution and reformulate our initial value problem in terms of the reference solution and the difference between the unknowns (5) which arise from the reference solution and the solution which we want to construct. By suitable choice of the remaining gauge source functions this can be done in such a way that the reference solution corresponds to vanishing data for our new initial value problem. The general stability properties of solutions for our new hyperbolic systems (Kato (1975)) and some particular properties of the regular conformal field equations allow one then to deduce the theorem.

The result stated above, which can be generalized in various directions, shows that the idea of a smooth conformal boundary is quite natural for solutions to Einstein's equations with positive cosmological constant.

6. The global conformal structure of solutions: Case $\lambda = 0$

There are a number of results which have been obtained in the case of vanishing cosmological constant but for reasons which will be discussed below the results are not as complete yet as in the previous case.

The formal expansions which are known as "Bondi type expansions" analyze on a formal level the "asymptotic characteristic initial value problem" where data are given on $\mathcal{I}^+$ (these data represent the "radiation field") and on an outgoing null hypersurface N which intersects $\mathcal{I}^+$ in a smooth space-like 2-sphere. If these data are analytic they determine in fact an analytic solution of the vacuum field equations near $N\cap\mathcal{I}^+$ which has a smooth "piece of a conformal boundary" (Friedrich (1982)). It is not necessary to make such strong smoothness assumptions.

EXISTENCE FOR ASYMPTOTIC CHARACTERISTIC INITIAL VALUE PROBLEM (Rendall (1988)):

Characteristic initial data on N and $\mathcal{I}^+$ of Class C^∞ for the regular conformal field equations determine a unique solution of Einstein's vacuum field equations of class C^∞ near $N\cap\mathcal{I}^+$ which has a smooth conformal boundary in the future.

Rendall analyzed characteristic initial value problems in much greater generality and discussed also weaker smoothness requirements.

While the existence result above shows, that Bondi-type expansions in general refer to "real" solutions of the Einstein equations, it does not tell us much about the global questions we wanted to answer. The following discussion is centered about a certain formulation of such a question, which may be considered as our "guiding problem."

Do there exist non-trivial "purely radiative space-times," i.e., smooth solutions of Einstein's vacuum field equations which have smooth, complete (in both senses discussed above) $\mathcal{I}^-$, $\mathcal{I}^+$ and regular i^-, i^+ ?

Since there would be no singularities which could be considered as the sources of the fields but the latter would be determined uniquely by the radiation field on $\mathcal{I}^- \cup i^-$ (say), such a space-time would have an unambiguous interpretation as representing pure gravitational radiation which comes in from infinity, interacts with itself in a nonlinear but smooth manner and escapes to infinity again. If we could give an answer to this question and, in case it were positive, even characterise those solutions in terms of standard Cauchy data, we would not only gain a deeper understanding of the notions of "gravitational radiation" compatible with Einstein's equations but also of the equations which govern the propagation of this radiation.

One could try to answer the question by finding an explicit solution which satisfies all requirements. Bičák and Schmidt (1989) have given a survey of boost-rotation-symmetric solutions which satisfy a part of the conditions by having a piece of a smooth, conformal boundary, regular i^-, i^+ and non-vanishing radiation fields, but all these solutions also have singularities. This is not surprising since Ashtekar and Xanthopoulos (1978) have shown that a purely radiative space-time admits at most a 1-dimensional space of Killing fields, unless it is Minkowski space. This makes the task of finding explicit solutions of the desired type quite difficult.

A general study of our problem can be tried in various ways. From the point of view of the physical interpretation of the solutions it would probably be most natural to solve the *"pure radiation problem"* where the radiation field is prescribed as initial data on the cone $\mathcal{I}^- \cup i^-$. As a first step one would like to prove a semi-global existence result by showing that these data determine a solution of Einstein's equations "near" the "initial surface" $\mathcal{I}^- \cup i^-$. In terms of the regular conformal field equations this would be a local existence result and it turns out that the decisive difficulty here is less the "semi-globality" than the presence of the vertex at i^-. Because of this more care must be taken in formulating regularity conditions on the data (which can be done) and for the solution. It follows that the radiation field determines a unique formal expansion type solution and that any solution for which the conformal fields (5) are of class C^1 is determined by the radiation field (Friedrich (1986a)). In the analytic case estimates can be given on the expansion coefficients which imply the existence of an analytic solution for these data near i^-.

The standard Cauchy problem for asymptotically flat initial data on R^3 offers another possibility for analyzing our question. In this approach, one may distinguish two subproblems. The "i^0-problem" asks which standard Cauchy data develop into space-times which admit "pieces of $\mathcal{I}^-$, $\mathcal{I}^+$ near i^0." Here i^0 is thought of as a point in the conformally extended space-time. The solution of this problem requires a much deeper analysis of the equations and the solutions than has been discussed up to now; the reason for this being that the fields (5) are partly non-smooth and partly even unbounded at i^0 unless the mass vanishes, i.e. unless the data represent flat space. If the i^0-problem admits non-trivial Cauchy data, one could construct in the corresponding solutions "hyperboloidal hypersurfaces;" i.e. space-like hypersurfaces which intersect $\mathcal{I}^+$ (say) in a space-like 2-sphere. I shall call the initial data for the regular conformal field equations implied on such a hypersurface "hyperboloidal initial data." The "hyperboloidal initial value problem" then asks which hyperboloidal initial data develop into solutions with smooth future null infinity $\mathcal{I}^+$ and regular future time-like infinity i^+. The analysis of the hyperboloidal problem yields the

REDUCTION OF THE PROBLEM OF THE EXISTENCE OF PURELY RA-DIATIVE SPACE-TIMES TO THE i^0-PROBLEM (Friedrich (1983, 1986c, 1988b)):

Smooth hyperboloidal initial data develop into solutions of Einstein's vacuum field equations $Ric[\tilde{g}] = 0$ which possess a smooth piece of $\mathcal{I}^+$ near the boundary of the initial surface. Hyperboloidal initial data sufficiently close to Minkowskian hyperboloidal initial data evolve into solutions with smooth conformal boundary $\mathcal{I}^+$ and regular time-like infinity i^+. Similar results hold for the Einstein-Maxwell-Yang-Mills equations.

I think it is quite remarkable that, provided a "smallness" assumption is made on the data to ensure that the lifetime of the solution is long enough, the regularity of i^+ is forced on the solution by the regular conformal field equations. The same property of these equations also has interesting consequences for solutions having a Cauchy surface with topology different from that of R^3 and it is likely that a deeper analysis of the regular conformal field equations will produce further such properties.

The behavior of the solutions near i^+ can again be described conveniently in terms of conformal geodesics (though an expansion of the field near i^+ is in the present case not of the same simplicity as (14) because of the more complicated structure of $\mathcal{I}^+$ (Friedrich and Schmidt (1987)). Recently Cutler (1989) discussed solutions with regular i^+ in more detail and generality. As an application of the result above the following has been obtained:

THE EXISTENCE OF SMOOTH SOLUTIONS OF THE EINSTEIN-MAXWELL EQUATIONS WITH SMOOTH COMPLETE $\mathcal{I}^-$, $\mathcal{I}^+$ AND REGULAR i^-, i^+ (Cutler and Wald (1989)):

There exist smooth time-symmetric initial data on R^3 for the Einstein-Maxwell equations with a conformally flat metric, vanishing electric field, and an axisymmetric magnetic field B with compact support which are Schwarzschild with positive mass near space-like infinity. For B chosen sufficiently small but $\neq 0$ these data develop into a solution with a smooth and complete conformal boundary, regular future and past time-like infinity and positive ADM-mass.

This is the first result on the existence of solutions of the field equations with the desired asymptotic behavior. Though it does not give us purely radiative vacuum solutions, it opens interesting possibilities for generalizations.

A more consequent use of the regular conformal field than that made in the results above may open still other ways to look for purely radiative space-times. One may ask, whether there exist smooth Cauchy data for the regular conformal field equations such that on the initial hypersurface there is a point i with the following property. At i the equations $\Omega = 0$, $d\Omega = 0$ hold; the 4-dimensional Hessian of Ω is proportional to the nonphysical metric at i, and Ω is positive on the rest of the initial hypersurface. I shall call such data "radiative initial data" for the following reason. If we evolve the data with the regular conformal field equations we find that $\Omega = 0$ but $d\Omega \neq 0$ on the future light cone of i, Ω is positive in the future of this cone and the metric obtained there yields after conformal rescaling a solution of Einstein's vacuum field equations, for which the future light cone of i represents past null infinity and i itself the regular past time-like infinity. One would like to think of this solution as being parts of purely radiative space times.

An analysis of the radiative initial data shows, however, that the conditions on Ω imply at i an infinite sequence of further conditions, called in the following "radiativity conditions," on some of the other conformal fields (5). It turns out that solution of the conformal constraint equations which satisfy all the radiativity conditions can be constructed and one way to do it, which gives some of them quite explicitly, is by exploiting the results by Beig and Simon (1981) on the analyticity of static solutions at space-like infinity.

EXISTENCE OF RADIATIVE INITIAL DATA (Friedrich (1988a)):

To any asymptotically flat static initial data set (of sufficient smoothness) can be associated a radiative initial data set such that the corresponding 3-metrics are conformal to each other. These initial data sets determine unique solutions of Einstein's vacuum field equations which have a smooth past null infinity and a regular past time-like infinity.

It can be shown that among those solutions to Einstein's equations are many which have not more than one symmetry. Furthermore it may be pointed out, that there is no "smallness assumption" needed here. The solutions considered above are all analytic. To work out their structure in detail it may be useful to study the complete analytic extension of the conformal spacetime. The picture which arises then is as follows. Starting from the radiative Cauchy data the conformal space-time evolves as a solution to the regular conformal field equations. In general there will be several points, i_1, i_2 say, on the initial surface where the radiativity condition is satisfied and there will also be a singularity. In the future of the points i_1, i_2 the "physical space-times" will come into existence and the singularity will evolve also into the future. If one follows the analytic extension of this solution further into the future the decisive question one would like to answer is: *Will the singularity by necessity penetrate into the physical space-times or will the physical space-times just "touch" each other and the singularity "at some point" (which possibly represents space-like infinity for the physical solutions) and later develop a smooth future null infinity?* If the latter were true it would give quite a new twist to the point of view of Einstein (1941), who though that an asymptotically flat solution of the vacuum field equations should have a singularity unless it were trivial. Here we would see that, in the extended conformal space-time smooth purely radiative space-times coexist peacefully with singularities which were hidden from the physical space-times by a horizon which represented null infinity for those space-times.

7. Conclusion

We have seen that the conformal Einstein equations possess somewhat unexpected regularity and hyperbolicity properties. Some of the consequences of these properties on the conformal structure of the solution of Einstein's equations in the large have been worked out. In the case of a positive cosmological constant global and semi-global existence results and detailed information on the asymptotic structure of the solutions have been obtained and we have seen that the idea of the conformal boundary at infinity is in perfect harmony with the requirements of the field equations. In the case of a vanishing cosmological constant the problem of the global existence and smooth asymptotic behavior, which has been rephrased here as the question of the existence of purely radiative space-times has been reduced to the i^0-problem and a number of space-times which may eventually show the desired behavior have been shown to exist. The key open problem is the i^0-problem. Its solution should lead to a more complete picture of what an "isolated gravitating system" looks like, to a more concise notion of "radiation" in general relativity, and to a deeper insight into "how Einstein's equations work." This in turn should have consequences for the design of approximation procedures and on numerical relativity, where numerical calculations could then be performed in a finite picture but right up to the infinity of the physical space-time.

References

Ashtekar, A. (1984), in *Gen. Relat. Grav.*, ed. B. Bertotti et al., (Reidel, Amsterdam).

Ashtekar, A., and Xanthopoulos, B.C. (1978), *J. Math. Phys.* **19**, 2216.

Beig, R. (1981), *Acta Physica Austriaca* **53**, 249.

Beig, R., and Simon, W. (1980), *Commun. Math. Phys.* **78**, 75.

Bičák, J., and Schmidt, B. G. (1989), *Phys. Rev.* **D**, in press.

Blanchet, L., and Damour, T. (1986), *Phil. Trans. Roy. Soc. Lond.* **A320**, 379.

Bondi, H., van der Burg, M. G., and Metzner, A. W. K. (1962), *Proc. Roy. Soc. Lond.* **A269**, 21.

Choquet-Bruhat, Y., and York, J. W. (1980), in *General Relativity and Gravitation*, ed. A. Held (Plenum, New York).

Choquet-Bruhat, Y., and Novello, M. (1987), *C. R. Acad. Sci. Paris* **305**, Serie II, 155.

Christodoulou, D. (1986), *Commun. Math. Phys.* **105**, 337.

Cutler, C., and Wald, R. M. (1989), *Class. Quantum Grav.* **6**, 453.

Cutler, C. (1989), *Class. Quant. Grav.*, in press.

Einstein, A. (1941), *Univ. Nac. Tucuman, Revista* **A2**, 5.

Friedrich, H. (1979), in *Proceedings of the Third Gregynog Relativity Workshop*, ed. M. Walker, Max-Planck green rep. MPI-PAE/Astro, 204.

Friedrich, H. (1981*a*), *Proc. Roy. Soc. Lond.* **A375**, 169.

Friedrich, H. (1981*b*), *Proc. Roy. Soc. Lond.* **A378**, 401.

Friedrich, H. (1982), *Proc. Roy. Soc. Lond.* **A381**, 361.

Friedrich, H. (1983), *Commun. Math. Phys.* **91**, 445.

Friedrich, H. (1985), *Commun. Math. Phys.* **100**, 525.

Friedrich, H. (1986*a*), *Commun. Math. Phys.* **103**, 35.

Friedrich, H. (1986*b*), *J. G. P.* **3**, 101.

Friedrich, H. (1986*c*), *Commun. Math. Phys.* **107**, 587.

Friedrich, H. (1988*a*), *Commun. Math. Phys.* **119**, 51.

Friedrich, H., (1988*b*), Max-Planck green rep. MPA, 402.

Friedrich, H., and Schmidt, B. G. (1987), *Proc. Roy. Soc. Lond.* **A414**, 171.

Kato, T. (1975), *Arch. Ration. Mech. Anal.* **58**, 181.

Klainerman, S. (1988), in *Mathematics and Relativity*, ed. J. A. Isenberg (AMS, Providence).

Moncrief, V. (1987), *Class. Quantum Grav.* **4**, 1555.

Newman, E. T., and Penrose, R. (1962), *J. Math. Phys.* **3**, 566.

Penrose, R. (1963), *Phys. Rev. Lett.* **10**, 66.

Penrose, R. (1965), *Proc. Roy. Soc. Lond.* **A284**, 159.

Penrose, R. (1980), in *Essays in General Relativity*, ed. F. J. Tipler (Academic Press, New York).

Rendall, A. (1989), *Proc. Roy. Soc. Lond.* **A**, in press.

Sachs, R. K. (1962), *Proc. Roy. Soc. Lond.* **A270**, 103.

4

Progress in 3D
numerical relativity

Takashi Nakamura
National Laboratory for High Energy Physics,
Oho, Tsukuba-shi, Ibaraki-ken, 305, Japan

1. Inflationary supercomputing

There are many problems in numerical relativity. From a view point of asymptotic behavior of space-times they are divided into three categories as:

V: Vacuum space-times;

AF: Asymptotically flat space-times with matter;

ANF: Asymptotically non-flat space-times with matter.

On the other hand from a view point of dimension of the problem they are divided into another three categories as:

1D: A problem in which all the physical quantities depend on one spatial coordinate x^1 and the time t;

2D: A problem in which all the physical quantities depend on two spatial coordinates x^1 and x^2 and the time t;

3D: A problem in which all the physical quantities depend on three spatial coordinates x^1, x^2 and x^3 and the time t.

Every problem can be characterized according to these two classifications. For example, the two black hole collision calculated by Smarr[1] belongs to V-2D. There are also many numerical methods for each problem. They include the characteristic method, Regge calculus, finite element method, finite difference method, spectral method, particle method and so on. Moreover the choice of coordinate conditions is not usually unique for each method. Therefore there are a tremendous variety of possibilities for each author's problem. In reality there are up to 30 papers on numerical relativity in this conference which are reviewed by Centrella[2] in this volume.

One of the main factors, however, which determines progress in numerical relativity is the power of available computers. The speed of calculations and the size of memories have been increasing. Roughly speaking the maximum speed of super computers as a function of time t in the Christian Era can be expressed by

$$\text{Speed} = 10 \, \text{MFLOPS} \cdot \text{Exp}(H(t - 1970)), \qquad (1.1)$$

and

$$H = 0.3 \, \text{year}^{-1},$$

where MFLOPS means 10^6 floating decimal operations per second. Eq. (1.1) is similar to the expansion law of the de-Sitter universe with a Hubble constant $0.3 \, \text{year}^{-1}$. So the power of supercomputers is increasing in inflationary fashion as in the early universe and the characteristic time scale of the increase of the power of supercomputer is $H^{-1} = 3$ years. For $t = 1989$ the speed becomes $3 \, \text{GFLOPS}$. In reality there are several supercomputers with a peak speed of ~ 3 GFLOPS and a memory size of ~ 500 Mbytes, where 8 bytes = 1 word = 1 memory for a floating decimal.

What does the existence of such machines means to numerical relativity? Let us consider solving the Einstein equations for a perfect fluid under general initial conditions, that is, initial conditions with no symmetries. Then how many grids can one use with a 500 Mbyte machine? For such a general problem one needs ~ 50 variables including the auxiliary memories necessary for numerical computations. However with 500 Mbyte memories one can use $\sim 100^3$ grids. This number of grids is not so small since 2D numerical relativity started with $\sim 30^2 - 40^2$ grids.

Here I mainly report on recent progress in 3D numerical relativity. The main goal here is:

Construct **general** *general relativistic code and*

solve $G_{\mu\nu} = 8\pi T\mu\nu$ *for*

any *initial conditions as initial value problems.*

2. 3D codes

One of the major difficulties in constructing axially symmetric codes in numerical relativity is how to avoid coordinate singularities in cylindrical and spherical polar coordinates. Let us consider a point on the z-axis in cylindrical coordinates. This point is clearly a single point in the real space but is assigned to many points in the coordinate space, that is, to arbitrary values of ϕ. In order to guarantee that this is a single point in the real space, one can easily show that certain relations called regularity conditions should be satisfied by components of metric tensors such as

$$\gamma_{RR} = \gamma_{\phi\phi}/R^2 . \tag{2.1}$$

However in a naive finite difference method it is very difficult to guarantee the above relation numerically due to truncation errors. The violation of regularity conditions usually yields numerical instabilities. Physically the violation of regularity on the axis means that there is a true physical singularity. This situation is essentially the same in the Newtonian gravitational collapse problem. However in numerical relativity the difficulty becomes worse due to the second spatial derivative terms in the evolution equations of the extrinsic curvatures. One of the methods of avoiding this difficulty is to use the regularized variable defined by

$$g = (\gamma_{RR} - \gamma_{\phi\phi}/R^2)/R^2 . \tag{2.2}$$

If one uses g and γ_{RR} instead of γ_{RR} and $\gamma_{\phi\phi}$, regularity is automatically satisfied.

Now in non-axially symmetric cases, what are the regularity conditions? If one uses (x, y, z) coordinates, each coordinate point has a one to one relation to a point in the space-time and there are no problems concerning regularity. However if one uses spherical polar coordinates, for example, the transformation becomes as follows:

$$\gamma_{rr} = \gamma_{xx} \sin^2\theta \cos^2\phi + \gamma_{xy} \sin^2\theta \sin 2\phi + \gamma_{yy} \sin^2\theta \sin^2\phi$$

$$+ \gamma_{yz} \sin 2\theta \sin\phi + \gamma_{zz} \cos^2\theta + \gamma_{zx} \sin 2\theta \cos\phi .$$

Let us consider the origin $r = 0$, for example. As $\gamma_{xx}, \gamma_{yy}, \gamma_{zz}, \gamma_{xy}, \gamma_{yz}$ and γ_{zx} have definite independent values even at the origin, the above relation tells us that the metric tensor in spherical polar coordinates should depend on θ and ϕ even at the origin to guarantee the regularities. So it will be very difficult to find the regularized variables like Eq. (2.2). The only variables which behave well at $r = 0$ and on the axis seem to be $\gamma_{xx}, \gamma_{yy}, \gamma_{zz}, \gamma_{xy}, \gamma_{yz}$ and γ_{zx} although there is no proof for this statement.

Stark[3] is now extending his 2D code[4] using Bardeen-Piran coordinate conditions to 3D. He is using (r, θ, ϕ) coordinates with $80 \times 12 \times 16$ grids. His code at present is undergoing testing. He is trying to overcome the above difficulties. However it remains to be seen whether his approach is adequate or not. He is putting in a pressure reduced TOV polytrope with rotation added. With sufficient rotation numerical roundoff will be expected to break the axisymmetry and lead to the growth of the unstable non-axisymmetric modes.

The above consideration suggests that we had better use $\gamma_{xx}, \gamma_{yy}, \gamma_{zz}, \gamma_{xy}, \gamma_{yz}$ and γ_{zx} as basic variables. Then there is no merit in writing down the Einstein equations in (r, θ, ϕ) coordinates. However unless one has a supercomputer with memory greater than 200 Mbytes or so, it is impossible to use (x, y, z) coordinates as grids. This is a dilemma. We must use (r, θ, ϕ) as grids due to the lack of sufficient memory while we must adopt $\gamma_{xx}, \gamma_{yy}, \gamma_{zz}, \gamma_{xy}, \gamma_{yz}$ and γ_{zx} as basic variables to avoid coordinate singularities. A naive method of overcoming this difficulty may be to convert $\partial/\partial x, \partial/\partial y, \partial/\partial z$ to finite difference versions of $\partial/\partial r, \partial/\partial \theta, \partial/\partial \phi$. In reality I tried this method but encountered serious numerical instabilities on the axis due to zero-divided-by-zero type terms. So we need a more sophisticated method.

Every quantity Q in (x, y, z) coordinates can be expressed as

$$Q = \sum_{l,m} r^l Q_{lm}(r^2, t) Y_{lm}(\theta, \phi). \qquad (2.3)$$

One possible method is to use $Q_{lm}(r^2, t)$'s as the basic variables of the Einstein equations. The merit here is that we can get rid of the coordinate singularities using the analytic properties of $Y_{lm}(\theta, \phi)$. To estimate the effect of non-linear terms on the evolution of $Q_{lm}(r^2, t)$, the integral with respect to solid angle of the source term is needed. In reality Nakamura, Oohara and Kojima[5] solved 3D pure gravitational waves using this method and obtained good results. Marck and Bonazolla[6] are now proceeding further. They expand $Q_{lm}(r^2, t)$ by Tchebychef polynomials $T_n(r)$ as

$$Q = \sum_{n,l,m} a_{nlm}(t) T_n(r) Y_{lm}(\theta, \phi). \qquad (2.4)$$

In their method $a_{nlm}(t)$'s are the basic variables and they are now developing a 3D code.

On the other hand Wilson[7] is now considering binary neutron stars using 40^3 grids in (x, y, z) coordinates. In (x, y, z) coordinates there are no problems concerning regularities. Wilson assumes that the metric tensor γ_{ij} is equal to $g\delta_{ij}$. g is determined by solving the Hamiltonian constraint equation. The lapse (α) is determined from the maximal slicing condition. The shift β_i is determined by

$$(\beta_i^{|j})_{|j} + (\beta_{|i}^j)_{|j} - \frac{2}{3}(g\beta_{|j}^j)_{|i} = (2\alpha K_i^j)_{|j}. \qquad (2.5)$$

The form of the metric tensor assumed in Wilson's method is not consistent for fully general relativistic cases. However if one considers the semi-relativistic situation with low luminosity of the gravitational waves, Wilson's approach may be valid.

Nakamura and Oohara[8] are developing a fully general relativistic code in (x, y, z) coordinates. At present they are using $\alpha = 1$ and $\beta^i = 0$. Since there are no assumptions on the form of metric tensor in this case, the code construction is very complicated. It is very possible that many mistakes will be made somewhere in writing basic equations, finite difference versions of equations, typing FORTRAN source code and so on.

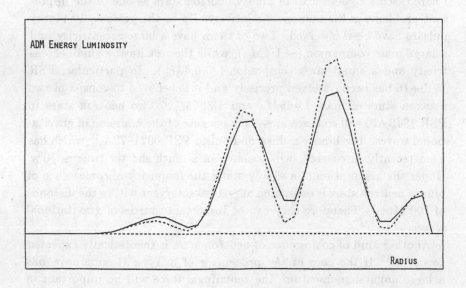

ADM ENERGY LUMINOSITY

RADIUS

Fig. 1. ADM energy luminosity as a function of distance from the center. The solid and dashed lines show numerical and analytic results, respectively.

To avoid these difficulties they use the algebraic software package RE-DUCE by which one can obtain FORTRAN source code directly without errors. In a sense a computer itself makes a program for numerical computations. This kind of technique is indispensable for development of 3D numerical relativity to make calculations free from errors. As a first demonstration of 3D code, the time evolution of the Teukolsky wave ($l = m = 2$) with 80^3 grids is shown in Figure 1.[8] The CPU time for the entire evolution, which has typically 3000 time steps is about three hours using a Hitachi S810. In Fig. 1 a solid line shows the computed ADM energy luminosity as a function of the distance from the center. Here ADM energy luminosity means the surface integral of the ADM energy flux at a constant radius. A dashed line shows the ADM energy luminosity estimated from the analytic solutions. Fig. 1 shows globally that solid lines agree with dashed ones. This means that the propagation of the gravitational waves are successfully simulated by this 3D metric code including all of the non-linear terms of the Einstein equations.

3. Coalescence of binary neutron stars

I here discuss coalescences of binary neutron stars as one of the important problems in 3D numerical relativity. Up to the present, 10 binary pulsars have been observed. Two of them have a large eccentricity and a large mass companion ($\sim 1.4 M_\odot$), while the rest have a small eccentricity and a small mass companion ($\leq 0.4 M_\odot$). In particular, PSR 1913+16 has been observed precisely and it is believed to consist of two neutron stars of mass $1.445 M_\odot$ and $1.384 M_\odot$.[9] Two neutron stars in PSR 1913+16 will coalesce in $\sim 10^8$y because of the emission of gravitational waves. The binary millisecond pulsar PSR 0021-72A,[10] which has been recently discovered, will coalesce in a much shorter time $\sim 10^6$y. Under the assumption of a steady state, the frequency of coalescence of binary neutron stars is estimated at ~ 10 events/year within the distance of 100 Mpc.[11] Therefore they can be important sources of gravitational waves.

Another kind of coalescence of neutron stars is theoretically expected to occur.[12] If the core of the progenitor of a Type II supernova has a large angular momentum, the centrifugal force will be important in some stage of the collapse into a final neutron star. The core radius where the centrifugal force is comparable to the gravitational force is proportional to the square of the angular momentum. When the size

of the core decreases to this radius, the core contracts principally along
the rotational axis and then a thin disk will be formed. Such a thin
disk is known to be gravitationally unstable irrespective of the equation
of state,[13] and it fragments into several pieces in a free fall time scale.
Each fragment looks like a neutron star and is called a proto neutron
star. Proto neutron stars will coalesce again to form a single neutron
star owing to the emission of gravitational waves. If the number of
fragments is two, the system is essentially the same as a binary neutron
star like PSR 1913+16 in the final coalescence stage. A time profile
of neutrino events,[12] a sub-millisecond pulsar[14] and the existence of a
Jupiter-like secondary[15] suggests that this kind of process really occurred
in SN 1987A. If a scenario like this applies to a large fraction of Type II
supernovae, the frequency in the events of burst emission of gravitational
waves within the distance of 10 Mpc increases to ~30 events/year. Then
it is urgent to estimate the amount, the wave pattern and the spectrum
of the gravitational waves emitted in coalescence of a binary neutron
star.

Since a fully general relativistic code including the matter does not
exist at present, we use a Newtonian 3D hydrodynamics code to assess
the total energy and the wave pattern of the gravitational waves as
well as the final destiny of coalescing binary neutron stars. To take
into account the effect of radiation reaction, we just add a radiation
reaction potential, which is expressed by the fifth time derivatives of the
quadrupole moment to the Newtonian potential. Since there are many
formalisms for the radiation reaction of the gravitational waves,[16] the
calculations in this article should be considered as a first step towards
solution of the coalescing problem.

4. Numerical results

The basic equations are the three-dimensional hydrodynamics equations.
The radiation reaction potential is expressed as

$$\psi_{react} = 0.2(\frac{d^5}{dt^5}D_{ij})x^i x^j . \tag{4.1}$$

To express a hard equation of state, we use a polytropic equation of
state with $\gamma=2$. We put in initially two spherical neutron stars of mass
M_0 and radius r_0 at $y = \pm r_0$ on the y-axis, which means that we start
calculations when two neutron stars just contact each other. As for the

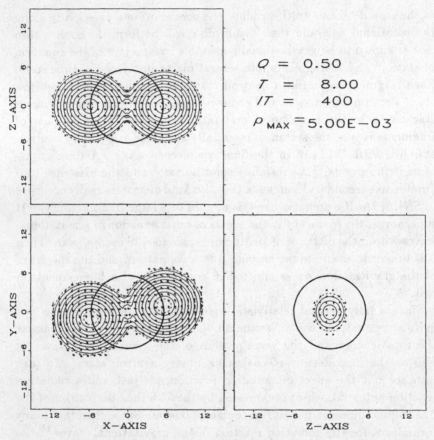

$$Q = 0.50$$
$$T = 8.00$$
$$IT = 400$$
$$\rho_{MAX} = 5.00E-03$$

Fig. 2(a)-2(m).   Density contour on the x-y, y-z and z-x planes for $q = 0.5$. The time t and time step number IT are indicated. Solid lines are drawn at steps of one tenth of the maximum density and the inner and the outer dashed lines indicate, respectively, 19/20 and 1/20 of the maximum density. Arrows indicate the velocity vectors of the matter. A fat line shows a circle of radius $4GM_0/c^2$ to show the size of a spherical black hole for comparison.

initial velocity we assume that

$$v_x = -y\Omega, \quad v_y = x\Omega \quad \text{and} \quad \Omega = 0.5\sqrt{\frac{GM_0}{r_0^3}}\, q, \qquad (4.2)$$

where q is a parameter used to specify the total angular momentum of the system. Thus the system is assumed to be rigidly rotating with respect to the origin $x = y = z = 0$. We take the units of

$$M = M_\odot, \quad L = \frac{GM_\odot}{c^2} = 1.5\,\text{km}, \quad T = \frac{GM_\odot}{c^3} = 5 \times 10^{-6}\,\text{sec}. \quad (4.3)$$

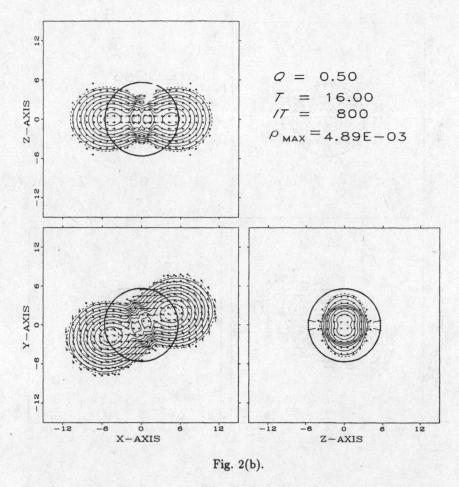

$$Q = 0.50$$
$$T = 16.00$$
$$IT = 800$$
$$\rho_{MAX} = 4.89E-03$$

Fig. 2(b).

To describe neutron star appropriately, we fix $r_0 = 6$ in our units, which means the radius of the neutron star is 9 km irrespective of mass M_0.

We will estimate the amount of gravitational radiation emitted using the quadrupole formula, which requires the third time derivatives of the quadrupole moment. Using the continuity equation, we first reduce the first time derivative of the quadrupole moment to[17]

$$\frac{d}{dt}Q_{ij} = \int \rho(x^i v^j + x^j v^i)dV\,, \qquad (4.4)$$

where Q_{ij} is defined by

$$Q_{ij} = \int \rho x^i x^j dV\,.$$

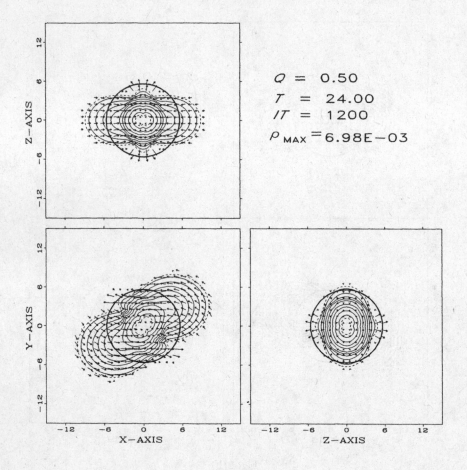

$$Q = 0.50$$
$$T = 24.00$$
$$IT = 1200$$
$$\rho_{MAX} = 6.98E-03$$

Fig. 2(c).

Next using the equation of motion[17] we can express the second time derivative of the quadrupole moment as

$$\frac{d^2}{dt^2}Q_{ij} = \int \left\{ 2\rho v^i v^j + 2P\delta_{ij} - \rho\left(\frac{\partial \psi}{\partial x^i}x^j + \frac{\partial \psi}{\partial x^j}x^i\right) \right\} dV . \qquad (4.5)$$

In Eq. (4.5), ψ is contained in the integrand but the integral has compact support because ρ is multiplied by the potential terms.

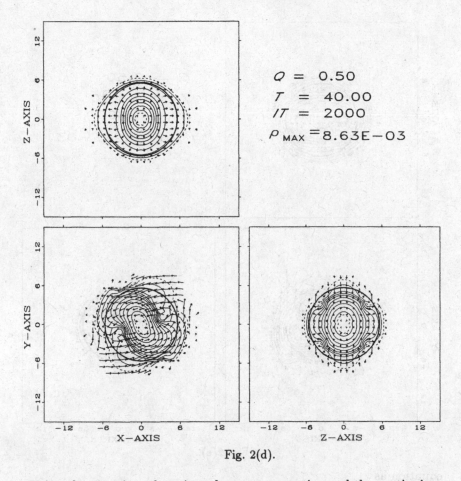

Fig. 2(d).

Using the equation of motion, the energy equation and the continuity equation, we rewrite the third time derivatives as

$$\frac{d^3}{dt^3}Q_{ij} = \int \left[2P(\frac{\partial v_i}{\partial x^j} + \frac{\partial v_j}{\partial x^i}) - 2(\gamma - 1)P\,\mathrm{div}(\mathbf{v})\delta_{ij} \right.$$
$$- 2\rho(v_i \frac{\partial \psi}{\partial x^j} + v_j \frac{\partial \psi}{\partial x^i}) + \mathrm{div}(\rho\mathbf{v})(x^i \frac{\partial \psi}{\partial x^j} + x^j \frac{\partial \psi}{\partial x^i})$$
$$\left. - \rho(x^i \frac{\partial \dot\psi}{\partial x^j} + x^j \frac{\partial \dot\psi}{\partial x^i}) \right] dV.$$

(4.6)

Except for $\dot\psi$, all the quantities with time derivatives are transformed to those only with spatial derivatives. So they are not noisy but have definite values. As for $\dot\psi$ one may perform the numerical time difference of ψ. But we can derive a new Poisson equation for $\dot\psi$ using the continuity

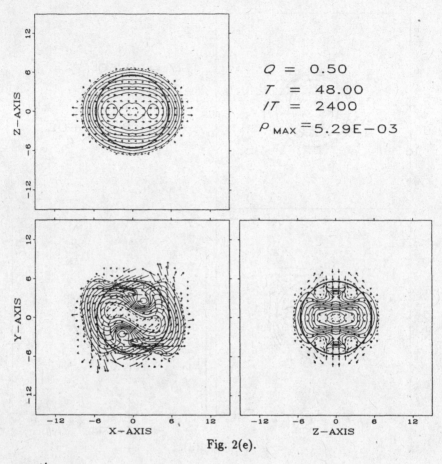

$$Q = 0.50$$
$$T = 48.00$$
$$IT = 2400$$
$$\rho_{MAX} = 5.29E-03$$

Fig. 2(e).

equation as

$$\Delta\dot{\psi} = -4\pi G \mathrm{div}(\rho\mathbf{v}). \tag{4.7}$$

Since the r.h.s of Eq. (4.7) has no time derivatives, $\dot{\psi}$ is solved for without noise using the same numerical code as for ψ.

We can repeat the above procedure to determine d^4Q_{ij}/dt^4. This time we must solve another poisson equation for $\ddot{\psi}$ which satisfies

$$\Delta\ddot{\psi} = 4\pi G \frac{\partial}{\partial x^i}\left[\frac{\partial \rho v^i v^l}{\partial x^l} + \frac{\partial P}{\partial x^i} + \rho\frac{\partial\psi}{\partial x^i}\right]. \tag{4.8}$$

We determine d^5Q_{ij}/dt^5 through a numerical time difference of the quantity d^4Q_{ij}/dt^4. So we can compute ψ_{react} through the integration with compact support and the numerical time difference. This may seem tricky at first. One may not believe that the fifth time derivatives can be

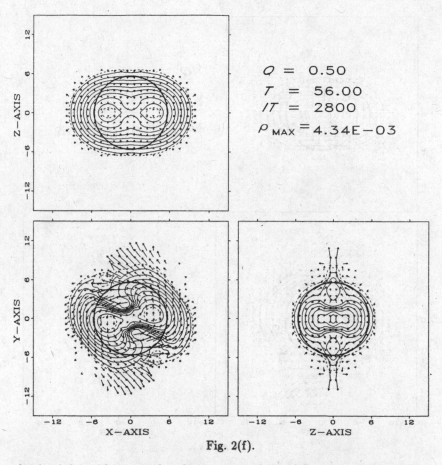

$Q = 0.50$

$T = 56.00$

$IT = 2800$

$\rho_{MAX} = 4.34E-03$

Fig. 2(f).

obtained from the second order accurate finite difference scheme. However in Oohara and Nakamura[18] we checked this method by computing the fifth time derivatives of the quadrupole moment for the collapse of the homogeneous ellipsoid using our 3D code. We compared the results with semi-analytic solutions obtained by solving the ordinary differential equations and confirmed that the results are satisfactory. One can obtain accurate estimates of the fifth time derivatives if one pays for more CPU time for two extra Poisson equations. A numerical scheme for hydrodynamics equations is the same as in Oohara and Nakamura.[18] We solve three Poisson equations Eqs. (2.5), (4.7), and (4.8) by an ICCG method described in Oohara and Nakamura[19] under appropriate boundary conditions.

We take a $141 \times 141 \times 141$ grid. We performed two simulations with $q = 0.35$ and 0.5. A typical CPU time per one time step is about 8

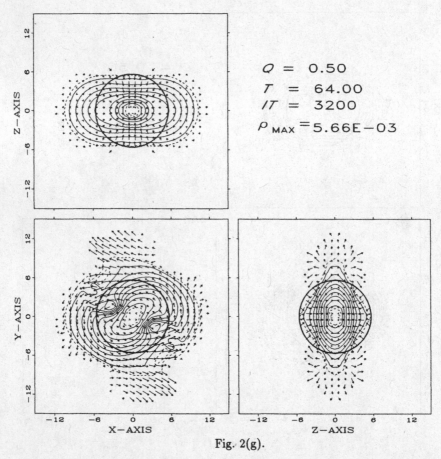

$Q = 0.50$
$T = 64.00$
$IT = 3200$
$\rho_{MAX} = 5.66E-03$

Fig. 2(g).

seconds using a Hitachi S820/80, because we must solve three Poisson equations at each time step. We performed the numerical calculation for each model until $t \sim 600$ in our units and the total CPU time needed for one model is typically 80 hours. We set $M_0 = 1.4 M_\odot$ for all models. Therefore the total mass is $2.8 M_\odot$, which means that each model looks like the final stage of a binary neutron star system such as PSR 1913+16, which is believed to consist of two neutron stars of mass $1.445 M_\odot$ and $1.384 M_\odot$.[9] We show in Figs. 2(a)-2(m) contours of density and velocity vectors on the x-y, y-z and z-x planes for $q = 0.5$. In each figure, solid lines are drawn at intervals of one tenth of the maximum density. Inner and outer dashed lines indicate, respectively, 19/20 and 1/20 of the maximum density. A fat line is a Schwarzschild radius of the black hole of mass $2M_0$. If the binary becomes the spherical black hole, all the matter will be swallowed inside this circle. Fig. 2(a) is the almost

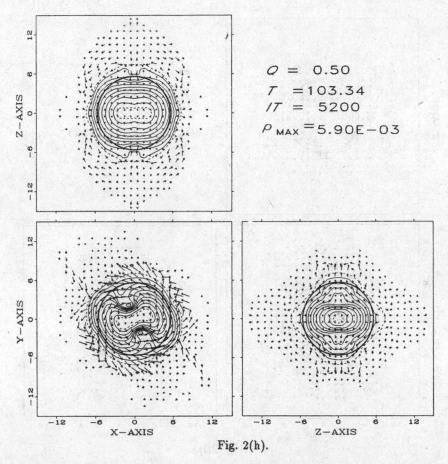

$$Q = 0.50$$
$$T = 103.34$$
$$IT = 5200$$
$$\rho_{MAX} = 5.90E-03$$

Fig. 2(h).

initial stage when two neutron stars just contact each other. At $t = 16$
(Fig. 2(b)) two stars rotate by about 30 degrees and coalescence begins
in the central region on account of the gravitational tidal force. The
views of y-z and z-x planes show that the expanding motion along z-axis
appears in the coalescing region. At $t = 24$ (Fig. 2(c)) the coalescence
seems to be completed in the central region temporarily. The rate of
contraction in the central region is greater than in the outer region and
then the angular velocity of the central part increases by the conservation
of angular momentum. Consequently the shape of the density contour
in the inner part rotates in advance of the outer part. In the views
of y-z and z-x planes, we see the expansion along the z-axis and the
contraction along x and y axis. At $t=40$ (Fig. 2(d)), however, the
central part begins to expand. The contraction in the previous stage
makes the centrifugal force increase and the centrifugal force overcomes

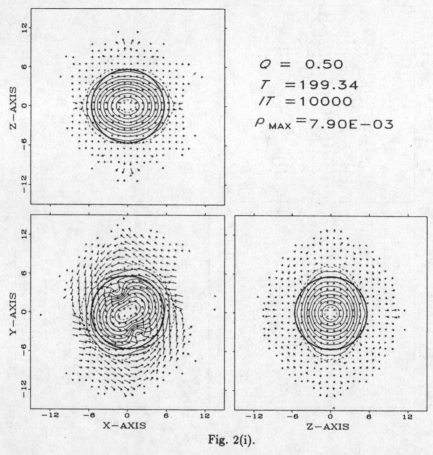

$$Q = 0.50$$
$$T = 199.34$$
$$IT = 10000$$
$$\rho_{MAX} = 7.90E-03$$

Fig. 2(i).

gravity during this stage. This expansion causes the angular velocity
to decrease. Then the contours of density have almost the same phase.
At the same time the expansion of velocity along the z-axis decreases.
The expansion of the central part continues at the time of Fig. 2(e).
In this stage the expanding motion along z-axis almost stops. Finally
each neutron star appears again at $t = 56$ (Fig. 2(f)). The motion
along the z axis reverses its direction. In the next stage at $t = 64$ (Fig.
2(g)) the coalescence temporarily ceases again because of the decrease
of the centrifugal force and we see the contraction along the z-axis and
expansion along the x-axis and y-axis. Then almost the same things as
in the previous stage proceed. At $t = 103$ (Fig. 2(h)), however, we see
the dumbbell shape coalesced binary, which is different from Fig. 2(f).
This kind of oscillation continues several times and at each oscillation
the dumbbell shape (Fig. 2(i)) becomes more and more faint. Finally

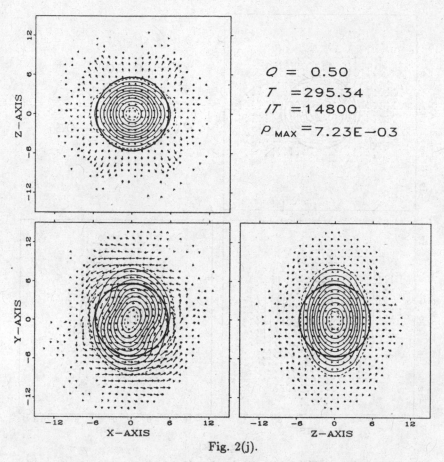

$$Q = 0.50$$
$$T = 295.34$$
$$IT = 14800$$
$$\rho_{MAX} = 7.23E{-}03$$

Fig. 2(j).

the system settles down to be a rotating bar shown as in Fig. 2(j) at $t \sim 300$. This is due to both the back reaction effect and the re-distribution of the angular momentum of each fluid element since the specific angular momentum is not conserved in a non-axially symmetric dissipative system. After this stage, the shape becomes more and more axially symmetric due to the loss of angular momentum and the energy by the gravitational waves (Figs. 2(k) to 2(m)). In the final stage of our numerical simulation (Fig. 2(m)) almost all the matter seems to be inside the Schwarzschild radius. Since the total angular momentum and the total energy of the system in the final stage are 1.45 and -0.527, respectively in our units, we expect the value of $q (\equiv J/M_{tot}^2 = a/m)$ of the final black hole is 0.28 ($= 1.45/(2.8 - 0.527)^2$). $-T/W$ is also as small as 0.035 where T and W are the kinetic and the gravitational energy, respectively. This is due to the loss of the angular momentum

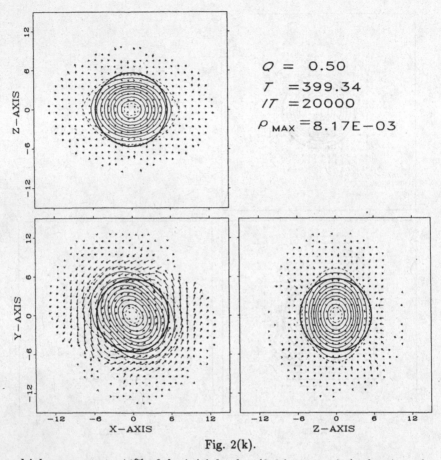

Fig. 2(k).

which amounts to 44% of the initial value (2.6 in our units), that is, only 56% of the initial angular momentum is left. Therefore the final black hole is far from the extreme Kerr black hole.

In Figure 3 we show the energy flux and the central density ρ_c as functions of time. The central density ρ_c begins to increase when the coalescence starts but the increase stops at $t \sim 35$ (Fig. 2(d)-(e)). At $t=64$ (Fig. 2(g)), ρ_c starts to increase again. After several such oscillations, the variation of ρ_c becomes rather regular up to $t \sim 400$. The energy flux shows behavior similar to ρ_c. The luminosity L for $t \geq 400$ empirically obeys the relation as

$$L \propto \mathrm{Exp}(-\frac{t}{100}). \qquad (4.9)$$

So $\tau_{damp} \sim 100$ is the characteristic time of the decrease of the luminosity as the system becomes more and more axially symmetric. This

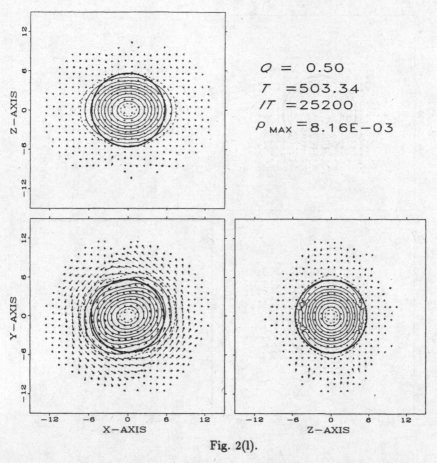

$Q = 0.50$
$T = 503.34$
$IT = 25200$
$\rho_{MAX} = 8.16E-03$

Fig. 2(l).

exponential decline of the luminosity can be seen in the other simulation ($q = 0.35$) with $\tau_{damp} \sim 120$. The total energies emitted are 0.063 and 0.043, for $q = 0.5$ and 0.35, respectively.

The wave form of the gravitational radiation is obtained from the second time derivative of the quadrupole moment. We show the wave form observed on the z-axis at 10 Mpc in Figure 4. The wave pattern consists of three parts; 1) the transient oscillation and rotation for $t \leq 1$ msec, 2) the regular pattern for 1 msec $\leq t \leq 2$ msec and 3) the damping oscillation for $t \geq 2$ msec. An important impression from the wave form is that the damping of the amplitude is not so large compared with the damping of the luminosity in Fig. 3. This is due to the fact that the luminosity is proportional to the square of the amplitude. Therefore from the observational point of view, one can expect the longer time observability of the low luminosity source for the fixed total radiated energy.

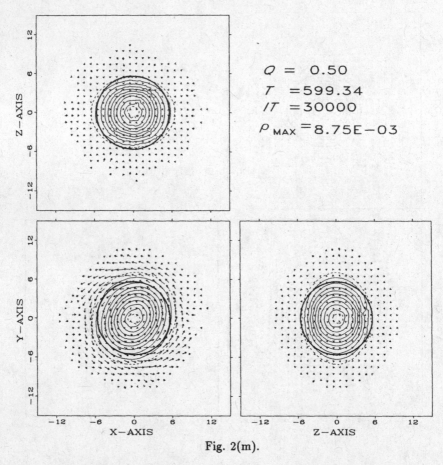

$$Q = 0.50$$
$$T = 599.34$$
$$IT = 30000$$
$$\rho_{MAX} = 8.75E-03$$

Fig. 2(m).

Fig. 4 shows that in reality we can expect bursts of gravitational waves of $h \sim 10^{-21}$ with a few milliseconds duration from the source in the Virgo cluster of galaxies.

The most accurate method of performing a simulation of the coalescence of binary neutron stars is to construct a fully general relativistic code, in which the back reaction is included exactly. So even if a very good formalism for the post-Newtonian hydrodynamics exists, it should be considered as only a prelude to fully general relativistic calculations. From this point of view it is urgent to combine the hydrodynamics code with the 3D metric code.

The results of §4 are based on a recent paper by Nakamura and Oohara.[20] The numerical calculations were performed on the HITAC S820/80 at the Data Handling Center of National Laboratory for High Energy Physics (KEK).

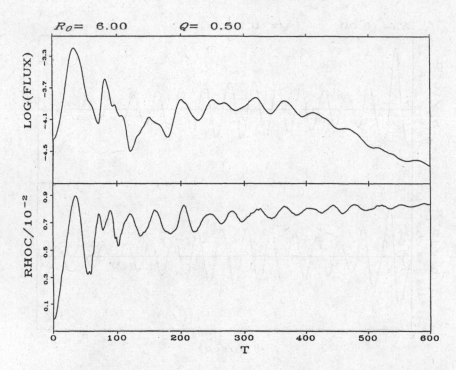

Fig. 3.   Luminosity and central density ρ_c as functions of time t for $q = 0.5$.

This work was supported in part by a Grant-in-Aid from the Scientific Research of Ministry of Education, Science and Culture (01306006, 01652509).

References

1. L. Smarr (1979), in *Sources of Gravitational Waves*, ed. L. Smarr (Cambridge University Press, Cambridge), p. 245.
2. J. Centrella, Chapter A7, this volume.
3. R. Stark (1989), in *Frontiers of Numerical Relativity*, eds. D. Hobill, C. Evans and S. Finn (Cambridge University Press, Cambridge).
4. R. F. Stark and T. Piran (1985), *Phys. Rev. Lett.* **55**, 891.
5. T. Nakamura, K. Oohara and Y. Kojima (1987), *Prog. Theor. Phys. Suppl.* **90**, 1.
6. J. Marck and S. Bonazolla (1989), in *Frontiers of Numerical Relativity*, eds. D. Hobill, C. Evans and S. Finn (Cambridge University Press, Cambridge).
7. J. Wilson (1989), in *Frontiers of Numerical Relativity*, eds. D. Hobill, C. Evans and S. Finn (Cambridge University Press, Cambridge).

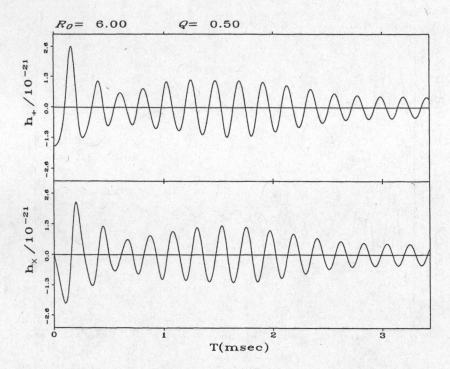

Fig. 4. Wave forms of h_+ and $h_\times$ observed on the z-axis at 10 Mpc for $q = 0.5$.

8. T. Nakamura and K. Oohara (1989), in *Frontiers of Numerical Relativity*, eds. D. Hobill, C. Evans and S. Finn (Cambridge University Press, Cambridge).

9. J. H. Taylor (1987), in *General Relativity and Gravitation 11*, ed. M. A. H. MacCallum (Cambridge University Press, Cambridge), p. 209.

10. J. G. Ables (1989), in *Proceedings of the Fifth Marcel Grossmann Meeting on General Relativity*, eds. D. G. Blair and M.J. Buckingham (World Scientific, Singapore), in press.

11. B. F. Schutz (1986), *Nature* **323**, 310.

12. T. Nakamura and M. Fukugita (1989), *Astrophys. J.* **337**, 466.

13. P. Goldreich and D. Lynden-Bell (1965), *Mon. Not. Roy. Astr. Soc.* **130**, 97.

14. J. Kristian et al. (1989), *Nature* **338**, 234.

15. T. Nakamura (1989), *Prog. Theor. Phys.* **81**, 1006.

16. For a review see T. Damour (1987), in *Gravitation in Astrophysics*, eds. B. Carter and J. B. Hartle, (Plenum, New York), p. 3.

17. L. S. Finn (1989), in *Frontiers in numerical relativity*, eds. C. R. Evans, L. S. Finn and D. W. Hobill (Cambridge University Press, Cambridge), p. 126.

18. K. Oohara and T. Nakamura (1989), *Prog. Theor. Phys.* **82(3)**.

19. K. Oohara and T. Nakamura (1989), *Prog. Theor. Phys.* 81, 360; K. Oohara and T. Nakamura (1989), in *Frontiers of Numerical Relativity*, eds. D. Hobill, C. Evans and S. Finn (Cambridge University Press, Cambridge), p. 74.
20. T. Nakamura and K. Oohara (1989), *Prog. Theor. Phys.* 82(6).

A1

Exact solutions and exact properties of Einstein's equations

Vincent Moncrief

Department of Physics and Department of Mathematics
Yale University, P.O. Box 6666,
New Haven, CT 06511 USA

The study of exact solutions and exact properties of Einstein's equations is a rather broad mathematical subfield of general relativity. Of the roughly 80 abstracts submitted to the symposium devoted to this topic, time limitations permitted only a small fraction to be presented orally. Table 1 lists the papers given at the two sessions of this symposium. The 16 presented papers fall roughly within the categories of "exact solutions," "gravitational energy," "mathematical results" and "symmetry properties of Einstein's equations" and are briefly discussed under those headings in the following. Since most of the oral presentations described extremely recent research results, they did not, for the most part, include references to published papers concerning these results. For this reason the attached reference list is extremely sketchy.

Exact solutions

Virtually all of the known exact solutions of Einstein's equations involve some significant element of idealization. One usually imposes a stringent geometrical symmetry upon the solutions to be considered and, in the non-vacuum case, simplifying assumptions upon the matter sources to be included. Goenner and Sippel discussed several classes of exact solutions which, though highly idealized in the geometrical sense (being in fact pp waves), were nevertheless more realistic in their material content. The sources included both Maxwell fields and a viscous, heat-conducting plasma subject to certain natural energy and entropy inequalities. Several classes of solutions were discussed which represented the generation of a gravitational wave by an electromagnetic wave and a temperature wave propagating in the viscous gas.

An elegant result for electrovacuum null fields of Petrov type D was discussed by Van den Bergh and Debever. They described their proof (which relied heavily upon the use of the "np" package in the symbolic algebra system Maple) that:

Einstein-Maxwell null fields of Petrov type D (with or without Λ term) have zero twist.

This result explains the otherwise mysterious absence of twisting solutions in the literature.

In another study of the Einstein-Maxwell equations Dagotto, et al., discussed the application of inverse scattering techniques to the construction of (single-pole) exact solutions. In the stationary case their approach reproduced the Kerr-Newman-NUT family of solutions whereas, in the nonstationary case, it led to a large class of solutions displaying solitonic waves.

Xanthopoulos described a surprising solution of the Einstein-Maxwell-stiff-perfect-fluid equations which appears to represent the collision of two mixtures of plane gravitational and electromagnetic waves (in an otherwise flat spacetime) which results, in the interaction region, in the production of a relativistic fluid (having $p = \rho$) in addition to the scattered gravitational and electromagnetic waves. This solution generalizes earlier work by Xanthopoulos and Chandrasekhar[1] which had shown that the collision of null dust may result in the formation of an extremely relativistic matter field.

The interpretation that such solutions actually represent, for example, the transformation of colliding clouds of null dust into other forms of matter was questioned by MacCallum, Feinstein and Senovilla. They argued that ambiguities in the temporal development of solutions (which can lead to the apparent transformations mentioned above) result from an incomplete physical specification of the problem. For several representative cases (including e.g., electromagnetic and massless scalar fields) they showed that an apparent ambiguity in the evolution disappears when the underlying physical equations are considered in more detail.

A detailed model of the problem of colliding clouds of null dust was discussed by Taub who showed that the stress energy tensor in the region of interaction must take the form

$$T_{\mu\nu} = \rho_1 U_\mu U_\nu + \rho_2 V_\mu V_\nu - 2Ce^{-\omega} g_{AB} \delta_\mu^A \delta_\nu^B$$

where $\{\rho_1, U_\mu\}$ and $\{\rho_2, V_\mu\}$ are the energy densities and (null) velocity fields of the two dust clouds respectively, where ω and g_{AB} are certain metric functions and where C is a function which characterizes transverse stresses in the colliding dust clouds. Taub showed that if C is specified (e.g., as a function of ρ_1 and ρ_2) then the evolution of the spacetime metric is always unambiguous. Different choices for C however can lead to quite different reactions between the clouds. If $C = 0$ for example then the clouds do not interact whereas, if $C^2 = \rho_1 \rho_2$ then they interact to produce a stiff perfect fluid as in the example of Xanthopoulos and Chandrasekhar.

The cloud of controversy between the various viewpoints of Xanthopoulos, MacCallum, et al., and Taub had apparently not settled, nor was the issue resolved to everyone's satisfaction, by the end of GR12.

Gravitational energy

The most elegant way to prove the positivity of the total energy of an asymptotically flat spacetime is due to Witten.[2] Witten's proof is somewhat mysterious, however, in that it entails the introduction of an auxiliary spinor field whose physical or geometrical significance is not immediately clear. Dimakis and Müller-Hoissen have shown however that the spinor field introduced in the proof of the theorem determines a certain orthonormal frame field with respect to which the integrand (on a spacelike hypersurface) of the usual energy expression (without spinors) becomes manifestly non-negative. This result goes a long way towards clarifying the role played by Witten's spinor in the proof of positivity of energy.

From the proof of energy positivity one knows that the unique global minimum of the ADM (Arnowitt-Deser-Misner) mass functional is flat, empty space. By considering the variation of the ADM mass with respect to a restricted class of perturbations, however, Reula has derived a set of tensorial equations for the gravitational initial data whose solutions he hopes to interpret as non-trivial initial data sets containing, on the initial hypersurface, no gravitational radiation. He has established the existence of solutions of his equations for a large class of material sources.

The significance of the energy momentum pseudotensors of Einstein and Landau and Lifschitz has been clarified by Frauendiener who showed that they are the components of pullbacks of certain natural 3-forms living on the bundle of linear frames. These forms are not tensorial -

a fact which helps to explain the behavior of the pseudotensors under coordinate transformations.

Mathematical results

The characteristic initial value formulation of Einstein's equations provides an interesting and useful alternative to the more conventional Cauchy initial value formulation. In particular the problem of constraints is greatly simplified and the characteristic approach is often more naturally suited to the study of radiation. Rendall discussed an elegant proof of the existence and uniqueness of solutions of Einstein's equations (in vacuum or with perfect fluid sources) when initial data is prescribed on two intersecting null hypersurfaces. To a certain extent Rendall's proof consists of showing how to reduce the characteristic problem to the ordinary Cauchy problem (in harmonic coordinates) by a method which seems previously to have been overlooked.

Some new results on the validity of Huygens' principle for type II spacetimes were discussed by McLenaghan, et al. Recall that for the (conformally invariant) scalar wave equation Huygens' principle is valid on a spacetime V_4 if for every Cauchy initial value problem and every event $x_0 \in V_4$ the solution at x_0 depends only on the data on an arbitrarily small neighborhood of the intersection of the lightcone through x_0 with the chosen Cauchy hypersurface. Analogous definitions can be given for Maxwell's equations and Weyl's neutrino equation. Huygens' principle is valid on any conformally flat spacetime and on any spacetime which is conformally related to an exact plane wave spacetime. The determination of any additional spacetimes for which Huygens' principle holds is still an open problem but partial results have been obtained based upon the Petrov classification of the Weyl tensor of the spacetime in question. In particular Carminati, Czapor and McLenaghan have proven that:

> *For every Petrov type II spacetime for which the conformally invariant scalar wave equation, Maxwell's equations or Weyl's equation satisfies Huygens' principle the principal null directions of the Weyl tensor are necessarily tangent to null congruences that are geodesic and shear free.*

If Einstein's equations are formulated on manifolds of the form $V_4 = \Sigma \times I\!R \times S^1$ where Σ is a compact two-manifold and the metric is assumed to have a spacelike Killing field tangent to the fibers of the circle bundle $V_4 \to \Sigma \times I\!R$ then one can project these equations (in the manner of

Kaluza, Klein and Jordan) to a set of field equations on the base manifold $S \times R$. These projected equations have the form of a set of 2+1 dimensional Einstein equations coupled to a harmonic map field and (for certain choices of Σ) some additional topological data. If one turns off the harmonic map field and the topological data then one is left with the problem of solving the pure vacuum Einstein equations on the 2+1 dimensional manifold $\Sigma \times R$. Even though such spacetimes are necessarily flat their global description may entail a nontrivial evolution of the "Teichmüller parameters" of Σ. Moncrief described the ADM reduction of Einstein's equations on $\Sigma \times R$ and showed how this procedure leads to a (finite dimensional) Hamiltonian system defined on the cotangent bundle of the Teichmüller space of Σ.

Symmetry properties of Einstein's equations

For vacuum metrics with one (spacelike) Killing field there is a well-known local reformulation of the Einstein equations in terms of a Lorentzian metric γ_{ab} on the quotient 3-space and the norm U and twist potential ψ of the assumed Killing vector. The resulting system includes a non-vacuum set of Einstein equations in the quotient 3-space with the norm and twist fields providing the "material" sources. Fackerell discussed several remarkable properties of these reduced field equations. In particular he showed that if one derives the integrability condition on the Ricci tensor $\hat{R}_{ab}$ of the 3-metric γ_{ab} which is needed to ensure that $\hat{R}_{ab}$ can be equated to a source of the appropriate type (constructed from the norm and twist potential), one finds that it implies the existence of a vector field K^a in the quotient 3-space for which $\pounds_K \hat{R}_{ab} = 0$ and $K^a \hat{R}_{ab} = 0$. This same vector field also leaves U and ψ invariant. Locally at least one can express the Killing field as $\frac{\partial}{\partial \phi}$ and the Ricci collineation vector K as $\frac{\partial}{\partial x^0}$ for suitably chosen coordinate functions ϕ and x^0. In addition one can (in the generic case when the norm and twist are functionally independent) take U and ψ as the two remaining coordinates on the original (4-dimensional) spacetime. In the special case for which K^a is a null vector in the quotient 3-space Fackerell showed that the metric is completely determined in terms of two new real potentials whose existence he established. Further progress along these lines seems very promising and potentially quite important for the further unraveling of the one-Killing-field Einstein equations.

To solve the Einstein equations explicitly one often (as in the case described above) imposes an isometry or perhaps a homothety upon the metrics to be considered. Isometries and homotheties leave the Einstein tensor $G_{\mu\nu}$ invariant and hence are rather natural symmetries to consider. Proper (i.e., non-homothetic) conformal symmetries, on the other hand, do not leave the Einstein tensor invariant and so place severe restrictions upon the spacetimes which can exhibit them and also satisfy the field equations. Sharma discussed some new results which generalize earlier work by Collinson and French,[3] Eardley, et al.,[4] and Garfinkle and Tian.[5] In particular he proved that:

> *Solutions of Einstein's equations with covariantly constant energy momentum tensor and a proper (non-homothetic) conformal symmetry are of type O or N.*

As is well-known, the local symmetry of covariance under diffeomorphisms of Einstein's equations is not represented in a straightforward manner in the standard phase space description of general relativity. Lee and Wald have studied this issue within a rather general framework. Starting with the space of all field configurations, $\mathcal{F}$, they construct on it a degenerate symplectic structure. Phase space, with a non-degenerate symplectic form, is obtained by passing to an appropriate quotient space. Within this setting have investigated the relationship of the Poisson bracket algebra of constraint functions on phase space to the algebra of "field dependent" local symmetries on $\mathcal{F}$, which can be projected to the constraint submanifold of phase space.

The splitting of spacetime into space plus time, which is inherent in the Hamiltonian approach to Einstein's equations, has been studied from another viewpoint by Jantzen and Carini. They introduce (into a region of spacetime) a parametrized reference system consisting of both a slicing (i.e., foliation) by a family of spacelike hypersurfaces and a "threading" of these hypersurfaces by a transversal congruence of timelike curves. Using spacetime frames and natural projections associated to the slicing and threading viewpoints they gave a precise definition of gravitoelectric and gravitomagnetic vector and tensor fields and clarified the significance of the corresponding decompositions of Maxwell's[6-8] equations, which had served as a model for the gravitational case.

Table 1.

Symposium on exact solutions and exact properties of Einstein's Equations

===

J. Frauendiener, *Geometric Description of Energy Momentum Pseudotensors*

J. Lee and R. Wald, *Constraints and Local Symmetries*

O. Reula (talk presented by J. Ehlers), *Initial Data Sets of Minimal Energy*

E. D. Fackerell, *Ricci Collineations and New Potentials for Vacuum Spacetimes with one Killing Vector*

A. Dimakis and F. Müller-Hoissen, *Note on the Positive Energy Theorem*

A. Feinstein, M. A. H. MacCallum and J. M. M. Senovilla, *On Ambiguous Evolution and Production of Matter in Spacetimes with Colliding Waves*

B. Xanthopoulos, *Transformation of Electromagnetic Energy into Matter in Collisions of Plane Waves*

A. H. Taub, *Collision of Impulsive Gravitational Waves Followed by Dust Clouds*

J. Carminati, S. R. Czapor and R. G. McLenaghan, *The Validity of Huygens' Principle on Type II Spacetimes*

A. Dagotto, R. Gleiser, G. Gonzalez and J. Pullin, *Exact Solutions to the Einstein-Maxwell Equations with Alekseev's Inverse Scattering Technique*

H. F. M. Goenner and R. Sippel, *Exact pp-Wave Solutions of the Einstein-Maxwell Equations Generated by a Viscous Heat-Conducting Plasma*

A. D. Rendall, *The Characteristic Initial Value Problem for the Einstein Equations*

N. Van den Bergh and R. Debever, *Twisting Einstein-Maxwell Null Fields of Petrov Type D*

R. Sharma, *Matter-Symmetric Solutions of Einstein's Field Equations with Conformal Symmetries*

R. T. Jantzen, *Gravito-What Fields?*

V. Moncrief, *How Solvable is Witten's 2+1 Dimensional Topological Gravity Theory?*

===

References

1. S. Chandrasekhar and B. C. Xanthopoulos (1985), *Proc. Roy. Soc. Lond.* **A402**, 37, and (1986), *Proc. Roy. Soc. Lond.* **A403**, 189.
2. E. Witten (1981), *Commun. Math. Phys.* **80**, 381. See also J. Nester (1981), *Phys. Lett.* **83A**, 241; and A. Dimakis and F. Müller-Hoissen (1985), *J. Math. Phys.* **26**, 1040, App. A.
3. C. D. Collison and D. C. French (1967), *J. Math. Phys.* **8**, 701.
4. D. Eardley, J. Isenberg, J. Marsden and V. Moncrief (1986), *Commun. Math. Phys.* **106**, 137.
5. D. Garfinkle and Q. Tian (1987), *Class. Quant. Grav.* **4**, 137.
6. K. S. Thorne and D. MacDonald (1982), *Mon. Not. Roy. Astr. Soc.* **198**, 339.
7. P. Benvenuti (1960), *Ann. Scuola Normale Sup. Pisa*, Ser. III **14**, 171.
8. G.F.R. Ellis (1973), in *Cargese Lectures in Physics*, Vol. 6, ed. E. Schatzman (Gordon and Breach, New York). See also L.D. Landau and E. M. Lifshitz (1975), *The Classical Theory of Fields* (Pergamon Press, New York).

A2

Spinors, twistors, and complex methods

N. M. J. Woodhouse
Wadham College
Oxford OX1 3PN, England

Five of the talks in the workshop concerned the use of twistor theory or related complex methods to solve nonlinear equations, including Einstein's equations; the sixth, by Y. Ne'eman, addressed the issue of the definition of spinors on a general manifold.

E. T. Newman described joint work with L. J. Mason and J. Ivancovitch on the splitting problem for the anti-self-dual (ASD) Yang-Mills equation on Minkowski space. In Ward's (1977) construction, solutions are obtained from a general patching matrix $F(L, \overline{M}, \zeta)$ (a holomorphic function of the complex variables $L, \overline{M}$ and ζ) by finding splitting matrices $G_0(x, y, z, t, \zeta)$ and $G_\infty(x, y, z, t, \zeta)$ such that

$$F(t - z + \zeta(x - iy), x + iy + \zeta(t + z), \zeta) = G_\infty^{-1} G_0, \qquad (1)$$

with G_0 holomorphic in ζ in a neighbourhood U_0 of $\zeta = 0$ and G_∞ holomorphic in ζ in a neighbourhood U_∞ of $\zeta = \infty$. Eq. (1) is required to hold on the annular intersection $U_0 \cap U_\infty$.

Although there is no general procedure for solving such a Riemann-Hilbert problem, G_0 and G_∞ can be constructed explicitly when F is upper triangular. The approach of Newman and his coworkers is to try to reduce F to triangular form by making a 'gauge transformation'

$$F \longmapsto FS$$

where S is holomorphic (and nonsingular) on U_∞; F will be upper triangular if $F_{21}S_{11} + F_{22}S_{21} = 0$. Their idea is to approximate F by a sequence $F_{(m)}$ ($m \in Z$) of patching matrices for which S can be found,

93

with $F_{(m)} \to F$ in the limit $m \to \infty$. The matrices $F_{(m)}$ are constructed by truncating the Laurent series of F_{21}/F_{22} to a finite number of terms. There is an extension of the procedure for $SL(n, \mathcal{C})$ gauge fields.

I. Hauser described his and F. J. Ernst's method for constructing the metric in the scattering region of two colliding plane waves by solving a certain type of Hilbert problem (Hauser and Ernst (1989)). The metric is written in the form

$$ds^2 = h_{ab} dx^a dx^b + 2g_{uv} du dv \quad (a, b = 1, 2)$$

where the coefficients depend only on the nonignorable coordinates u and v (the wave fronts are $u = 0$ and $v = 0$). If $\rho = \sqrt{\det h}$ where h is the 2×2 matrix with entries h_{ab}, then the vacuum field equations reduce to

$$\partial_u(\rho h^{-1} \partial_v h) + \partial_v(\rho h^{-1} \partial_u h) = 0 \tag{2}$$

together with an equation for the gradient of g_{uv} which is automatically integrable.

The function ρ satisfies the wave equation $\rho_{uv} = 0$, and so must be of the form $\rho = \frac{1}{2}(s(v) - r(u))$ for some single-variable functions r and s. Given u and v, r and s are uniquely determined by the conditions $r(0) = -1$ and $s(0) = 1$.

The key to the method is the reinterpretation of Eq. (2) as the integrability condition for the existence of a matrix-valued potential $P(u, v, \tau)$, determined by

$$\partial_u P = \frac{1}{2}(\tau - r)^{-1}(\rho k^{-1} - \mathrm{i})\partial_u(k)P \,,$$

$$\partial_v P = -\frac{1}{2}(\tau - s)^{-1}(\rho k^{-1} + \mathrm{i})\partial_v(k)P \,,$$

where

$$k = h \begin{pmatrix} 0 & 1 \\ -1 & 0 \end{pmatrix}$$

and τ is a complex parameter. For fixed u and v, $P(u, v, \tau)$ is holomorphic in τ on the whole Riemann sphere, less the intervals $[-1, r]$ and $[s, 1]$ on the real axis. At $\tau = \infty$, $P = I$. Moreover, $P(u, v, \tau)P(0, v, \tau)^{-1}$ is holomorphic on $\mathcal{C}\backslash[-1, r]$ and $P(u, v, \tau)P(u, 0, \tau)^{-1}$ is holomorphic on $\mathbb{C}\backslash[s,1]$. Thus $P(u,v,\tau)$ and hence the metric can be recovered from $P(0,v,\tau)$ and $P(u,0,\tau)$ by solving a Hilbert-type splitting problem. The matrices $P(0,v,\tau)$ and $P(u,0,\tau)$ are determined by the initial data.

L. J. Mason tackled the same problem, and described a similar solution which emerges from twistor theory. First he introduced new coordinates $U = -r$ and $V = s$ in the u,v plane (they are analogous to Weyl coordinates in stationary axisymmetric solutions). Then $\rho = \frac{1}{2}(U + V)$ and Eq. (2) coincides with a reduction of the ASD Yang-Mills equations for a gauge field on Minkowski space invariant under a spacelike boost and an orthogonal translation (by adapting Ward's (1983) treatment of the stationary axisymmetric case). Under a reduced form of Ward's correspondence (Ward (1977)), such gauge fields are associated with holomorphic bundles on a non-Hausdorff Riemann surface (Woodhouse and Mason (1988)).

The local trivializations of the bundle corresponding to h are given by matrix-valued solutions Q of the linear system ('Lax pair')

$$\partial_U Q = -\frac{1}{2}\left(1 \pm \sqrt{\frac{\gamma - V}{\gamma + U}}\right) h^{-1}\partial_U(h)Q$$

$$\partial_V Q = -\frac{1}{2}\left(1 \pm \sqrt{\frac{\gamma + U}{\gamma - V}}\right) h^{-1}\partial_V(h)Q$$

where γ is a complex parameter. The integrability condition is again Eq. (2), but with u and v replaced by U and V. It is the square roots that give rise to the non-Hausdorff nature of the Riemann surface.

Mason's method for determining the metric in the scattering region is first to find the patching matrices for the bundle in terms of the initial data of the colliding waves; and then to determine h in the interaction region by an extension of Ward's splitting procedure. The construction is clearly very closely related to that of Hauser and Ernst and it would be interesting to make the connection explicit. The twistor approach seems to be well adapted to the analysis of the space-time singularity that can occur at $U + V = 0$.

G. A. J. Sparling reported on joint work with L. J. Mason on the twistor interpretation of the soliton hierarchies associated with the Korteweg-de Vries and nonlinear Schrödinger equations. Their method is based on the observation that both equations are reductions of the ASD Yang-Mills equations in a space-time with signature $++--$ under the group generated by a timelike translation and an orthogonal null translation (Mason and Sparling (1989)).

In the KdV case, for example, one writes the metric in the form $ds^2 = du\,dv - dy\,dt$ and the Yang-Mills potential as

$$A = \frac{1}{2} \begin{pmatrix} \alpha & \beta \\ \gamma & -\alpha \end{pmatrix}.$$

The ASD condition on A is that the curvature of $F = dA - A \wedge A$ should be a combination of $du \wedge dy$, $dv \wedge dt$ and $du \wedge dv + dt \wedge dy$ alone. If

$$\alpha = q\,du + \frac{1}{2}(q_{xx} - 2qq_x)dt\,, \qquad \beta = du - q_x dt\,,$$

$$\gamma = (q_x - q^2)du + \frac{1}{4}(q_{xxx} - 4qq_{xx} - 2q_x^2 + 4q^2 q_x)dt + dy\,,$$

where $x = u-v$ and q is a function of x and t, then the ASD condition implies that $\phi = -q_x$ is a solution of the KdV equation $4\phi_t = \phi_{xxx} + 12\phi\phi_x$. There is a similar statement for the nonlinear Schrödinger equation.

Under Ward's correspondence, an ASD Yang-Mills field which is invariant under a timelike translation corresponds to a holomorphic bundle over $Z(2)$, the complex line bundle of Chern class 2 over $\mathbb{CP}_1$ (i.e. $Z(2)$ is the holomorphic tangent bundle). The bundles corresponding to solutions of the KdV and nonlinear Schrödinger equations have additional symmetry under the group generated by the null translation, which acts on $Z(2)$ leaving one fibre fixed.

By replacing $Z(2)$ by $Z(n)$ (which is the line bundle of Chern number n over $\mathbb{CP}_1$), one obtains systems of partial differential equations in n independent variables containing the KdV equation or nonlinear Schrödinger equation as the first term. On letting $n \to \infty$, these become the hierarchies discovered by Gardner, Greene, Kruskal and Miura (1967) in the case of the KdV equation and by Zakharov and Shabat (1974) in the case of the nonlinear Schrödinger equation. The equations in the hierarchies can also be realized as reductions of the ASD Yang-Mills equations.

C. R. Lebrun described an extension of Penrose's nonlinear graviton construction (Penrose (1976)). Under mild convexity assumptions, the space of complex null geodesics in a four-dimensional complex conformal manifold M is a five-dimensional complex contact manifold $\mathcal{N}$, which encodes the conformal geometry of M.

When M is Minkowski space, $\mathcal{N}$ is the hypersurface $Z^\alpha W_\alpha = 0$ in $PT \times PT^* = CP_3 \times CP_3^*$ (the product of projective twistor space with its dual). The idea in the general case is to try to reconstruct this embedding as a sequence of formal neighbourhoods, or truncated Taylor expansions, in the spirit of the analysis of the flat-space Yang-Mills equations by Isenberg, Yasskin and Green (1978) and Witten (1978). This is always possible up to the fourth order. The existence of the fifth formal neighbourhood is equivalent to the vanishing of the Bach tensor

$$B_{ab} = (\nabla^c \nabla^d + \frac{1}{2} R^{cd}) C_{abcd} \qquad (a, b, \cdots = 0, 1, 2, 3);$$

and of the sixth formal neighbourhood to the vanishing of the Eastwood-Dighton tensor

$$E_{abc} = C^{ef}{}_{cb} \nabla^d C^*_{afed} - C^{*ef}{}_{cb} \nabla^d C_{afed};$$

(these results, now rigorously established, were conjectured by Baston and Mason (1987)).

Provided that C_{abcd} is algebraically general, the vanishing of B_{ab} and E_{abc} is equivalent to the metric being conformal to an Einstein metric (Kozameh, Newman and Tod (1985); Baston and Mason (1987)).

Y. Ne'eman drew attention to the inaccurate statement that is made in some textbooks on general relativity that there do not exist 'spinor' representations of $GL(n, R)$ and the diffeomorphism groups of R^n, so that it is not possible to define spinors on a general manifold without a metric. This is only true if one restricts attention to finite-dimensional representations and spinors with a finite number of components. Such representations *do* exist if one admits 'spinors' with an infinite number of components (the double-covering groups $\overline{GL}(n, R)$ are groups of infinite matrices). The representations in the case $n = 3$ were constructed by D. W. Joseph and Y. Ne'eman and catalogued by Dj. Šijački; and in the case $n = 4$, the multiplicity-free representations were constructed and catalogued by Y. Ne'eman and Dj. Šijački. See Šijački (1975), Šijački and Ne'eman (1985), and papers cited therein. It is possible to use such spinor representations to construct infinite component 'manifields' which can be required to satisfy generally covariant field equations; see, for example, Mickelsson (1983).

References

Baston, R. J. and Mason, L. J. (1987), Conformal gravity, the Einstein equations and spaces of complex null geodesics, *Class. Quant. Grav.* 4, 815-26.

Gardner, C. S., Greene, J. M., Kruskal, M. D. and Miura (1967), Methods for solving the Korteweg-de Vries equation, *Phys. Rev. Lett.* 19, 1095-7.

Hauser, I. and Ernst, F. J. (1989), Initial value problem for colliding gravitational plane waves, I. *J. Math. Phys.* 30, 872-887; part II will be published in the same journal.

Isenberg, J., Yasskin, P. B. and Green, P. S. (1978), Non-self-dual gauge fields, *Phys. Lett.* 78B, 462-4.

Kozameh, C., Newman, E. T. and Tod, K. P. (1985), Conformal Einstein spaces, *Gen. Relat. Grav.* 17, 343-52.

Mason, L. J. and Sparling, G. A. J. (1989), Nonlinear Schrödinger and Korteweg-de Vries are reductions of self-dual Yang-Mills, *Phys. Lett.* 137, 29-33.

Mickelsson, J. (1983), On $\overline{GL}(4, R)$-covariant extensions of the Dirac equation, *Commun. Math. Phys.* 88, 551-67.

Penrose, R. (1976), Nonlinear gravitons and curved twistor theory, *Gen. Relat. Grav.* 7, 31-52.

Šijački, Dj. (1975), The unitary irreducible representations of $\overline{SL}(3, R)$, *J. Math. Phys.* 16, 298-311.

Šijački, Dj. and Ne'eman, Y. (1985), Algebra and physics of the unitary multiplicity-free representations of $\overline{SL}(4, R)$, *J. Math. Phys.* 26, 2457-2464.

Ward, R. S. (1977), On self-dual gauge fields, *Phys. Lett.* 61A, 81-2.

Ward, R. S. (1983), Stationary axisymmetric space-times: a new approach, *Gen. Relat. Grav.* 15, 105-9.

Witten, E. (1978), An interpretation of classical Yang-Mills theory, *Phys Lett.* 77B, 394-8.

Woodhouse, N. M. J. and Mason, L. J. (1988), The Geroch group and non-Hausdorff twistor spaces, *Nonlinearity* 1, 73-114.

Zakharov, V. E. and Shabat, A. B. (1974), A scheme for integrating the nonlinear equations of mathematical physics by the method of the inverse scattering problem I., *Functional Analysis and its Applications* 8, 226-35.

A3

Alternative gravity theories

Mauro Francaviglia
Istituto di Fisica Matematica "J.-L. Lagrange",
Università di Torino,
Via C. Alberto 10,
10123 Torino, Italia

1. Introduction

In this short report I will try to condense in a few pages some of the
topics which, among those presented at the Workshop A3, I felt to be
the most suited for being reviewed, although serious limitations of space
will prevent me from mentioning other interesting issues which were also
raised at the Conference.

The large number of contributions received (over 70) and the good
quality of a number of them shows that, although general relativity
is a well-established discipline (both on theoretical and experimental
grounds), still the theory deserves efforts aimed at producing alternative or more general frameworks for investigating the classical properties
of gravity. These are either devoted to produce alternative viewpoints
or interpretations of standard general relativity, or at constructing, discussing and proposing experimental tests for alternative (a priori non-
equivalent) descriptions of the dynamics of the gravitational field and
its interaction (or unification) with external matter fields.

As is well known, classical alternative theories of gravitation can be
roughly classified as follows:

(**A**) Theories based on a still 4-dimensional picture, under the assumption that the dynamics of the gravitational field is more com-

plicated (at least a priori) than Einstein's one. This goal may be achieved either by: (i) assuming the field to be a metric $g_{\mu\nu}$ in space-time and allowing the Lagrangian to be more general than Hilbert's one $g^{\mu\nu} R_{\mu\nu} \sqrt{g}$; or: (ii) by assuming that the gravitational dynamics is hidden in the combination of a linear connection $\Gamma^{\mu}_{\nu\lambda}$ and a metric $g_{\mu\nu}$ (which might possibly be absent from the Lagrangian L), possibly mediated by tetrads, spin-connections or group variables coming from the action of the linear group or the Poincaré group.

(B) Theories based on higher-dimensional pictures, i.e., on the assumption that space-time is replaced by a higher-dimensional manifold (usually a G−bundle over space-time, where G is some physically meaningful Lie group). The evolution in this direction brings into the realm of supergravity and strings, which were however out of the domain of this Workshop and will be touched upon in other reports in these *Proceedings*.

In the sequel I shall briefly mention some of the contributions at this Workshop, dividing them in accordance with the previous classification. A last section will be devoted to mentioning some further topics which could hardly be classified in the aforementioned scheme.

2. Four-dimensional theories

Among 4-dimensional theories a renewed interest has grown about so-called *higher derivative gravity*, i.e, theories based on a metric Lagrangian containing curvature terms other than the linear Hilbert's term $R\sqrt{g}$.

M. Ferraris reported about a joint work with G. Magnano (and myself), aimed at investigating problems of dynamical equivalence between various non-linear theories and general relativity (with matter). The dynamical equivalence is here meant as a full correspondence among the solutions of the original 4th-order equations and a set of new equations which are generated via a Legendre transformation mechanism[1] and turn out to be of the second order in a new metric and some additional fields. In many cases of concrete physical interest (e.g., whenever L depends only on the Ricci tensor) the theory is essentially transformed into general relativity coupled with a spin-two and a spin-zero field (as in Stelle's[2] linear analysis of $L = (aR^2 + bR_{\mu\nu}R^{\mu\nu})\sqrt{g}$). For details see Refs. 3, 4, and 5 where also the case of an explicit dependence of the Lagrangian on Weyl's tensor was considered. Important arguments were raised by C. H. Brans, who discusses the operational definition of a "physical metric,"[6,7] and by F. Occhionero, who pointed out the interest of higher order theories for cosmology (see also Ref. 8).

A remarkable analysis on various aspects of higher order gravity was presented by F. Müller-Hoissen, who addressed the study of the Lovelock action in dimensions possible higher than four. He shows that the dimensional reduction of this action leads to a generalized Einstein-Yang-Mills action in lower dimension, with the possible addition of scalar fields and/or sigma models. The interesting feature is that field equations, which have second order (instead of fourth) in the full bundle, preserve this character when reduced to standard space-time in four dimensions. A nice review can be found in Ref. 9 (and references quoted therein).

The Lovelock Lagrangian was considered also by C. V. Vishveshwara et al.,[10] who studied scalar perturbations of spherically symmetric black-hole solutions of these theories, together with their stability properties.

J. Wheeler discussed the idea that a Weyl geometry could be used to model quantum mechanics. By postulating that an object undergoes any displacement with a probability which is inversely proportional to the change in length it experiences as a result, one is led directly to a Wiener path integral as the main object of interest in a Weyl geometry. This path integral has conformal invariance in exact analogy to the phase invariance of the Feynman path integral. A nontrivial check on the formulation is suggested by the classical limit given by the vanishing of the first variation of the path integral itself. This variation leads to a special form of the Weyl vector, which agrees precisely with the form required by the combined axiomatic approaches to general relativity proposed by Ehlers, Pirani and Schild and by Audretsch, Gahler and Straumann. In the special classical limit the length changes induced by the nonvanishing Weyl field are not observable, because the paths of classical objects are restricted. The Weyl field itself can be measured in this limit, and is given by a certain function of the velocity and the magnetic field.[11,12]

In the above framework the two metrics originally introduced by Dirac were used. A different perspective was addressed by J. Wood and G. Papini, who try instead to eliminate the original objection to Weyl's work, namely that the theory lacks fundamental standards of length. These are reintroduced by means of the constraint that the field $|\beta|$ equals the gradient of a class of multivalued functions which describe regions of space-time in which the physical properties of the particles are dif-

ferent from those outside. Lengths have then an absolute value along
world-lines linking these regions (as in all physical situations where electromagnetic measurements are made). In this way scales are restored to
a conformally invariant theory. The form of the action is then borrowed
from Dirac, and extended by making the field β a complex function of
real variables (so that there are here two scalar fields). In their work[13]
they suggest that the theory has a quantum mechanical content in the
form of a Klein-Gordon equation with conformal terms and Caianiello's
maximal acceleration.[14]

In the domain of metric-affine pictures for gravity, one should
also mention some interesting contributions by L. Garcia de Andrade
(who considered nonlinear electrodynamics in curved space-times with
torsion[15]) and by R. J. Petty (who considers torsion as translational
holonomy per unit area and suggests interpreting the intrinsic angular
momentum density as a continuum approximation of space-time dislo-
cations with time-like Burger's vectors; see Ref. 16).

3.   Five-dimensional theories

In the domain of five-dimensional theories a rather interesting contribution was given by D. Brill,[17] who discussed an important explicit
example of Kaluza-Klein solutions having negative energy. According
to some suggestions due to E. Witten,[18] the positivity of total gravitational energy could be violated in five-dimensional theories with special
topologies. Using the appropriate ADM surface integral, Brill shows in
fact that the lower limit of the total mass-energy of asymptotically flat
initial geometries is negative for space-like sections having the topology
$R^2 \times S^2$ of a "bubble solution." Moreover, it is shown how this lower
limit, which can be arbitrarily negative, is explicitly related to the size
of the bubble, but it does not apparently show "sensitivity" on the scale
at which compactification occurs.

One should also mention an interesting (although preliminary) contribution by P. Macedo,[19] who considered the problem of electrical neutrality of large bodies in the context of Jordan-Thiry theory. Assuming
that the "total scalar charge" of a distribution of particles is usually
positive (also for electrically neutral bodies) he suggested limits for the
scalar field coupling constant and raised the question of the scale to
which "electrical neutrality" should apply.

4. Miscellaneous

Under this last heading I will mention some further contributions which do not properly fall into the previous two categories.

In the domain of scalar-tensor theories J. Skea[20] addressed the problem of determining the critical points of the cosmological Bergmann-Wagoner field equations, showing explicitly that the stability of all the critical points depends substantially on the type of perturbation envisaged. In particular, one of these critical points is the analogue of the de Sitter solution and has an unstable behavior.

A final mention should be given to H. Terazawa (pre-geometric theories of gravity, i.e., theories in which Einstein's gravity is reproduced as an effective theory of low energy: see Ref. 21 for a review), to J. Bekenstein (theory of "phase coupling gravitation," where the missing mass of extragalactic astrophysics is suggested to be reflecting a non-Newtonian behavior of gravity at large scales; see Ref. 22) and to V. Tapia (an interesting approach to "General relativity à la string," i.e., via the embedding method in a higher dimensional flat space; see Ref. 23).

References

1. Magnano, G., Ferraris, M. and Francaviglia, M. (1990), *J. Math Phys.*, to appear.
2. Stelle, K. S. (1977), *Phys. Rev.* **D16**, 953; and (1978), *Gen. Relat. Grav.* **9(4)**, 353.
3. Magnano, G., Ferraris, M. and Francaviglia, M. (1987), *Gen. Relat. Grav.* **19(5)**, 465.
4. Magnano, G., Ferraris, M. and Francaviglia, M. (1988), *Class. Quant. Grav.* **5(6)**, L95.
5. Magnano, G., Ferraris, M. and Francaviglia, M. (1989), *Class. Quant. Grav.*, to appear.
6. Brans, C. H. (1988), *Class. Quant. Grav.* **5(12)**, L197.
7. Magnano, G., Ferraris, M. and Francaviglia, M., *Class. Quant. Grav.*, submitted.
8. Maeda, K. (1988), *Phys. Rev.* **37(4)**, 858.
9. Müller-Hoissen, F., preprint GOEDT-TP 55/89 (August 1989); Müller-Hoissen F. and Sippel, R. (1988), *Class. Quant. Grav.* **5**, 1473.
10. Iyer, B. R., Iyer, S. and Vishveshwara, C. B. (1989), in *Abstracts of contributed papers, GR-12*, Boulder, CO, USA, Vol. I., 173.
11. Hochberg, D. and Wheeler, J. T., preprint BA-89-33.
12. Hochberg, D. and Wheeler, J. T. (1989), Scale-invariant metric from conformal gauge theory, preprint.

13. Wood, W. R., Papini, G. and Cai, Y. Q. (1989), Quantum aspects of theories of the Weyl-Dirac type, preprint.
14. Caianiello, E. R. (1981), *Lett. Nuovo Cimento* **32**, 65.
15. Garcia de Andrade, L. C. (1989), On the nonlinear electrodynamics in Riemann-Cartan space-time, preprint.
16. Petti, R. J. (1986), *Gen. Relat. Grav.* **18(5)**, 441.
17. Brill, D. and Pfister, H. (1989), States of negative total energy in Kaluza-Klein theoɪy, preprint.
18. Witten, E. (1982), *Nucl. Phys.* **B195**, 481.
19. Macedo, P. (1989), Electric Neutrality and the Jordan-Thiry scalar field, preprint.
20. Skea, J. E. F. and Burd, A. B. (1989) Cosmological models in the Bergmann-Wagoner theory of gravity, preprint.
21. Terazawa, H. (1989), in *Proc. 5th Marcell Grossman meeting on recent developments in theoretical and experimental general relativity, gravitation and relativistic field theories, Perth* (1988), eds. R. Ruffini et al. (World Scientific, Singapore).
22. Bekenstein, J. D. (1988), in *General relativity and relativistic astrophysics*, eds. A. Coley, C. Dyer, and T. Tupper (World Scientific, Singapore).
23. Tapia, V. (1989), *Class. Quant. Grav.* **6**, L49.

A4

Asymptotia, singularities, and global structure

B. G. Schmidt
Max-Planck-Institute für Astrophysik,
Karl-Schwarzschild Str. 1,
D-8046 Garching bei München, Federal Republic of Germany

Let me organize this report by comparing some of the issues with the achievements described in the contributed papers.

Two important questions of **"Asymptopia"** are: (1) Relations between the sources and the asymptotic field of space-time, and (2) The existence and smoothness property of solutions of the field equations admitting a null infinity in the sense of Penrose. In view of its importance it is regrettable that not a single paper addressed the first question. Apparently it can still only be treated in the context of approximation methods. (Compare the workshops A5, A6.) Concerning the second question there is still no proof or counter example known.

There were however two contributions dealing with existence questions of solutions with certain asymptotic properties. Choquet-Bruhat demonstrated the existence of global solutions of the Yang-Mills Higgs equations of Anti-de Sitter space-time, under the condition that there is no radiation at timelike infinity. Reula showed - via implicit function theorem techniques - that near the Schwarzschild solution there does exist the expected number of stationary solutions of the vacuum field equations, with well defined Geroch-Hansen multipole moments. Up to now this was only known for the Weyl solutions, hence in the axisymmetric case.

The further contributions under this heading consisted mainly in extensions or refinements of already known results or approaches to certain questions. Let me mention two examples: Bičak and Schmidt extended the investigation of the global structure of boost-rotationally symmetric

105

vacuum spacetimes. Kozameh and Newman described further properties of "light cone cuts of null infinity."

About **"Singularities"**: we know from the early "singularity theorems" that many physically relevant space-times are geodesically incomplete. Later results made it plausible that the incompleteness will generically be a curvature singularity. In these investigations the explicit form of the field equations is not used. Therefore we know almost nothing about the specific nature of the singularities of solutions of Einsteins equations. This situation was not changed by the papers presented!

Some known theorems were generalized, singularities of exact solutions were further investigated. Krienle for example generalized a theorem by Hawking and Penrose. De Felice and Bradley analysed the curvature near the Kerr singularity.

"Global Structure" is in indeed a wide field! The Cosmic censorship hypothesis; causal structure; distance between Lorentz metrics; colliding plane waves; localisation of gravitational energy were discussed amongst other themes. New was an attempt by Klinkhammer, Morris, Thorne and Yurtsever to use wormhole space-times to get some hints about relations between causality violations and quantum fields.

Altogether, a lot of work is done in detailed questions but no really new idea emerged. The fundamental questions seem still to be too complicated to be attacked with the mathematical tools available.

A5

Radiative spacetimes and approximation methods

Thibault Damour
Institut des Hautes Etudes Scientifiques
91440 Bures sur Yvette, France
and D.A.R.C., Observatoire de Paris - CNRS
92195 Meudon Cedex, France

1. Introduction

The workshop consisted of an introductory overview and of seven specialized talks. The talks were selected among twenty-five submitted
abstracts and were presented in an order consistent with the three subsequently discussed categories, i.e. from rigourous *mathematical results
about* approximation methods, to *definition of* approximation methods,
to *physical results* obtained *by* approximation methods. This order was
chosen to emphasize the following easily forgotten fact. Ideally, physics
should connect the mathematical axioms defining our theories to the
results of observations or experiments by means of a tight chain of deductions. However, in practice, this chain of deductions often contains
gaps, and one of the main sources of gaps lies in the use of mathematically ill justified approximation methods as substitutes for the exact
theories. In order to bridge the latter gap one needs, on the one hand,
some clear algorithmic definition of the approximation methods used
together with a formal study of the structure of its successive terms,
and, on the other hand, mathematical theorems proving that the formal
sequence defined by the "approximation method" is either convergent,
or asymptotic, to some exact solution. Progress on both aspects has

recently been obtained, and has been reported, or quoted, during the workshop.

Talks presented at the workshop fell into three categories. Talks on *mathematical results about approximation methods* showed instances where the conceptually important gap between mathematics and the use of approximation methods by physicists can be closed, or narrowed. Talks on the *definition and formal properties of approximation methods* presented well-defined formal algorithms which have been studied and developed up to higher orders of approximation than hitherto known. Moreover, these and other talks presented recent progress in the *application of approximation methods to radiative spacetimes*. Altogether, the field seems to have undergone an appreciable progress since GR-11, as witnessed, for instance, by the derivation of the higher-order corrections to the "standard quadrupole equations," by means of new approximation methods which are partly backed up by rigourous mathematical results.

2.   Mathematical results about approximation methods

In his talk, B. G. Schmidt started by quoting the recent theorem about the existence of c^{-1}- smooth solutions of the initial value constraints obtained by J. Ehlers and M. Lottermoser (1989). This result constitutes a solid first step towards a mathematical grasp of the so-called "post-Newtonian approximation methods." However, because of the difficulty of controlling the effect of the time evolution on the c^{-1}- dependence, one is still far from having any mathematical justification of the use of the post-Newtonian approximation schemes. By contrast, Schmidt reported recent results (Damour and Schmidt (1989)), based on a theorem of Rendall (1989), which prove quite generally that the formal power series generated by "post-Minkowskian approximation methods" (i.e. formal non-linearity expansions) are (locally, in vacuum) the Taylor series of some exact, parameter dependent, solution of Einstein's equations. In other words, post-Minkowskian expansions are asymptotic to G-dependent exact solutions. Moreover, by combining the techniques used in the proof with theorems of H. Friedrich (1989) one proves the existence, outside some spatially compact domain, of many (G-smooth) exact solutions admitting a regular conformal compactification at future null infinity.

J.W. Barrett and R.M. Williams presented interesting recent results about the convergence properties of the Regge calculus, an "approximation method" which can be considered as being both of fundamental (quantum gravity ?) and practical (numerical relativity) importance. As emphasized, among other things by Barrett, one would like to have, rather than a definition of the convergence of a sequence of a piecewise flat Riemannian manifold to some supposedly given smooth curved Riemannian manifold (as used e.g. in the previous theorems of Cheeger et al. (1984), a *criterion*, *à la Cauchy*, for the convergence of a sequence of piecewise flat Riemannian manifolds which is defined intrinsically as a restriction on the behaviour of the members of the sequence relative to each other, without referring itself to the curved limit. In the case of the "linearised Regge calculus" (corresponding to the linearised approximation to General Relativity), the following results, among others, have been obtained: a Cauchy-type definition of the convergence of a sequence of linearised Regge configurations has been given (through the notion of Cauchy convergence of a corresponding sequence of curvature distributions) (Barrett (1988)), the existence of such convergent sequences has been proven (Barrett and Williams, (1988)) and the properties of the limits have been studied, notably the fact that solutions of the linearised Regge equations converge to smooth solutions of the linearised Einstein equations (Barrett (1988)). Let us hope that these results can be generalised to the pseudo-Riemannian case, and to the full, non-linear, Regge calculus.

3. Definition and formal properties of approximation methods

The chairman reported on recent work done in collaboration with L. Blanchet about the definition and study of a formal algorithm which combines a multipolar post-Minkowskian expansion scheme (for a generic radiative gravitational field outside a bounded source) with a post-Newtonian expansion scheme (for the field in the source and its near-zone). This scheme possesses nice structural features which allow one both to prove general results about its structure, and to implement it explicitly (Blanchet and Damour (1988)). For instance, this scheme gives a general proof of the necessary breakdown of all (pure) post-Newtonian schemes at some finite c^{-n} order, by exhibiting the irreducible appearance of "hereditary terms" (depending on the full past-history of the material source) in the 4PN approximation of the near-zone metric. G. Schäfer presented a new approximation method for General Relativity which combines the advantages of the post-Newtonian and of the post-Minkowskian (non-linearity) expansions not through matching, as the previously mentioned method, but through the use of an ADM-type reduced Hamiltonian approach. In this approach the use of adapted variables, and gauge conditions, allows a clean separation between elliptic and hyperbolic equations (without having to make any formal near-zone expansion). When the latter, ill–justifiable, expansion is introduced this scheme leads to well-defined iterations up to the 3.5 PN level included.

Brumberg and Kopejkin (1989), and Kopejkin (1988), have recently set up an approximation method which implements in a very explicit manner the idea of matching several post-Newtonian expansions having different spatial domains of validity. Their work brings progress in the problem of the motion of N slowly-moving weakly self-gravitating bodies, and in the operational definition of relativistic reference frames.

Concerning the formal properties of approximation methods, one can also mention the work reported in an abstract by J. Ehlers and A.R. Prasanna, in which an example of a breakdown of the WKB method is exhibited (the "matter mode" appearing at the level of the eikonal equation, but admitting no transport equation for its amplitude, even at the lowest order).

4. Application of approximation methods to radiative space-times

The application of the first method mentioned in the previous section has allowed one to go beyond the "standard quadrupole equations," which have been the central focus of the corresponding workshops of the previous two GR meetings (Ehlers and Walker (1984), Schutz (1987)). Note that we use here the terminology "quadrupole equations," instead of the more familiar terminology "quadrupole formulas," to emphasize that all the "quadrupole moments" appearing in such equations must be fully specified, either as functionals of the matter variables, or as directly, or indirectly, observable quantities, so that some specific physical consequences can be effectively drawn from the corresponding "quadrupole equation" (the "quadrupole formula" recently studied by Kundu (1989), which generalizes an intermediate result of Damour (1986), is somewhat disappointing in this respect, though, in principle, it is a well specified "equation", because it relates an easily measurable radiative quadrupole moment to another "field quadrupole moment" which is physically very difficult to observe in the far-wave-zone gravitational field).

Concerning the "radiation-reaction quadrupole equation" it has been possible to compute explicitly its modification due to the "hereditary" influence of the past-history of the source (a fractional correction of the order of $O(c^{-3})$) (Blanchet and Damour (1988)). Concerning the "far-field quadrupole equation" (relating the radiative quadrupole moment to the structure and motion of the material source) one has derived both the $O(c^{-2})$, and the $O(c^{-3})$ corrections to the standard Einstein-Landau-Lifshitz result (Blanchet and Damour (1989)). These terms have been obtained without making use of ill-justified retardation expansions, and are given by integrals having a spatially compact support. Moreover, the dependence of the O (c^{-3}) "tail" term over the remote past history of the system can help to delineate the boundary between the spacetimes admitting a smooth *scri* at future null infinity (studied by Friedrich (1989)), and those that do not peel "as expected" (generic case of Christodoulou and Klainerman (1989)).

J.W. Guinn presented recent progress in the understanding of the radiative decay modes of Schwarzschild black holes ("quasi normal modes") obtained by pushing the WKB approximation of the Regge-Wheeler-Zerilli radial equation to very high orders (Guinn et al. (1989)). The results were compared with those obtained by Leaver (1986) by a completely different approach.

Finally, the talk of C.W. Lincoln presented some intriguing features of the motion and gravitational wave generation of coalescing binary systems (Lincoln (1989)). These features (essentially a richer than expected spectrum of the emitted radiation), if confirmed, would be of interest for the future astronomy of gravitational waves.

References

Barrett, J. W. (1988), *Class. Quant. Grav.* **5**, 1187.

Barrett, J. W. and Williams, R. M. (1988), *Class. Quant. Grav.* **5**, 1543.

Blanchet, L. and Damour, T. (1988), *Phys. Rev.* **D37**, 1410.

Blanchet, L. and Damour, T. (1989), *Ann. Inst. H. Poincaré* **50**, in press; and to be submitted.

Brumberg, V. A. and Kopejkin, S. M. (1989), *Nuov. Cim.* **103B**, 63.

Cheeger, J., Müller, W. and Schrader, R. (1984), *Commun. Math. Phys.* **92**, 405-454.

Christodoulou, D. and Klainerman, S. (1989), to be published.

Damour, T. (1986), in *Gravitational Collapse and Relativity*, eds. H. Sato and T. Nakamura, (World Scientific, Singapore), p. 63.

Damour, T. and Schmidt, B. G. (1989), GR-12 abstract (A5:03).

Ehlers, J. and Lottermoser, M. (1989), in preparation.

Ehlers, J. And Prasanna, A.R. (1989), GR-12 abstract (A5:07).

Ehlers, J. and Walker, M. (1984), in *General Relativity and Gravitation*, ed. B. Bertotti et al., (Reidel, Dordrecht), p. 125.

Friedrich, H. (1989), GR-12 plenary talk; Chapter 3 in these proceedings.

Guinn, J. W., Will, C. M., Schutz, B. F. and Kojima, Y. (1989), to be submitted; see also GR-12 abstract (A5:09).

Kopejkin, S. M. (1988), *Cel. Mech.* **44**, 87.

Kundu, P. K. (1989), GR-12 abstract (A5:14).

Leaver, E. W. (1986) *Phys. Rev.* **D34**, 384.

Lincoln, C. W. (1989), GR-12 abstract (A5:15).

Rendall, A. (1989), Green report MPA 438, Garching.

Schäfer, G. (1989), GR-12 abstract (A5:21).

Schutz, B. F. (1987) in *General Relativity and Gravitation*, ed. M. A. H. MacCallum, (Cambridge University Press, Cambridge), p. 369.

A6

Algebraic computing

M. A. H. MacCallum
Astronomy Unit, School of Mathematical Sciences,
Queen Mary and Westfield College,
Mile End Road, London E1 4NS, U.K.

1. Introductory survey

The implementation of a new computer algebra system is time consuming: designers of general purpose algebra systems usually say it takes about 50 man-years to create a mature and fully functional system. Hence the range of available systems and their capabilities changes little between one GR meeting and the next, despite which there have been significant changes in the period since the last report.

I do not believe there is a single "best" system (though like everybody else I am biased towards the systems I actually use), and in particular one should be extremely cautious about any claims about comparative efficiency of systems.[1] These introductory remarks therefore aim to give a very brief survey of capabilities of the principal available systems and highlight one or two trends. The most recent full survey of computer algebra in relativity (as far as I know) is in Ref. 2, while very full descriptions of the Maple,[3] REDUCE[4] and SHEEP applications[5] will appear in a forthcoming lecture notes volume.

The oldest of the still current general purpose systems are REDUCE and Macsyma. REDUCE is a highly portable system available on a very wide range of machines and is sufficiently cheap to have become widespread in most parts of the world. Its main disadvantage until recently has been the rather small range of auxiliary packages for applied mathematics, but this is improving rapidly with the availability of con-

113

tributed programs through electronic mail (to reduce-netlib@rand.org: an initial mail should contain the one line 'send index').

Macsyma has for many years been the leader in terms of auxiliary packages. It is a very large system (e.g. the running image for a Sun is 13 Mb), widely used in the U.S. and Europe. There are two versions, a commercial product sold by Symbolics, Inc., and a "public domain" version sold by the U.S. Department of Energy.

The small system MuMATH, first distributed as MuMATH-79 for Z80-based microcomputers, and rewritten as MuMATH-83 for the 16-bit IBM PCs, has not been widely used in relativity, perhaps because it seems to use too small a data space to be able to cope with the size of the Einstein equations. In recent years Dyer and Harper of the University of Toronto have created a tensor package MuTensor using MuMATH, which was described at GR-11 and shown to be surprisingly effective. At GR-12 there was no new paper on this package but it was referred to in the papers on Maple (see below). (It should be noted that MuMATH is no longer distributed and has been replaced by DERIVE, which is as yet not programmable.)

More important for relativity are the new systems Maple and Mathematica, both written in C rather than the Lisp which underlies the older systems. (C's object-oriented extension C++ has been used directly for some purposes, as described in the session.) Maple is fast growing to rival Macsyma in its applications packages, and its structure makes it an excellent system for teaching purposes. Although I personally have some criticisms of the design (e.g. the inconveniently unpredictable orderings of the output, the lack of automatic simplifications and the lack of a compiler for Maple code) it is a very good system.

Mathematica is still new: I doubt if it has yet had its 50 man-years of development. It is being very aggressively marketed and its features seem to be designed with the physicist in mind (to a degree where some mathematicians have complained about the system in much the same way they complain about physicists' mathematics in general). Its unique strengths lie in its integration of numerical and graphical facilities with algebraic ones, and its very nice user interface: in these respects it puts other systems, whose progress in these areas, while not non-existent, has been leisurely, to shame. Workers on other computer algebra systems have been hesitant to give an opinion of the algebraic part after the rather acrimonious controversy[6] about the previous algebra system SMP also designed by Mathematica's principal designer, Stephen Wolfram.

The other general purpose systems are little used for relativity, though it may be that, by GR-13, IBM will have released its large and powerful, and very high-level, system SCRATCHPAD II as a commercial product. Moreover, of the specialized systems which were in use among relativists, Camal and Ortocartan have undergone little or no development in recent years, and Schoonschip remains confined to a small number of people working with quantum field theory problems. (As machine facilities have grown the value of other small specialized systems has declined and most of them have become defunct.) Hence the only other system discussed at the GR-12 session was SHEEP, with its extensions CLASSI and STENSOR.

A major change already affecting gravity theorists' use of computer algebra is that home computers have the power of mainframes of 10 years ago, and the software has been transferred to these cheaper but equally powerful machines. I refer of course to machines with 32-bit central processors and memory of at least 1 Mb expandable to 4Mb or beyond: the software has been available on "workstations" or departmental minicomputers somewhat longer.

To be specific, one can now run REDUCE on an Acorn Archimedes or R140, or an Amiga; REDUCE and Maple on an Atari ST; and REDUCE, Maple or Mathematica on an Apple Macintosh. These and Macsyma also run on an IBM PS/2, and a version of SCRATCHPAD will run on that machine too. These algebra systems are all general purpose. Of the specialized systems for relativity, SHEEP and its extensions are the only ones known to me to run on any of those machines (namely on an Atari ST), but Schoonschip also runs on some personal computers.

The programs available for relativists have developed somewhat since GR-11. The main trend is for increased abstraction: more and more programs handle differential forms (e.g. EXCALC, Maple's difforms, Macsyma's cartan) or tensors with symbolic indices (e.g. STENSOR, Math-Tensor, EXCALC) and the "indicial" manipulators can all interface to component-wise computations.

To relativists, algebraic computing is primarily a tool, rather than an end in itself, although interesting and new problems can arise in designing algorithms for relativity. For this reason the number of papers presented to specialized sessions is likely to remain small even if the number of researchers using the tool is rising rapidly (see for example the references to its use in papers on numerical relativity both at GR-12 and GR-11,[7] and elsewhere.[9]) REDUCE developments were not covered by the 11 contributions (9 orally presented and 2 posters) and I will now briefly describe these.

The exterior calculus program EXCALC based on REDUCE is now distributed with REDUCE. For some details of its capabilities, which can be summarized as the ability to handle differential forms in a notation very close to that of a textbook, see Ref. 8. Other REDUCE programs for relativity are described in Refs. 2 and 4. One very useful recent development is a symbiotic system containing REDUCE and SHEEP (or CLASSI or STENSOR) developed by Dr. J. E. F. Skea in my department.[5] The resulting RCLASSI and RTENSOR systems were used in papers described below, though the program is not immediately portable to all REDUCE implementations.

Presentations were given connected with the other systems. Moreover, demonstrations and displays of the systems Macsyma, Maple, Math-Tensor (based on Mathematica), and SHEEP/CLASSI/STENSOR were given in addition to the oral session. (Thanks are due to Drs. Åman, Christensen, Hörnfeldt, Fee, McLenaghan, Parker, Petti and others for providing these, and to the University of Colorado for the loan of equipment.)

2. Papers presented

The first presentation after my introductory survey was "Lobachevski geometry, relativity and object-oriented programming" by Prof. Joseph Dreitlein of Colorado. He described how C++ allows abstract data types to be defined and hence permits a good form of function handling. It is extensible and efficient, and has good capabilities for graphics interfaces, e.g. with the system Iris. The applications he had made included relativistic kinematics and the geometry of Lobachevski space. In particular he mentioned the problems of geodesic motion on a homogeneous space with identifications made in order to compactify the space. In the 2-dimensional case the space might then have genus ≥ 2 and the motion could be chaotic.

The second talk scheduled was "Algebraic computing in gravitation theory" by A. P. Doohovskoy of the Massachusetts Institute of Technology, reporting work using Macsyma. Unfortunately the author did not attend. Instead Dr. Richard Petti, the current head of the Symbolics Macsyma group, gave a short review of the status of relativity programs in Macsyma. The Ctensor and Itensor programs for component and indicial manipulation have recently been corrected and improved. In particular both have been extended to provide handling of frame fields and of affine torsion and nonmetricity. New programs, Cartan for exterior calculus and Atensor for tensor algebra, are now available. Cartan provides handling of Lie derivatives and contractions of vectors with forms as well as the basic exterior algebra and differentiation. Atensor covers multilinear operators, symplectic, Clifford and Grassmannian algebras, and Lie enveloping algebras.

Prof. Ray McLenaghan (Waterloo) then spoke on "General relativity calculations in Maple" (a related paper "Computation of the general Ricci tensor in five dimensions" by Fee, McLenaghan and Pavelle was in the poster session), which also served to introduce the paper "Conversion of spinor equations to dyad form in Maple" by Czapor, McLenaghan and Carminati, which was more fully described by Dr. Czapor. Prof. McLenaghan reviewed the current capabilities of the Maple programs for relativity which provide the following: computation of connection and curvature in coordinate, general moving frame and complex null bases; determination of the Petrov type of the Weyl tensor; computer-aided integration of the Einstein equations in Newman-Penrose and orthonormal bases; computation of integrability conditions; classification of Bianchi group type from structure constants; and some differential form capabilities. A Maple version of the MuTensor package is also available.

In the design of the Maple packages great attention has been paid to algorithm choice and specification.[3] The poster session paper illustrated this: Maple's calculation of the 5-dimensional Ricci tensor had been much faster than the earlier computation by L. Hörnfeldt of the University of Stockholm using STENSOR. (A discrepancy in the numbers of terms found is probably due to the differing representations of the two systems.)

In another session Carminati, Czapor and McLenaghan showed how one of the advanced capabilities of Maple, the Gröbner basis package which provides a systematic method of reduction of sets of polynomial equations (similar capabilities exist in REDUCE and Macsyma), could be used to investigate the integrability conditions of Huygens' principle.

Dr. Czapor's talk concerned the new Maple program which can take spinor equations in indicial form and generate the corresponding dyad equations, i.e. write out all components of the equations using Newman-Penrose notation. Some details of the implementation, such as the representation of spinors as "dynamic tables" in Maple, the procedures which had to be defined, and the miscellaneous utilities required, were given.

The next talks by Dr. Christensen and Prof. Parker concerned their new system MathTensor, which they aim to market commercially as an extension to Mathematica. (This, incidentally, is a new development in computer algebra for relativity, the programs for which, though not the underlying systems, have up-to-now been free.). The title was "Math-Tensor. A system for the manipulation of tensors by computer:" with subtitles "An overview" and "Examples of applications." The overview was given by Prof. Parker, who described the structure of the system as a kernel of functions and rules, the most basic being simplification functions capable of coping with long expressions containing products and contractions of Riemann and other tensors, to which a knowledge base for specific tensors was added. This made the system easy to extend and suitable for pedadogic purposes. The system recognises the summation convention, and 'knows' for example that commuting covariant derivatives leads to terms involving the Riemann tensor. Dr. Christensen's talk gave some detailed examples, in particular of the first and second variations of the gravitational field actions and of the use in computing the "Hamidew coefficients" used in curved space quantum field theory. This system is still under very active development.

Michael Bradley (Stockholm and Turin) described a new SHEEP application in his work on "New cosmological solitonic solutions generated by using SHEEP," work done with Curir and Francaviglia of Turin. They had developed a program, using CLASSI and RCLASSI, capable of performing the Belinskii-Zakharov algorithm for generating new solutions of the vacuum Einstein equations from old. The input consisted of the seed metric and B-Z potential, and some parameters of the solitons. The output could be given in a form suitable for use in a Fortran program to give graphical representations of the behaviour of the metrics, and some of these were presented. (The only paper in this workshop on the classification techniques for exact solutions embodied in CLASSI was "Cylindrically symmetric solutions and the computer data base of exact solutions" by Collins and d'Inverno of Southampton University, which was presented in the poster session. Activity in this area continues to make progress and some further details are given in Ref. 5.

The final two papers concerned the STENSOR system written by Lars Hörnfeldt which provides indicial tensor capabilities for SHEEP. Alex Harvey (City University of New York) presented his joint paper with Hörnfeldt, "Calculation of the Geheniau-Debever scalars by means of

STENSOR." Prof. Harvey gave a brief outline of STENSOR's capabilities for manipulating complicated expressions involving indexed objects, in particular for using textbook type definitions of one tensorial object in terms of another, before illustrating this with the example of the scalar invariants of the Riemann tensor calculated in a geodesic coordinate system.

Finally Abraham Giannopoulos (Imperial College, London) presented his work with Varsha Daftardar on "STENSOR direct evaluation of the Ashtekar variables for any metric." The Ashtekar variables are described in Prof. Ashtekar's plenary lecture (this volume), and the talk concerned a program which can take as input the three-dimensional metric of a slice in spacetime induced by the four-dimensional metric, and a triad, compute such quantities as the Gauss law, Hamiltonian and momentum constraints, and evolution equations and check that they are satisfied by given sets of components (presumably it could also be used in computer-aided integration). This work used RTENSOR.

References

1. M. A. H. MacCallum (1989), Comments on the performance of algebra systems in general relativity; and a recent paper by Nielsen and Pedersen (1989), *ACM SIGSAM Bulletin* **23** 22-25.
2. M. A. H. MacCallum (1989), Symbolic computation in relativity theory, in *EUROCAL 87 (Proceedings of the European Conference on Computer Algebra, Leipzig, 1987*, ed. J. H. Davenport, Lecture Notes in Computer Science (Springer-Verlag, Berlin and Heidelberg).
3. R. G. McLenaghan (1990), Maple applications to general relativity, to appear in *Proceedings of the first Brazilian School on computer algebra*, ed. M. J. Rebouças (Oxford University Press, Oxford).
4. J. D. McCrea (1990), REDUCE in general relativity and Poincaré gauge theory, to appear in *Proceedings of the first Brazilian School on computer algebra*, ed. M. J. Rebouças (Oxford University Press, Oxford).
5. M. A. H. MacCallum and J. E. F. Skea (1990), SHEEP: a computer algebra system for general relativity, to appear in *Proceedings of the first Brazilian School on computer algebra*, ed. M. J. Rebouças (Oxford University Press, Oxford).
6. M. Monagan, G. Gonnet and B. Char (1986), Symbolic mathematical computation, *Communications of the ACM* **29**, 680-682.
7. T. Nakamura (1987), Numerical relativity, in *General Relativity and Gravitation: Proceedings of the 11th International Conference*, ed. M. A. H. MacCallum (Cambridge University Press, London and New York).
8. E. Schrüfer, F. Hehl and J. D. McCrea, (1987), Application of the REDUCE package EXCALC to the Poincare gauge field theory, *Gen. Relat. Grav.* **19**, 197.

9. C. Evans, L. S. Finn and D. Hobill (1989), eds., *Frontiers in numerical relativity*, (Cambridge University Press, London and New York).

A7

Numerical relativity

Joan M. Centrella
Dept. of Physics and Atmospheric Science,
Drexel University, Philadelphia, PA 19104 USA

In this report the current status of numerical relativity is presented. Progress in the field is discussed through reviews of work on gravitational radiation and cosmic strings.

Numerical relativity is a young and vital area, growing in many dimensions. This field has its roots in the effort to calculate the gravitational wave emission from astrophysical sources, and it continues to be energized by the promising new developments in gravitational wave detectors. For example, there are a number of new code-building efforts; several of these are aimed at constructing 3-D codes while others are concerned with putting more realistic physics into 1-D and 2-D codes. There is increased activity in combining analytic and numerical techniques to calculate the gravitational radiation produced. And researchers continue to emphasize the need for improved numerical analysis to insure the accuracy and stability of the codes.

There is also increasing diversity in the types of problems being treated numerically, from cosmic strings to inflation, along with the more traditional stellar collapses and black hole collisions. And new people are joining the community, both those just beginning their research careers as well as more experienced researchers crossing over from other areas. The interaction of these new recruits with the "seasoned veterans" of numerical relativity is producing many new ideas and approaches.

This continued blossoming of numerical relativity is fueled by the growth of computing resources. On the hardware side, supercomputers are becoming increasingly available to the academic community and improvements in workstations are bringing increased power to the scientist's desktop. Software advances are also making a significant impact as numerical relativists continue to embrace visualization and learn to make greater use of symbol manipulation programs. Overall, this improvement in computing resources is a *sine qua non* for the progress that is so clearly visible in the field of numerical relativity.

During the workshop, 6 papers were presented orally, 4 on the topic of gravitational radiation and 2 on cosmic strings. These are listed in Table 1. Summaries of the individual oral papers follow; in cases of multiple authorship, the superscript * indicates the presenter.

Slicing the Schwarzschild Spacetime and *New Results on Black Holes and Brill Waves in Axisymmetric Spacetimes*, D. Bernstein, D. Hobill*, and L. Smarr: A major thrust of this group's work is to develop accurate numerical methods for 2-D axisymmetric calculations of black hole spacetimes that generate gravitational waves. To this end, the Schwarzschild solution was used as a testing ground for methods that will later be used in 2-D, since its properties are well understood and many of them can be calculated analytically. A series of analytic code tests was generated, based on the static, geodesic, and maximal time slicings of Schwarzschild. Codes were written to calculate these properties using various finite difference schemes (Bernstein et al. (1989)) and the results were compared. Among the findings is that, for the time evolution of the central lapse function on maximal slices using second order methods, the Euler, leapfrog, MacCormack, and Brailovskaya methods are more accurate than the Lax-Wendroff method. Improvement is possible in all methods by increasing the number of grid zones.

Also, embedding diagrams of the 3-metric in flat Euclidean space were prepared. A videotape showing the time evolution of the embedded surfaces for different slicing conditions was presented.

The rest of this group's presentation concerned new results for a vacuum 2-D axisymmetric calculation. The gravitational radiation produced by an oscillating black hole was studied using Brill wave initial data with throat boundary conditions. Previous workers in this area had set the shift vector to zero, a gauge condition that produced a numerical instability on the axis (L. Smarr, private communication; J. Thornburg, private communication). By using shift potentials (Evans et al. (1986)) this problem was eliminated. The time evolution of the Newman-Penrose shear was shown to be a well-behaved quadrupolar wave in the calculation. The gravitational wave amplitudes were extracted by using the Abrahams-Evans method (Abrahams and Evans (1988); Abrahams (1989)) of analytically matching the first order multipole moments with the numerical solution in the linear near-zone. The measured quadrupole moment wavelength agreed with the theoretical $\ell = 2$ quasi-normal mode wavelength to within 1%.

Finally, the problem of maximally slicing the Schwarzschild solution was run on the 2-D code and the results compared with those run on a 1-D code. For 100 radial zones, the code results agreed. However, further runs with the 1-D code demonstrated that 1000 zones are needed to reproduce the analytic result accurately for a metric component out to $t = 100$ M, as the lapse collapses and the grid is pulled into the black hole. A 2-D code with this resolution would require about 1000 hours on a Cray X-MP. Adaptive gridding, in which the zones are automatically distributed where they are needed (Evans (1986)), was suggested as a solution to this problem.

In all, the careful, systematic approach to code development and testing taken by this group exemplifies the new generation of work in numerical relativity.

Close Black Hole Binary Systems, S. Detweiler* and J. Blackburn: This work is concerned with binary black hole systems in the regime in which the secular effect of radiation reaction is long compared to the dynamical timescale (Blackburn and Detweiler (1989)). In this case, the geometry is approximately time independent when observed from a frame of reference rotating with the black holes. Analytically, a Killing vector field is imposed on the geometry to represent the time symmetry in the rotating frame. This requires standing wave boundary conditions to keep the holes in steady circular orbits, with no loss of energy or angular momentum. A parametrized sequence of such geometries then gives an approximation to the dynamical evolution of a binary black hole system in this regime.

The 3 + 1 formalism is used. Specializing to the case of vacuum and assuming that the metric and extrinsic curvature do not change in the rotating frame, the constraint and evolution equations are used to derive a set of "steady state" equations (Detweiler (1989)). With these, a variational principle for the total mass of the system is derived. Simply put, it is found that for fixed masses and angular momenta of the holes, the mass of the system is an extremum at solutions of the steady state equations.

In this application the holes are chosen to have no spin angular momenta and to have fixed masses at their infinities; the total angular momentum is also fixed. Only one free parameter, the separation of the holes, is allowed to vary while looking for an extremum of the mass of the system. The variational principle is seeded by constructing parametrized geometries that reduce to known analytic solutions in appropriate limits. The resulting equations are solved numerically. These methods have been tested in the limits of circular orbits of a test particle around a Schwarzschild black hole and in the post-Newtonian limit of two equal mass black holes.

This approach produces a sequence of steady state geometries, parametrized by the separation of the holes, that gives an approximation to the evolution of the binary system. The gravitational radiation emitted is estimated from the binding energy of the system. For the case of equal mass black holes it is found that 3% of the initial mass of the holes is emitted as gravitational radiation before the frequency of the radiation equals the orbital frequency, at which point the steady state approximation breaks down. The frequency of the emitted radiation is less than that of the quadrupole normal mode of the final black hole.

The gravitational radiation can be used to determine the rate of evolution along the sequence. Given the entire sequence of solutions, the frequency and amplitude of the radiation can be determined as a function of time. In all, this work is quite original and very interesting in its own right, and provides an important test-bed calculation for 3-D codes studying the binary black hole problem in the future.

Gravitational Radiation from Stellar Collapse to a Black Hole: Odd- and Even-Parity Perturbations, E. Seidel: In this work, odd- and even-parity perturbations of stellar collapse models are studied numerically, with the emphasis on the gravitational radiation produced during the collapse to a black hole. As such, it extends earlier work on perturbations of dust collapse (Cunningham et al. (1978a,b)); 1980) and perturbations of realistic core bounce models (Seidel et al. (1988)).

The stellar models are produced by a May-White stellar collapse code using a polytropic equation of state. While the coordinate conditions in this type of code eventually break down as the black hole forms, in many cases it is still possible to follow the collapse far enough to obtain useful results on the radiation. A wide range of collapse models was studied, ranging from stars without pressure (i.e. dust collapse) to stars which collapse and then bounce at radii near the Schwarzschild radius.

Preliminary results indicate that while the quasi-normal modes of the black hole generally dominate the emission spectrum, some collapse models show significantly damped and distorted waveforms. Models with greater pressure collapse more slowly and emit lower amplitude ringing radiation. Furthermore, those models which develop shocks that approach the surface of the star as it crosses the peak of the scattering potential can produce an initial transient waveform unlike that which would be expected from purely quasi-normal mode excitation. Following this initial transient, the waveform settles into a more recognizable quasi-normal waveform, although its amplitude may be reduced (when compared to models which collapse without any significant hydrodynamic effect). This work indicates that while quasi-normal ringing is a general feature of nonspherical collapse to a black hole, the details of the collapse can affect both the amplitude and shape of the emitted waveform.

An interesting calculation in its own right, this work also provides an important test-bed problem for calibrating full numerical relativity codes studying stellar collapse.

A View From Null-Infinity: Nonlinear Effects on Radiation, R. Gomez,* J. Winicour, and R. Isaacson: This group is pioneering the use of techniques for constructing spacetimes numerically based on the characteristic initial value problem, in which spacetime is foliated by a set of characteristic hypersurfaces (Gomez and Winicour (1989); Gomez et al. (1989)). This is radically different from the standard $3 + 1$ approach based on the evolution of Cauchy hypersurfaces. Since the radiation fields approach infinity along characteristic hypersurfaces, the extraction of radiation is greatly simplified in comparison with the Cauchy evolution. In addition, compactification methods are used to describe the asymptotic properties of the radiation, thus mapping an infinite radial domain into a finite coordinate region and allowing asymptotic radiative properties to be formulated rigorously. Here this approach is used to evolve scalar fields obeying linear and nonlinear wave equations.

Null coordinates employing retarded time are used, with the initial data being given over a characteristic surface at some fixed retarded time. The scalar wave equation is then converted into an integral equation using an identity based on bicharacteristics. This is then discretized, resulting in an evolution algorithm for marching out to null infinity along successive null cones. This algorithm is calibrated by comparison with analytic solutions and is shown to possess global second order convergence and to satisfy a conservation law. Its stability requirements are also addressed.

This code was then used to study the evolution of both linear and nonlinear scalar waves. Common initial data, restricted for simplicity to spherical symmetry, was chosen for all runs in order to reveal those qualitative properties generic to evolution under different potentials. The linear waves showed no tails, with the entire evolution governed strictly by propagation effects. Solutions with nonlinear potentials showed very interesting features associated with backscattering, the formation of kink-like configurations, and radiative tail decay. In all cases these features were evident in a natural way using the null hypersurface approach, whereas they would have been difficult at best to extract from conventional Cauchy evolutions.

These results are very encouraging. The basic algorithm is directly applicable to other hyperbolic systems such as the Einstein equations, with the addition of radial differential equations. This work has already begun (Isaacson et al. (1985)), and promises to bring powerful techniques to bear on very important problems in gravitational radiation.

Cosmic String Evolution - A Numerical Simulation, B. Allen* and E. Shellard: In this work a network of intercommuting cosmic strings is evolved numerically in a homogeneous radiation-dominated expanding background universe (Allen and Shellard (1989)). In certain particle physics models these strings appear at a very early phase transition in the universe. Long strings cross to form loops, which lose energy through gravitational radiation and can serve as the seeds for the formation of structure. For the model to succeed, a sufficient number of loops must form so that the string network will lose energy and not dominate the energy density of the universe. Previous work has shown that the energy density of the network decreases and approaches a small constant fraction of the total. But there is disagreement over the value of this constant and the means by which this "scaling" solution is approached. This work presents new independent numerical simulations which resolve some of the differences between previous efforts.

The basic difficulty with simulating string networks is that string velocity and direction discontinuities, called "kinks," form when two strings intercommute. To evolve the kinks accurately a different algorithm, drawn from the literature on hyperbolic conservation methods and designed to handle shocks, is used. Careful attention is paid to locating and evolving string segments that cross, with the probability of reconnection chosen as unity. The initial conditions are a random walk with a given correlation length, with non-zero velocities assigned to the points along the string and random orientations given to the strings. Energy conservation and other consistency checks were monitored during the runs.

A number of runs were done varying the initial string density in the network for a fixed set of initial conditions. The energy density in long strings approaches a constant value in a range that agrees with the earlier work of Bennett and Bouchet (1988a,b) but disagrees with that of Albrecht and Turok (1985,1989). Qualitatively, many loops and kinks form. Loops are mainly created on scales much smaller than the horizon and generally fragment further. Video animations of the evolving string system were shown; they point to long string intercommutations and the large number of oppositely propagating kinks as the sources of the small loops. These runs seem to exhibit even more small scale structure than the earlier results of Bennett and Bouchet (1988a,b). Important quantitative questions regarding the production of loops and their final scaling distribution remain, and this code is well-suited to tackling these issues directly.

Electromagnetic Charge and Current Density Distribution Resulting from Superconducting String Collisions, P. Laguna* and R. Matzner: In this work the interaction of two superconducting bosonic cosmic strings is studied numerically to determine the conditions under which the strings intercommute, i.e. reconnect when they cross (Laguna and Matzner (1989)). These strings carry both a string sector "magnetic" field and a current which can sustain superconductivity below a certain critical value. The intercommutation probability is a very important parameter in this scenario since, as with nonconducting cosmic strings, loops must form and radiate their energy away to avoid the domination of the energy of the universe by infinitely long strings. Cosmic string reconnection has been studied extensively for the nonconducting case (Matzner (1988)), and those techniques and codes have been adapted to handle the superconducting case here.

A 3-D code has been written which solves the equations for the evolution of superconducting cosmic strings in flat space, neglecting their gravitational effects. This code has been extensively tested. A videotape showing the collision of two strings carrying 85% of the critical current and approaching at a velocity of 0.75c with a crossing angle of 90° was presented. Two cases were considered: collisions in which the current initially flows in the same direction as the string sector magnetic field in both strings, and collisions in which the direction of the current is reversed in one of the strings.

The results closely resemble the nonconducting simulations, with new features arising due to the currents. When the current in each string is parallel to the string sector magnetic field, the superconducting current is still present along each reconnected string. Point charges appear at and move with the kinks, decreasing in strength as the kinks evolve. In the other case, in which the direction of the current is reversed in one of the strings, the reconnection allows no continuous path for the current to follow. Cancellation of the current begins at the point of collision and travels out with the kinks. Point charges form at the kinks, and charge accumulation develops in the region where the current cancels. In all cases, an expanding spherical front of electromagnetic radiation develops at the point of intercommutation. This bubble could provide a mechanism for sweeping away the surrounding plasma and thus driving the formation of large scale structure.

In addition to the oral papers, 15 papers were presented as posters; these are listed in Table 2. The diversity of topics and approaches used in these papers further testifies to the vitality of this young field.

I thank all of those who participated in the Numerical Relativity Workshop at GR12 and thereby made it a success. In particular, I thank the authors of the oral papers, who provided me with extended abstracts, preprints, copies of view graphs, and very useful conversations, all of which contributed substantially to the preparation of this review.

References

Abrahams, A. (1989), in *Frontiers of Numerical Relativity*, eds. C. Evans, L. Finn, and D. Hobill (Cambridge University Press, Cambridge), p. 110.

Abrahams, A. and Evans, C. (1988), *Phys. Rev.* **D37**, 318.

Albrecht, A. and Turok, N. (1985), *Phys. Rev. Lett.* **54**, 1868.

Albrecht, A. and Turok, N. (1989), preprint.

Allen, B. and Shellard, B. (1989), preprint.

Bennett, D. and Bouchet, F. (1988*a*), *Phys. Rev. Lett.* **60**, 257.

Bennett, D. and Bouchet, F. (1988*b*), in *Cosmic Strings: The Current Status*, eds. F. Accetta and L. Krause (World Scientific, Singapore), p. 74.

Bernstein, D., Hobill, D., and Smarr, L. (1989), in *Frontiers of Numerical Relativity*, eds. C. Evans, L. Finn and D. Hobill (Cambridge University Press, Cambridge), p. 57.

Blackburn, J. and Detweiler, S. (1989), preprint.

Cunningham, C., Price, R., and Moncrief, V. (1978*a*), *Ap. J.* **224**, 643.

Cunningham, C., Price, R., and Moncrief, V. (1978*b*), *Ap. J.* **230**, 870.
Cunningham, C., Price, R., and Moncrief, V. (1980), *Ap. J.* **236**, 674.
Detweiler, S. (1989), in *Frontiers of Numerical Relativity*, eds. C. Evans, L. Finn and D. Hobill (Cambridge University Press, Cambridge), p. 43.
Evans,C., Smarr, L., and Wilson, J. (1986), in *Astrophysical Radiation Hydrodynamics*, eds. K.-H. Winkler and M. Norman (Reidel, Dordrecht), p. 491.
Evans, C. (1986), in *Dynamical Spacetimes and Numerical Relativity*, ed. J. Centrella (Cambridge University Press, Cambridge), p. 3.
Gomez, R. and Winicour, J. (1989), in *Frontiers of Numerical Relativity*, eds. C. Evans, L. Finn and D. Hobill (Cambridge University Press, Cambridge), p. 385.
Gomez, R., Winicour, J., and Isaacson, R. (1989), preprint.
Isaacson, R., Welling, J., and Winicour, J. (1985), *J. Math. Phys.* **24**, 1824.
Laguna, P. and Matzner, R. (1989), *Phys. Rev.* **D** (to be published).
Matzner, R. (1988), *Computers in Physics* **2**, no. 5, 51.
Seidel, E., Myra, E., and Moore, T. (1988), *Phys. Rev.* **D38**, 2349.

Table 1.

Oral papers; * indicates presenter.

D. Bernstein, D. Hobill* and L. Smarr, *Slicing Schwarzschild Spacetime and New Results on Black Holes and Brill Waves in Axisymmetric Spacetimes*

S. Detweiler* and J. Blackburn, *Close Black Hole Binary Systems*

E. Seidel, *Gravitational Radiation from Stellar Collapse to a Black Hole: Odd- and Even- Parity Perturbations*

R. Gomez*, J. Winicour and R. Isaacson, *A View From Null-Infinity: Nonlinear Effects on Radiation*

B. Allen* and E. Shellard, *Cosmic String Evolution-A Numerical Simulation,*

P. Laguna* and R. Matzner, *Electromagnetic Charge and Current Density Distribution Resulting from Superconducting String Collisions*

Table 2.

Poster papers.

B. K. Berger and R. Church, *Monte Carlo Simulations of Quantized Cosmologies*

D. Bernstein, D. Hobill and L. Smarr, *Finite Difference Schemes for Numerical Relativity*

N. Bishop, C. Clarke and R. d'Inverno, *The Characteristic Initial Value Problem on a Transputer Array*

G. Cook, *Black Hole Initial Value Problems and Apparent Horizons*

G. Duncan, *Head-On Collision of a Black Hole and a Star*

D. Goldwirth, *A Spherical Relativistic Code for Inhomogeneous Inflation*

A. Kheyfets, W. Miller and W. Zurek, *3-D Covariant Hydrodynamics on a Curved Background*

Y. Kojima and B. Schutz, *3-D Hydrodynamics Code with Newtonian Gravity*

N. LaFave, *The Mathematics of Simplicial Lattices*

P. Laguna and D. Garfinkle, *Gravitational Properties of Long Straight Cosmic Strings*

J. McCracken and W. Miller, *Geometric Computing in GR: An $S^3 x S^3$ Null-Strut Calculus Algorithm*

A. Mezzacappa, R. Matzner and S. Bludman, *Improved Neutrino Transport in Stellar Collapse and Neutrino Spectra from Black Hole Birth*

P. Schinder and S. Bludman, *General Relativistic Neutrino Transport and Hydrodynamics in Spherically Symmetric Space-Time*

E. Seidel and W-M. Suen, *Collapse of a Self-Gravitating, Complex Scalar Field*

J. Thornburg, *The Time Evolution of Black Hole Spacetimes*

Part B.

Relativistic astrophysics, early universe, and classical cosmology

5

Gravitational lenses: theory and interpretation

R. T. Blandford
Theoretical Astrophysics,
California Institute of Technology,
Pasadena, CA 91125 USA

The theory of imaging of cosmologically distant point and extended sources, specifically quasars and galaxies by an intervening mass is described. Particular attention is paid to formalisms which allow one to understand the qualitative principles governing image formation. The importance of caustics is emphasized, particularly their role in the formation of highly magnified images. The prospects for measuring the Hubble constant and the cosmological density parameter are reviewed.

1. Introduction

The history of gravitational lensing is one in which general relativists can take some pride. The basic effect was anticipated by many researchers including Einstein (1936); Refsdal (1964); Press and Gunn (1973); Bourrassa and Kantowski (1975); long before the discovery by Walsh, Carswell and Weymann (1981) of the first convincing example of multiple imaging of a background quasar by an intervening galaxy. This is perhaps not too surprising since gravitational lensing is an almost trivial consequence of the general theory. What was surprising was how rich a field the elementary geometrical optics of gravitational lensing has become when stimulated by the observational discoveries reviewed here by Bernard Fort (Chapter 6, this volume). I intend to review some of the theoretical approaches to gravitational lenses that have been developed over the past ten years emphasizing those that are directly relevant to interpreting the observations.

Gravitational lenses have been heralded as important astronomical tools; specifically they are probes of the dark matter found in the outer parts of galaxies, rich clusters of galaxies and perhaps also the universe at large. It is a happy thought that we can use photons to probe the gravitational potential of a galaxy or cluster in much the same way that we use beams of relativistic electrons to explore an atomic nucleus. Lenses have also been welcomed as instruments for performing basic cosmography, specifically for measuring the Hubble constant and the deceleration parameter of the universe by monitoring the source variability in two or more of the images and by observing sources at two or more redshifts with the same lens. As will be discussed below, there are practical difficulties in implementing these measurements. A third astronomical use of gravitational lenses is as giant, though optically flawed, telescopes. Suitably located sources can be magnified by large factors, perhaps exceeding several hundred, and this allows us to see intrinsically fainter sources as well as examine their internal structure and spectra. With the discovery of arcs and rings, this capability, which was anticipated many years ago by Zwicky, is starting to be realized. Finally, by exploiting the stellar granularity in galactic lenses, it may be possible to probe regions within a quasar on scales of microarcseconds, finer than the angular resolution likely to be achieved otherwise within the foreseeable future.

Gravitational lenses therefore hold a strong interest for astronomers because they can satisfy their desire for larger telescopes and at the same time may provide a powerful tool for accomplishing some of the most important tasks in extragalactic astronomy. Their present interest to relativists is less obvious. They are surely not useful tests of the general theory; the measurement of the "Shapiro effect" and (less accurately) the solar light deflection (Shapiro, Chapter 12, this volume) already furnishes a superior determination of the post-Newtonian parameter γ, than cosmologically distant lenses are ever likely to provide. The discovery of "exotic" lenses, such as cosmic strings or isolated supermassive black holes could conceivably allow tests of strong field relativity. Unfortunately, there is no compelling evidence for the existence of lenses of this type; instead, there exist discouraging astrophysical limits on their space density.

Gravitational lenses have also not been very useful so far for cosmology. It is proving extremely difficult to extract useful quantitative information like the Hubble constant or the density parameter even within the framework of the simplest Friedmann world models. So, although most astronomers working on lenses are optimistic about the future, the existing examples can be, and indeed are, discussed using quasi-Newtonian methods. Photons are treated as particles moving in a Newtonian gravitational field with a ratio of gravitational to inertial mass of two and the universe is effectively flat with object and image distances given by the lens and source redshifts.

However, there are subtleties in the physics. These come in the handling of the geometrical optics and it is perhaps here that the relativist might recognise the primitive application of familiar techniques. For this reason, I shall emphasize methods for handling gravitational lens optics at the expense of both formal relativity and the astrophysical details of individual cases.

Recent reviews of gravitational lensing containing an extensive bibliography, can be found in the conference proceedings edited by Moran, Hewitt and Lo (1989); Kaiser and Lasenby (1979); the lecture notes of Blandford and Kochanek (1987); forthcoming conference proceedings edited by Mellier and Soucail (1990) and a forthcoming monograph by Schneider and Ehlers (1990).

2. Multiple imaging of point-like sources

Orders of magnitude

In order to fix ideas we should give some orders of magnitude. As is well known, the deflection of a ray at the solar limb is

$$\alpha = \frac{4M_\odot}{R_\odot} = 1.75''$$

($c = G = 1$ throughout). In general the deflection due to a self-gravitating body is proportional to the square of the escape velocity over the speed of light. For a galaxy or a cluster of galaxies, modeled as an isothermal sphere,

$$\alpha = 4\pi\sigma^2 \simeq 2'' \left(\frac{\sigma}{300 \text{ km s}^{-1}} \right).$$

Galaxies, with one dimensional velocity dispersions of $\sigma \sim 300$ km s^{-1} split images by one or two arc seconds; clusters with velocity dispersions typically three times larger can produce splitting as large as half a minute of arc. Multiple imaging clearly requires that the lens be as large as the impact parameter. Alternatively the surface density of mass within the lens must exceed $\sim 1/4\pi D$, where D is the distance to the source. For cosmologically distant sources, this can be evaluated as ~ 1 g cm^{-2}, a useful rule of thumb.

Fermat's principle

A convenient and elegant way to understand the formation of images, introduced by Nityananda, is to use Fermat's principle (cf. Schneider (1985); Blandford and Narayan (1986)). Let us consider a single lens plane and null geodesics connecting points on this plane (labelled by angular coordinates $r \ll 1$ on the sky as seen by the observer) to both the source and the observer. In the weak field limit, the metric near the lens can be written

$$ds^2 = -(1 + 2\phi)dt^2 + (1 - 2\phi)dx^i dx_i + O(\phi^2),$$

where ϕ is the Newtonian potential. Now let us compute the travel time along the two geodesics relative to the travel time along a single geodesic in the absence of the lens.

$$t = (1 + z_L)\left(\frac{D_{OS}D_{OL}}{2D_{LS}}(\mathbf{r} - \rho)^2 - 2\int \phi dz\right), \qquad (1).$$

where ρ labels the position of the source (in the absence of the lens). The distances D_{ij} are angular diameter distances that connect the observer (O), lens (L) and source (S). In a Friedmann cosmology, these can be evaluated as

$$D_{ij} = \frac{2}{\Omega_0^2 H_0}\frac{(G_i G_j + \Omega_0 - 1)(G_j - G_i)}{(1 + z_i)(1 + z_j)^2},$$

where $G_i = (1 + \Omega_0 z_i)^{1/2}$, z_i is the redshift and H_0 is the Hubble constant. The potential integral is recognized as the two dimensional or surface potential. It is convenient to introduce a scaled time so that

$$\tau = \frac{D_{LS}t}{(1 + z_L)D_{OL}D_{OS}} = \frac{1}{2}(\mathbf{r} - \rho)^2 - \psi(r),$$

where $\psi = 2D_{LS}\int \phi dz/D_{OL}D_{OS}$ satisfies a two dimensional Poisson equation

$$\nabla^2 \psi = \frac{2\Sigma}{\Sigma_{crit}}.$$

The differentiation is with respect to $\mathbf{r}$. $\Sigma_{crit} = D_{OS}/4\pi D_{OL}D_{LS}$ is the critical surface density for the formation of multiple images.

Now, by Fermat's principle, images of the source will be located where the travel time τ is stationary, i.e. where

$$\mathbf{r} = \boldsymbol{\rho} - \nabla\psi. \qquad (2)$$

The second term on the right hand side of this vector equation can be recognized as D_{LS}/D_{OS} times the deflection by the lens, and the equation can be immediately interpreted geometrically. This equation can be inverted to describe the one to many mapping from the source plane onto the image plane.

If we consider plotting contours of τ, then, in the absence of a lens, there will be a single minimum corresponding to the rectilinear propagation of light. (See Figure 1.) Adding a small mass displaces this minimum slightly away from the mass, corresponding to deflection of the ray. When a sufficiently large mass is introduced into the lens plane, extra stationary points will be formed in pairs, comprising a saddle and either a maximum or a minimum. We therefore see that a transparent, non-singular lens should form an odd number of images in total.

It is also possible to use this arrival time contour map to understand the magnification of images. As surface brightness is conserved through the lens, the flux magnification M is given by the ratio of the angular area of the image to that of the source (assumed much smaller than any structure in the lens). In other words,

$$M^{-1} = ||\frac{\partial\boldsymbol{\rho}}{\partial\mathbf{r}}|| = ||\nabla_i\nabla_j\tau||.$$

The magnification is therefore the reciprocal of the Gaussian curvature of the time delay surface, evaluated at the stationary points. Considering the arrival time contours, we see that the surface is quite flat when two images almost merge as in Q1115+080. Conversely, maxima located in the centers of lensing galaxies are likely to be highly curved, reflecting the shape of the gravitational potential well. The associated images will be highly deamplified and probably not detected at all and this can account for the fact that in practice, an even number of images is generally observed.

If we treat $\nabla_i\nabla_j$ as a tensor, then it also describes the shear of the image induced by the lens. Saddles therefore correspond to negative parity images and maxima to images that have been inverted twice and consequently have positive parity. In Q0957+561, the two images A, B have radio counterparts that have been resolved using Very Long Baseline Interferometry (VLBI). These images have opposite parity as the gravitational lens model requires. Now the heights of the stationary points

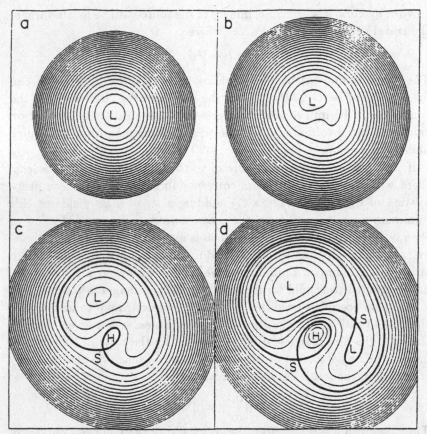

Fig. 1. Virtual arrival time contours for single elliptical lens of increasing strength. Images are formed at the stationary points, lows, saddles and highs. The times at the stationary points measure the time delays that would be measured by an observer monitoring a variable source. It can be seen that only certain combinations of arrival time order and parity are allowed. a) No lens. The single image is formed at the minimum of the paraboloidal time surface, corresponding to the rectilinear propagation of light. b) A weak lens deflects this minimum. c) An even stronger lens is able to create three images, a low, a saddle, and a high, for which the separatrix has the shape of a limaçon. The image locations correspond to those observed in Q0957+561. d) A stronger lens creates an additional low and saddle, with an extra separatrix in the shape of a lemniscate. The saddle and the low that are close to merging are very bright. This image arrangement is similar to that observed in Q1115+080. If the high and the other saddle were to merge then we would be left with an alternative three image topology to case c).

are just the additional time delays and so if the source varies then the order in which the images vary is associated with a particular ordering of the parities. In order to understand this we should draw the possible topologies by nesting the separatrix contours that pass through the saddle points. We find that there are two possible three-image topologies, known as a lemniscate and a limaçon, six five-image topologies and so on. (See Fig. 1.)

If massive black holes or cosmic strings really exist, then their arrival time surfaces are limiting cases of the limaçon and lemniscate respectively. The two images have opposite parity in the case of black holes and the same parity in the case of strings.

Using this approach it is also possible to prove that minima must be magnified and that an image known to be located on a maximum must be formed by a ray that has passed through at least a critical density of matter.

Elliptical lenses

Naturally occurring gravitational lenses differ from simple Gaussian lenses in that the deflection is generally not directly proportional to the displacement from the optic axis. This allows multiple images to form. In addition, natural gravitational lenses are generally non-circular (excepting isolated point masses). The breaking of the circular symmetry has a profound effect on the disposition and amplification of the images (cf. Narayan and Grossman in Moran et al. (1989)).

Transparent elliptical lenses form either one, three or five images of a point source. At high amplification, five image configurations have the largest cross section and seem to predominate among bright, multiply-imaged quasars. See Figure 2.

Caustics

If we consider a highly elliptical lens that generates "lemniscate" arrival time contours and allow the source to move in the source plane, then the saddle will approach one of the minima until it merges and both images increase in flux and then disappear. The locus of the merging images in the image plane is known as the *critical line*; the associated locus of the source is the *caustic*. Caustic lines divide the source plane into regions of image multiplicity. In this case, there are three images formed when the source lies within the caustic curve and only one image when it is located outside. Caustics are examples of catastrophes-generic singularities of Lagrangian mappings. There are two possibilities in a plane, *folds* involving the merging of two images and *cusps*, where three images are involved. Complex caustic networks can be created when the lens comprises several masses.

There are some simple asymptotic scaling laws associated with these two types of catastrophe. For example, if we Taylor-expand the reciprocal of the magnification about the critical curve, then the leading term is linear in the separation of the images from the critical curve. Now, the cross section for forming a pair of images both with magnification in excess of M is proportional to the relevant area of the source plane

$$\sigma(> M) \propto \int d^2\rho = \int d^2 r M^{-1} \propto M^{-2}.$$

Under quite general conditions, the fraction of bright images will diminish asymptotically as the inverse square of the magnification. Now, it is observed that the number of quasars more luminous than some luminosity L increase $\propto L^{-2.5}$, for bright quasars. Therefore the majority of bright, multiply-imaged quasars should be highly magnified, as indeed appears to be the case. This phenomenon goes by the name of *amplification bias*. It explains qualitatively why surveys of the very brightest quasars have been conspicuously successful in finding multiply imaged cases. However, when we look at fainter quasars, we find that the integral luminosity function only steepens $\propto L^{-1.5}$ and so amplification bias should not be important for fainter quasars (as also appears to be the case).

When we examine the cusp catastrophe, we find that the cross section for forming three images all magnified by more than M decreases as $M^{-2.5}$ and so cusp images ought not to show much amplification bias. Bright single images, as well as triple images can be formed when the source is located close to a cusp. When $M \gg 1$, we find that the cross section for forming a magnified single image is sixteen times the cross section for forming three images all magnified by more than M (Blandford, Kochanek and Soucail, in preparation). Another example of a scaling law, applicable when there are three bright images near a cusp, is that the amplifications of any pair are in inverse proportion to their measured separations from the third image. This might find an application if a supernova exploded in a triply-imaged galaxy located close to a cusp (Kovner and Paczyński (1988)).

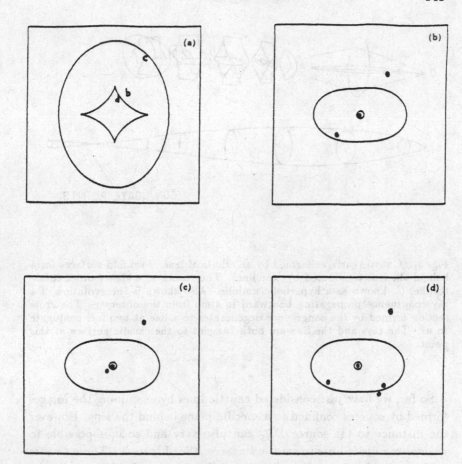

Fig. 2. Image arrangements produced by an elliptical gravitational lens. a) Source plane. The caustics divide the source plane into regions where 1, 3 or 5 images are produced. In this case, the inner caustic has four cusps and a source within it will generate five images. As the ellipticity in the lens is reduced, this caustic shrinks to a point. The outer caustic separates the three image region from the single image region. b) Critical curves and images in the image plane. When the source is located at b in the source plane, three images are formed. The brightest two are almost diametrically opposed. c) When the source is located at c, the brightest two images are said to be "allied" and straddle the smaller radial critical curve. d) A source at d produces five images in total. In this example, the source is close to the caustic and the two brightest images are separated tangentially on either side of the outer critical curve.

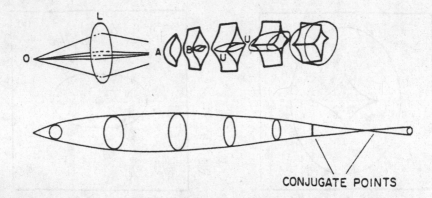

Fig. 3.   Caustic surfaces formed by an elliptical lens. Two fold surfaces form behind the lens terminated by cusp lines. They touch at a higher order catastrophe, U, known as a hyperbolic umbilic. Also shown is the evolution of a ray congruence propagating backward in time from the observer. The cross section formed by the congruence degenerates to a line at two foci conjugate to us. The rays and the line are both tangent to the caustic surface at this point.

So far, we have just considered caustic lines by examining the images formed by sources confined to a specific plane behind the lens. However the distance to the source D_{OS} can also vary and so it is possible to generalise caustic lines to caustic surfaces. (See Figure 3.) These caustic surfaces divide space up into regions where 1, 3 and 5 images can be formed. The caustic surface comprises fold surfaces that meet at cusp lines. Cusp lines can meet at higher order catastrophe points. In the case of a simple elliptical lens, these catastrophes are of hyperbolic umbilic type. More generally, they can also be of swallowtail or elliptic umbilic form. If we think of space as being filled with possible sources, then it is those that are just within the caustic surfaces that are highly amplified and create distinctive image arrangements.

There is a quite different, though no less illuminating way to visualise these caustic surfaces. (See Fig. 3.) This involves using the optical scalar equations to trace null geodesic congruences backward in time from the observer, through the lens to the source region. The expansion θ and the (complex) shear σ evolve according to

$$\dot{\theta} + \theta^2 + |\sigma|^2 = -R,$$

$$\dot{\sigma} + 2\theta\sigma = F,$$

where the dot signifies differentiation with respect to the affine parameter, essentially $\lambda = \int dt/(1+z)$, R is the Ricci scalar and the complex number F, which is related to the Weyl tensor, describes the gravitational field due to matter lying outside the congruence. If we track the cross-sectional area of a congruence then it will increase, reach a maximum, and then decrease to zero at a conjugate point where the cross section collapses to a line. The congruence is tangent to the caustic surfaces at a conjugate point and the caustic surface is formed as an envelope by the rays as they propagate backwards from the observer.

Microlensing

Soon after the discovery of the first gravitational lens, it was realized that the granularity in the potential caused by the stars might create additional sub- or micro- images of a sufficiently compact source. Although these microimages would be far too small to resolve, they might nevertheless cause a significant change to the integrated flux associated with the macroimage. The most dramatic changes occur when the source crosses a caustic formed by the interaction of the stellar and the background lens gravitational fields. As the caustic is crossed the observed flux will rise to a value limited by the angular size of the source. Furthermore, as there will almost certainly be some relative transverse motion between the stars and the line to the source, variability of the images will be induced. Now if a star, of mass M is able to form microimages, then the impact parameter of the ray is $b \sim (GMD/c^2)^{1/2}$ which is of order 10^{16} cm for solar type stars at cosmological distances. The probability that the flux is altered significantly by microlensing is called the optical depth $\tau \sim \Sigma_*/\Sigma_{crit}$, for $\tau \lesssim 1$. Here Σ_* is the surface density in stars. If the maximum amplification will be limited to $\sim (b/s)^{1/2}$, the time scale for the flux to vary due to microlensing is $t_{var} \sim b/v_\perp$, where $v_\perp$ is the relative transverse velocity. Typically this is ~ 10 yr. However, as the high magnifications are attributable to caustic crossings, we again find that the probability for a magnification in excess of M scales $\propto M^{-2}$.

The details of microlensing actually depend upon whether or not the macro image formed by the large scale gravitational field is a minimum, a saddle or a maximum. Conversely, if microlensing fluctuations are found, then they may provide another way to distinguish the image parities. When the macroimage is a maximum, then the star can completely

completely "swallow" it (e.g., Chang and Refsdal (1981)). Microlensing also depends upon the optical depth. When the optical depth is small $\Sigma_* \ll \Sigma_{crit}$, then only the single nearest star need be considered and there will only be occasional incidences of rapidly changing luminosity. When $\Sigma_* \sim \Sigma_{crit}$, several stars act nonlinearly to create an irregular caustic network in the source plane and a few subimages can be observed at any time so that the most dramatic changes in flux should be seen. However, when $\Sigma_* \gg \Sigma_{crit}$, (as can occur for images formed in dense galactic nuclei), there are so many subimages that the overall fluctuations are reduced. In this case, it is more useful to compute a point spread function which has the form of a central Gaussian core and an intensity $\propto \theta^{-4}$ at large distances.

The phenomenon of microlensing which should certainly occur under suitable conditions, has been frequently invoked to account for features of the observations of gravitational lenses. Notably there have been claims of observing a statistical excess of bright galaxies around nearby, bright galaxies. (Such claims are highly controversial.) Conversely, and more reliably there appears to be a marginally significant excess of cosmologically distant galaxies close to a complete sample of high redshift quasars. (About four times as many galaxies are observed as would be expected in random superpositions.) As most of these quasars are of intermediate luminosity, amplification bias caused by either the smoothed out galaxy or the constituent stars, can account for no more than a factor ~ 2 of enhancement. A larger sample of quasars must be examined before it can be concluded that there is an extra excess to be explained. However, it is clear that it may be possible to set some limits on either the sizes of the continuum-emitting region within the quasar sources or on the nature of the dark matter that dominates the mass density in the outer parts of galaxies. Two candidates for this dark matter are black holes of mass in excess of that of the sun and consequently unobservably slow variation, and low mass stars or Jupiters, for which variation should generally be seen.

Microlensing has also been invoked to account for the observed evolution of quasars, by postulating that most of the bright objects are highly amplified. However, this is now thought not to be a significant effect as the shape of luminosity function that would be produced does not correspond to that which is observed.

Microlensing is hard to avoid in the case of the gravitational lens 2237+035 where four images are observed around the center of a nearby spiral galaxy. Here, we know that most of the mass is stellar and can calculate that its surface density is about half the critical value. Provided that the source is sufficiently compact, independent large changes in luminosity should be seen in all four images with timescales of about a year. These can be distinguished from variations intrinsic to the source for which the relative delays should be typically about a week. Recently, all four images have been clearly resolved under conditions of excellent seeing and there are preliminary reports of brightness changes in two of the images (Hewett et. al. (1989), preprint).

3.   Arcs and rings - imaging of extended sources

Image formation by nearly-circular lenses

Although analysis of the location and fluxes of images of point sources can tell us quite a lot about the surface potential of the lens, we learn much more from the images of extended objects. This is because we can now sample the potential over an extended area rather than just at isolated points. In addition, the concern about microlensing changing the amplifications independently disappears. The spectacular discoveries of arcs and rings have transformed this into reality (Soucail et al. (1987); Lynds and Petrosian (1987); Hewett et al. (1988)).

In order to understand how these images are formed, we should first consider image formation by a single circular, though non-linear lens. Provided that the bending angle increases no more rapidly than linearly with distance from the optic axis, there will be a single radius where a source located on the optic axis can form a ring image. This is known as the Einstein ring. The average surface density interior to this ring equals the critical density. Now let us trace a radial *ribbon* of rays backwards from the observer to the source plane. As the lens is circular, a short radial segment on the image plane, cutting the Einstein ring, maps onto a short radial *spoke* on the source plane passing through the optic axis. We can map many such ribbons onto the source plane, each ribbon creating a spoke which passes through the optic axis when there is circular symmetry. (See Figure 4.)

The image arrangement from this type of lens is structurally unstable and if we introduce some small non-circularity, then the imaging changes. (See Figure 5.) For an elliptical perturbation, the ribbons will be deflected slightly tangentially as they cross the lens plane, giving similar transverse displacements to the spokes in the source plane. If we now superpose many spokes, their envelope forms a four-cusped astroid

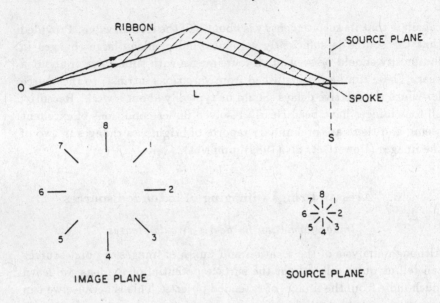

Fig. 4. Ribbons of rays propagating backwards from us past a circularly symmetric lens. The ribbons, labelled 1-8, intersect the source plane in radial spokes.

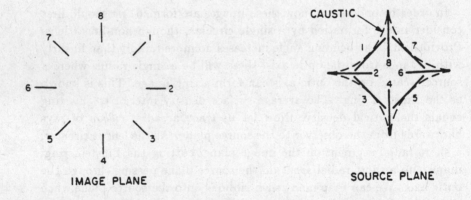

Fig. 5. As for Fig. 4, except that a small elliptical perturbation is introduced into the lens. The spokes in the source plane no longer pass through a common point and instead combine to form an astroid-shaped caustic envelope.

curve. This is the caustic curve because any source point that lies on it maps onto two merging images near the Einstein ring. This motivates

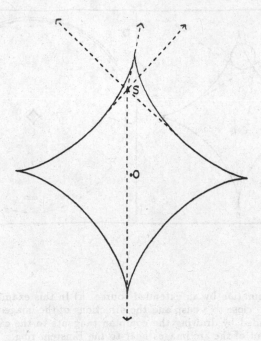

Fig. 6. Directions on the sky of images formed by a point source, S, located within an astroid caustic. The directions are tangent to the caustic.

an approximate geometrical construction for locating the images formed by a point source. (See Figure 6.) Firstly compute the caustic curve on the sky and then draw all the tangents to the caustic and extend them to the Einstein ring. When the point lies within an astroid caustic there will be four tangents and four images; when it lies outside, there will be two tangents and images. Source points on the tangent line close to the caustic will be translated along the line to the critical curve, usually with some enlargement. If the source lies close to, and within, a cusp, then three nearby images will be created.

This construction can be used with extended sources as well. (See Figure 7.) It is simply necessary to construct the common tangents to the caustic and the source to determine the extent of the arc image. When the source overlaps the interior of a cusp, a large arc will be created. Those source points that lie within the cusp will be triply-imaged, those that lie outside will be singly imaged. However, the tangents to the opposite cusp will only extend over a limited angle and only a small

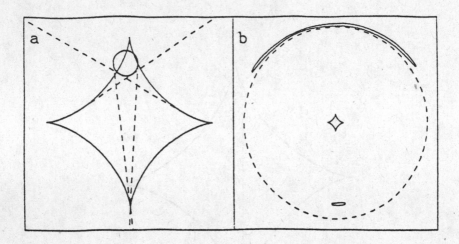

Fig. 7. Arc formation by an extended source. a) In this example, a circular source is located close to a cusp and the directions of the images of the source points are bounded by drawing the common tangents to the caustic and the source. b) Extent of the arc images near to the Einstein ring.

counter arc will be created. Sources that are not near to cusps create shorter arcs.

These features are just what is seen in the observations of cluster arcs, like that in Abell 370. The sources are believed to be mostly high redshift blue galaxies that have been counted most recently by Tyson, Valdes and Werk (1988), preprint. They are on average about ten arc seconds apart and this gives a sufficient sky density to make arcs common in those clusters that have a high enough central surface density to create a central astroid (or similar) caustic.

The same explanation suffices for a qualitative understanding of the two known radio rings (MG1131+0456, MG1634+1346). Here partial or complete ring images of background radio sources with angular diameters of a few arc seconds, are formed of background radio sources by a radio-quiet, intervening galaxy lens.

Modeling of lenses

As we have remarked, the images of extended sources constrain the lens models far more than the location and the amplification of individual images. It is therefore important to demonstrate that the observed arcs can indeed be reproduced by cluster lenses in which the mass has two components, one tracing the observed galaxies and the other representing the dark matter. This last is constrained so that the velocity dispersion of the galaxies has the observed value. It is found that arcs similar in shape and length to those observed can be reproduced in this manner, thereby strengthening our belief in the gravitational lens explanation.

A more sophisticated technique that we (Kochanek, Blandford, Lawrence and Narayan (1989)) have found to be very useful for the radio rings is to parametrize the lens model, using $\lesssim 10$ parameters, e.g. the ellipticity and orientation of the potential well. We then use the fact that the intensities of the images of multiply-imaged source points should be identical. We can compute the dispersion in the intensity for these multiple image points for any suitable lens model. We then vary the lens parameters so as to minimize this dispersion. In this way, we can derive a best-fitting lens and source model. We find that the resulting lens is capable of reproducing the radio structure observed at a different frequency as well as the polarization distribution (the polarization is parallel-propagated through the lens). This procedure produces a quite common source structure. The method works fairly well for the radio rings because the lenses are relatively isolated. It should also be applicable to the cluster arcs if and when they are observed with sufficient angular resolution.

4. Cosmological applications

Hubble constant

As was first recognized by Refsdal (1964), the measurement of the time lag between source variation as observed in two or more images provides a direct determination of the Hubble constant. Specifically, the combination $D_L D_{LS}/D_S$ which appears in Eq. (1) is inversely proportional to the Hubble constant. For the double quasar Q0957+561, the B image is anticipated to vary between one and one and a half years after the A image. Unfortunately, it has proved difficult to measure this quantity uncontroversially so far, despite the fact that the two best determinations (~ 1.1 yr, Schild, Vanderiest, private communication) agree within their stated errors. However, even when the source varies enough for an accurate measurement of the delay, the derived Hubble constant will remain controversial because of a lack of confidence in the lens model.

Part of the reason is that we know that the mass in clusters is mainly "dark" and we cannot determine its distribution accurately. One simple way to see how this influences the modeling is to observe that a large, uniform density, circular sheet of matter, centered on the image creates a jump in the expansion θ which is indistinguishable from a change in the distance to the source. Another problem is that deflection is produced by the component of gravitational acceleration perpendicular to the line of sight, whereas the galaxy velocities are controlled more by the parallel component of the gravitational acceleration. It is generally assumed in modeling that the cluster is spherically symmetric, whereas observed clusters are quite elliptical. There is probably a selection bias in favor of clusters that are elongated along our line of sight.

Finally, even if all these difficulties are overcome in some source, the contribution of additional mass along the line of sight will be unknown. It is usually hoped that the probability of having two lenses along a line of sight is small enough to be discounted. However, there are strong selection effects in favor of such an alignment. Stochastic perturbations induced by the many high redshift galaxies as well as fluctuations in the distribution of dark matter have also to be taken into account.

Perhaps the best hope for measuring the Hubble constant lies with the radio rings. Here, the lens model can be determined highly redundantly, especially if the critical curve can be traced on the sky using high resolution VLBI images.

Cosmological density parameter

In principle, measuring the relative values of D_{LS}/D_{OS} for two sources at different redshifts yields a value for the cosmological density parameter Ω_0 (assuming that a Friedmann-Robertson-Walker world model is adequate). However, in practice this quantity is highly insensitive to Ω_0. It is therefore more useful to use multi-arc cluster lenses to measure the masses in outer parts of rich clusters essentially independent of the cosmology.

Microwave background

One of the strongest constraints in observational cosmology derives from the present failure to observe fluctuations in the temperature of the microwave background on angular scales from $\sim$ 6 arcsec to degrees. These fluctuations should be imposed by density fluctuations at the last scattering surface, presumably during the epoch of recombination. This failure is already quite inconsistent with several theories for the origin of large scale structure. If and when microwave background fluctuations are measured, it may be possible to see the imprint of the intervening matter.

Several methods have been used to study light propagation in inhomogeneous cosmology (e.g. Dyer and Oattes (1988)). Firstly the universe can be imagined to expand on average at the same rate as a homogeneous Friedmann-Robertson-Walker model of the same average density. This gives a relationship between affine parameter and redshift. Then mass in the universe can be removed and added in the form of clumps with the requisite density correlations (usually none). Optical scalar equations are then used to trace rays through this universe. A second, related technique is to use "Swiss cheese" cosmologies in which matter is concentrated into the center of spherically symmetric holes. Provided that we ignore interactions between adjacent holes, the propagation of rays outside the holes is unaffected. However, this approach has the shortcoming that it ignores shear. Thirdly, a large cylinder can be hollowed out around the ray and the interior matter can be imagined to be concentrated into randomly located clumps and Monte Carlo methods used to trace rays. This can have the deficiency of ignoring the tidal effects of perturbations at the boundaries of the cylinder. The relationship between these different approximations is still a matter of controversy (e.g. Schneider and Weiss (1988)).

5. Conclusion

Gravitational lenses are now a major research field in extragalactic astronomy. We know of about seven probable and an additional seven possible multiple-imaged quasars. The number count of clusters exhibiting arcs (or "arclets") has now climbed to roughly eight. There are two radio rings and more are confidently anticipated.

Although gravitational lenses have neither proved useful in providing definitive answers to the important cosmological measurements of the Hubble constant and density parameter nor elucidated the nature of dark matter, these issues are starting to be significant in detailed lens modeling and some non-standard possibilities (e.g., $10^{15} M_\odot$ black holes) can already be ruled out. It is perfectly reasonable to hope that further observations of gravitational lenses, interpreted using elementary geometrical optics, will furnish serious answers to some of these questions in time for GR-13.

I thank the Harvard-Smithsonian Center for Astrophysics for hospitality. I am grateful to the Guggenheim Foundation, the Smithsonian Institution and the National Science Foundation (AST 86-15325) for financial support.

References

Blandford, R. D. and Kochanek, C. S. (1986), in *Dark Matter in the Universe*, eds. J. Bahcall, T. Piran and S. Weinberg (World Scientific, Singapore).

Blandford, R. D. and Kovner, I. (1988), *Phys. Rev.* **A38**, 4028.

Blandford, R. D. and Narayan, R. (1986), *Astrophys. J.* **310**, 568.

Bourassa, R. R. and Kantowski, R. (1975), *Astrophys. J.* **195**, 13.

Chang, K. and Refsdal, S. (1984), *Astr. Astrophys.* **132**, 168.

Dyer, C. C. and Oattes, L. M. (1988), *Astrophys. J.* **326**, 50.

Einstein, A. (1936), *Science* **84**, 506.

Hewitt, J. N., Turner, E. L., Schneider, D. P., Burke, B. F., Langston, G. I. and Lawrence, C. R. (1988), *Nature* **333**, 537.

Kaiser, H. and Lasenby, A. N. (1988), eds., *The Post-Recombination Universe*, (Kluwer Academic Press, Boston).

Kochanek, C. S., Blandford, R. D., Lawrence, C. R. and Narayan, R. (1989), *Mon. Not. R. Astr. Soc.* **238**, 43.

Kovner, I. and Paczyński, B. (1988), *Astrophys. J. (Lett.)* **335**, L9.

Lynds, R. and Petrosian, V. (1988), *Bull. Amer. Astr. Soc.* **18**, 1014.

Moran, J. N., Hewitt, J. and Lo, K.-Y. (1989), eds., *Gravitational lenses*, (Springer-Verlag, Berlin).

Press, W. H. and Gunn, J. E. (1973), *Astrophys. J.* **185**, 397.

Refsdal, S. (1964), *Mon. Not. R. Astr. Soc.* **128**, 295.

Schneider, P. (1985), *Astr. Astrophys.* **143**, 413.

Schneider, P. and Weiss, A. (1988), *Astrophys. J.* **327**, 526.

Soucail, G., Fort, B., Mellier, Y. and Picat, J.-P. (1987), *Astr. Astrophys.* **172**, 414.

Walsh, D., Carswell, R. F. and Weymann, R. J. (1979), *Nature* **179**, 381.

Zwicky, F. (1937), *Phys. Rev. Lett.* **51**, 290.

6

Observations of gravitational lenses

Bernard Fort

Observatoire Midi-Pyrénées,
14 avenue Edouard Belin,
31400 Toulouse, France

Introduction

During the last two years a burst of results has come from radio and optical surveys of "galaxy lenses" (where the main deflector is a galaxy). Even if this kind of work were better known and had already been reviewed for this assembly a few years ago, I cannot pass up the main results which have presently emerged. This will be the first part of the presentation.

On the other hand, in September 1985 we pointed out a very strange blue ring-like structure on a Charge-Coupled Device (CCD) image of the cluster of galaxies Abell 370 (Soucail et al. (1987a)). After ups and downs and a persistent observational quest, this turned out to be the Einstein arcs discovery (Soucail et al. (1988); Lynds and Petrosian (1989)), an important observational step in this particular field since the first observation of the double Quasi-Stellar Object (QSO) 0957 + 561 in 1979 (Walsh et al. (1979)).

Following this discovery, new observational results have shown that many rich clusters of galaxies can produce numerous arclets: tangentially distorted images of an extremely faint galaxy population probably located at redshift larger than 1 (Fort et al. (1989); Tyson (1989)). This new class of gravitational lenses proves to be an important observational topic (Mellier (1989a); Fort (1989a)). This story will be the second part of this presentation.

The classification between galaxy lenses and cluster lenses is somewhat arbitrary. The lens of few multiple QSOs, including the famous double QSO, is a combination of a galaxy plus a nearby cluster which helps to increase the image separation. In this way, within high resolution images of the cluster A2218 having several arcs, we noticed the first complete "optical" Einstein ring around a galaxy member acting as the primary deflector (Fort (1989)). Nevertheless this classification was chosen because it allows us to present an overview of what is going on for the classic works and to stress what I consider as an "avant-garde" subject to be explored.

Some parts of this presentation will be more tentative: the gravitational signature of cosmic strings and our guesses about possible next steps in gravitational lens observations.

Finally, I would like to say that this presentation was addressed to physicists more concerned with new trends than a complete review of observations. This remark particularly holds for the numerous studies of multiple QSOs.

2. Galaxy lenses

The VLA Survey

Soon after the discovery of the first double QSO in 1979 it became clear that the search for such lenses could not be done at random. In 1981, Burke and collaborators (Lawrence et al. (1984)) decided to use the VLA (Very Large Array) Radio Survey of about 4,000 radio sources to find new gravitational lenses. The survey was completed in 1986 and the results are still being analyzed (Hewitt et al. (1988a)).

The idea was that it is unusual for compact radio sources to display more than one compact component, and that multiple radio components with optical counterparts having the same redshift and the same spectra have a chance of being gravitationally lensed.

The survey provided new lenses (Table 1) and a few individual cases were extremely interesting for modeling. Some of them have been purposely reobserved with VLBI (Very Long Baseline Interferometry) in order to reconstruct high resolution images and to search for deamplified third images: 2016 + 112 (Langston et al. (1989)) and 0957 + 561 (Porcas et al. (1989)). Gorenstein et al. (1989) beautifully reconstructed the image of the double QSO 0957 + 561 with a spatial resolution of

Table 1. Galaxy lenses reasonably proved in July 1989. Additional candidates may be found in Surdej (1989).

Lens	Maximum image separation	Flux Ratio brightest pairs	Number of images detected	References
0957 + 561	6.1″	1.3	2	Walsh et al. (1979)
1115 + 080	2.7″	1.1	4	Weymann et al. (1979)
2345 + 007	7.3″	4	2	Weedmann et al. (1982)
2016 + 112	3.4″	1	3	Lawrence et al. (1984)
1635 + 267	3.8″	4.4	2	Djorgovski and Spinrad (1984)
2237 + 031	0.9″	1.2	4	Huchra et al. (1985)
0142 - 100	2.2″	7.6	2	Surdej et al. (1987)
3C324	3″	1.7	2/3	Le Fèvre et al. (1987)
0023 + 171	5″	3	2	Hewitt et al. (1987)
1131 + 046	2.1″	-	radio ring	Hewitt et al. (1988)
1413 + 046	0.77″	1.1	4	Magain et al. (1988)
0414 + 117	1.3″	1/2	4	Hewitt et al. (1989)
1654 + 135	7″	1.4	double radio arc	Langston et al. (1989)
1120 + 019	6.5″	69	2/3	Meylan et al. (1989)

10 m.arc second (Figure 1). They show for the first time a predicted change in parity from one image to the other.

More recent and famous results coming from this Survey concern the discovery of the first "radio" Einstein Rings MG 1131 + 0456 (Hewitt et al. (1988b)), and MG 1654 + 1346 (Langston et al. (1989)).

MG 1131 + 0456 has a ring-like morphology with two components A and B. Apart from the unusual geometry, the radio properties are completely compatible with the normal characteristics of a radio source. The optical counter-part should correspond to the deflecting galaxy, located inside the ring, from which the redshift is not yet measured. Kochanek et al. (1989) show that the geometry is compatible with a radio source having a core and two lobes imaged by the elliptic potential of a foreground galaxy (Figs. 2(a)-2(b)).

The second possible Einstein ring just discovered, MG 1654 + 1346 may be explained in a very similar way. The optical data reveal a QSO (radio source) at $Z = 1.74$ located approximately in the middle of the position of a faint radio lobe and the Einstein double arcs image (gravitational images of the second lobe). Exactly between the two radio arcs lies a foreground R galaxy at $Z = 0.254$ which mostly acts as the deflector. A modeling is compatible with a deflecting galaxy having an

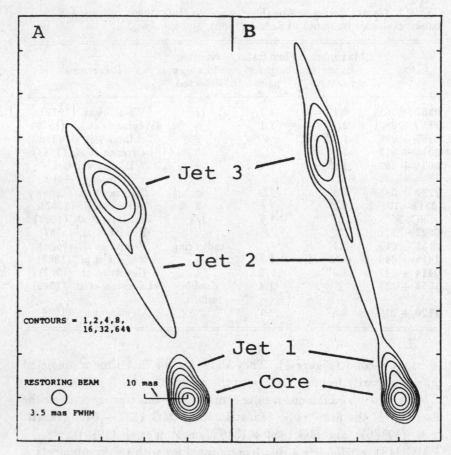

Fig. 1. The brightness distributions of 0957+561 A and B; 13 cm VLBI observations; 3.5 mas resolution. Note the evidence for the change in parity from one image to the others (from Gorenstein et al. (1989)).

equivalent velocity dispersion $\sigma = 220$ km/s and a mass to light ratio $M/L = 20$ compatible with the deflecting galaxy (Langston et al. (1989)).

In summary this extensive survey provided few new multiple QSOs and was surprising because there were almost none with angular separation below one arcsecond. This small number was used as an argument that the universe cannot be closed by typical galaxy masses around 10^{11} to 10^{12} solar masses (Hewitt (1986)). But it produces the first convincing gravitational images of an extended radio source by a galaxy lens, a particularly well designed case for modeling.

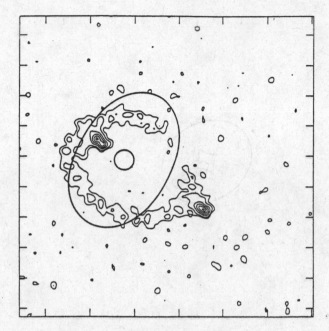

Fig. 2(a).   The 5 GHz map of the Einstein radio ring MG 1131+046

Optical Surveys

Many groups are completing optical surveys for multiple imaged QSOs (Crampton et al. (1988); Djorgovski and Meylan (1989); Hewett et al. (1985); Reboul et al. (1987); Surdej et al. (1987); Magain et al. (1988); Webster et al. (1988)). Below we will only present the two most extensive surveys.

R. Webster and collaborators have undertaken a large automated survey of gravitationally lensed QSOs. Wide field plates coming from the United Kingdom Schmidt Telescope (UKST) covering about 214 square degrees, were scanned with a high speed measuring machine. Different selection procedures like identification of a star profile, color indexes and finally, low resolution spectroscopy, were used to build up a fair sample of 296 QSOs with redshift below $Z = 3.3$ and a B magnitude from 16.5 to 18.75. At this point Webster et al. started to issue a list of possible lens candidates with all the doubles (but with separation greater than their 2 arc seconds resolution power). The probability of detecting them is almost 100%, when the flux ratio between the two components is not too large. To date, this survey gave 3 new possible gravitational lens candidates (Webster (1989)).

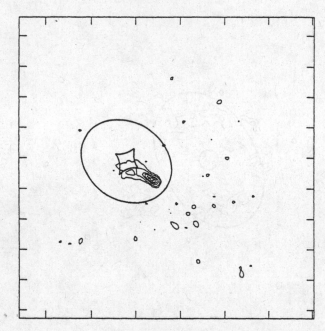

Fig. 2(b).   Shown is the calculated inversion of the source (from Kochanek et al. (1989)).

This ongoing survey is complementary to the radio VLA Survey. Its limitation is the impossibility of detecting close pairs below 2 arc second separation, but so far, the VLA Survey does not find such close pairs up to 0.4 arc second resolution.

In fact the most important result of this Survey is the detection of an excess number of galaxies superimposed on high redshift QSOs that we will discuss later (Webster et al. (1988)).

In the past few years another approach was followed by many observers to try to detect close pairs of lensed QSOs. It starts with the idea of a possible luminosity bias introduced by the gravitational lensing: the luminosity of some of the most distant bright QSOs or radiogalaxies (Le Fèvre et al. (1987); Surdej (1989)) may well be enhanced by the lensing of the smooth halo of an intervening galaxy and be associated with detectable multiple close images.

Surdej et al. (1988) started an optical Survey of 110 QSOs with an absolute magnitude larger than -29. From this sample they extracted 25 objects having fuzzy or elongated environments. Five of them were considered as possible lens images and two of them have been confirmed spectroscopically (Surdej et al. (1987); Magain et al. (1988); see Table 1).

Similarly, starting with a sample of the brightest 27 radio sources at large redshift, Le Fèvre et al. (1987) gave evidence of gravitational lens effects. The first case was 3C324 ($Z = 1.206$), whose spectrum could reveal a superimposed foreground galaxy at redshift $Z = 0.845$ (Le Fèvre et al. (1987)). They also obtained similar results on other bright radiosources with an intervening galaxy at lower redshift: 3C13; 3C208,1; 3C255A; 3C194 (Le Fèvre et al. (1988)).

For both for the Surdej and Le Fèvre works it is striking that the convincing cases found were obtained or confirmed (Nieto et al. (1988)) under excellent seeing conditions–better than 1 arc second. It is also worth noting that in the last year these two observational programs have merged within an European Southern Observatory (ESO) key programme.

All together, optical surveys have produced one or two "possible" new galaxy-lens candidates every year (Table 1). In fact the number of lenses that observers have actually detected is a matter of controversy with theoreticians because: 1) we cannot exclude physical pairs at the same redshift; 2) the deflecting galaxy is often not seen; 3) the geometrical configuration or flux ratio is sometimes irrelevant with reasonable modeling. Blandford (1989) raised this point and called for more strictness in assessing a lens discovery. Indeed Table 1 is affected by author bias, but it should represent the most probable lenses.

In summary the optical surveys mainly seem to have produced few but very interesting individual cases that we will discuss below for their implications on time delay measurements and microlensing effects, and they have also shown an overdensity of foreground galaxies near QSOs or radiosources. This effect of a luminosity bias introduced by gravitational magnification of intervening galaxies seems at present the most exciting output of these optical surveys.

Excess of foreground galaxies in the vicinity of a luminous distant object

In fact, Tyson (1986) was the first to report an excess of a factor 4 for galaxies up to an R magnitude $= 21$ within 30 arc seconds of a high-redshift QSO sample. But with a larger sample Webster (1988) found about the same overdensity value within 6 arc seconds from the QSOs with a sample of 285 objects at redshift $Z > 0.5$. Similarly, Fugmann

(1988) took deep CCD images of 12 distant flat spectrum QSOs ($Z >$ 1.7) and found a significant rise of galaxies ($R < 21.5$) near the QSO positions. There is also marginal statistical evidence of an excess of X-ray-emitting active galactic nuclei (AGN) at large redshift for an emitter close to foreground galaxies (Stocke et al. (1987)). All these results have prompted new theoretical analyses of the effect of microlensing or macrolensing as a possible important bias to the counts of QSOs and AGN at large redshift. Schneider (1989) shows that it is possible to explain an excess number of galaxies around high-redshift quasars as an effect of microlensing within the intervening galaxy. Narayan (1989) points out that one can also consider the magnification by smooth haloes of foreground galaxies. Neither can one exclude the magnification by intervening groups or clusters of galaxies which are still difficult to detect at redshift $Z > 0.5$. Kovner (1989) derives the maximum excess that can be produced by any ensemble of lenses and shows that the (galaxy) excess found by Webster and others is too large by about a factor of two if the magnification probability distribution is negligibly correlated with the intrinsic QSO flux. This means that there is a possible observational error in calculating the galaxy excess or that the background QSOs are already appreciably affected by lensing. In conclusion we now have evidence that some kind of lensing may actually bias the number counts and luminosity function of very distant objects.

QSO 2237 + 030: a good candidate for microlensing

This system is a QSO at $Z = 1.696$ lensed by the very central part of a nearby barred-spiral galaxy ($Z = 0.038$). Yee and collaborators (1988) took advantage of the excellent seeing conditions in Hawaii to demonstrate that the QSO is in fact split into 4 components (Figure 3). This lensed system was also observed and carefully modelled by Kayser et al. (1989a). They assumed that the dark matter follows the visible mass distribution of the galaxy with M/L constant. The remarkable consistency of the results with what we already know from the dynamical analysis of the galaxy center means that the lens model is "reliable." The time delay is so short (one day) that any variability of the QSOs will show up immediately in the four components. Inversely if a star from this very dense part of the galaxy rises to a brightening of one of the components it will be seen over a time scale of one year. For a long time theoreticians predicted the important possible effects of microlensing when interpreting the variability of Seyfert galaxies, BL Lac, AGN

(Chang and Refsdal (1979); Turner (1980); Setti et al. (1983)). This system is probably the best choice for finding an unquestionable proof of the possible effect of microlensing by stars (Kaiser and Refsdal (1989b)).

Time delay and microlensing for the double QSO 0957 + 561

A long time ago, it was predicted that the traveling time delay between the different gravitationally split images of variable objects could give a direct measurement of the Hubble constant H_0 if the lens system is known (e.g. Refsdal (1964)). Soon after the discovery of the double QSO several teams took up the challenge for this system and started photometric monitoring of the two images. Such observation needs a great amount of skill and tenacity because the intrinsic variability of 0957 + 561 is not larger than few hundreds to one tenth of magnitude on a time scale of several days to months (Keel (1982); Schild and Weekes (1984)). Reliable observations were done by several teams from 1980 to the present. (Lehar et al. (1988)) announced a first value of 500 days ± 100 days. Then Vanderriest et al. (1989), with more numerous data, succeeded in giving the time delay with an accuracy of 5%: 415 days ± 20 days. This result does correspond to a beautiful observing

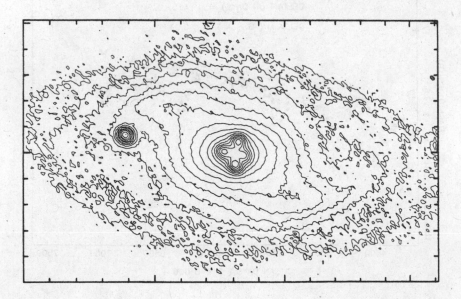

Fig. 3.   Contour plots of the lens system 2237+030 within the nuclear region of the corresponding Zwicky galaxy (from Yee (1988)).

performance (Figure 4). However large uncertainties in the calculation of H_0 still arise from imperfect modeling of the lens. The inferred value of H_0 (around 90 ± 50 km s^{-1}/Mpc) is compatible with other classical methods, but independent from any cosmic ladder. The observation of the third image, as it was done for QSO 2016 + 112 (Langston et al. (1989)) and the measurement of the velocity dispersion for the deflecting galaxy G1 and the associated cluster are very much needed to find H_0 with an accuracy of 20%.

This object deserves more attention because Vanderriest (1989) found marginal evidence of possible fluctuations of the relative amplification factor $a(t)$ between the two components which could be attributed to microlensing. This ratio is also very close to 1 in the visible part of the spectrum but has slightly different values at radio wavelengths or when using the emission lines. Such effects can be also understood in terms of microlensing if the region emitting at these different wavelengths does

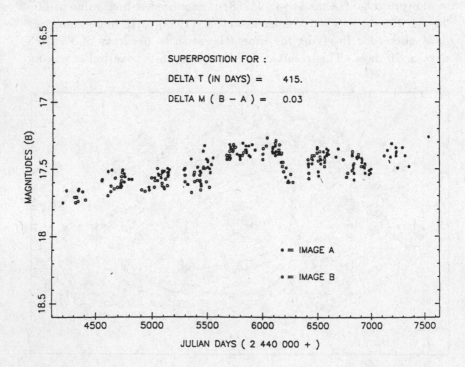

Fig. 4. Superposition of the two light curves A and B, but shifted by 415 days for B (from Vanderriest et al. (1989)).

not fall similarly within the corresponding Einstein radius of the microlens.

3.   The twin galaxy field 0249-186 and cosmic strings

The possible existence of cosmic strings and the possibility that they might have seeded the formation of galaxies and clusters at the early stage of the universe was an exciting topic at the beginning of the last decade (e.g. Vilenkin (1981)). So far the existence of these objects has not yet been proved. Using the millisecond timing argument of pulsars (e.g. Hogan and Rees (1984)) or the temperature fluctuations in the microwave background (Gott (1985)) the dimensionless mass per unit length was set to $\mu = 10^{-5}$. If cosmic strings or loops do exist with unit mass around 10^{-6} they should produce split images of background objects with separation of the order of arcseconds. This class of lensing will have a peculiar signature that we can summarize as follows: 1) There is no magnification of the two images stretched along the string; 2) the flux ratio shall be 1 for a pair of images; 3) The images have the same parity, recognizable with the HST (Hubble Space Telescope) if the sources have clear morphological properties; 4) The differential velocity on each side of the string should be of the order $8\pi G\mu/c^2 = 100$ km/s.

The serendipitous find of Cowie et al. (1987) of the twin-galaxies in the field 0249-186 is presently the only candidate for lensing by a cosmic string. They found within a 45×45 arc-second square field 5 pairs which almost have the properties we expect for a string loop having $\mu = 6 \times 10^{-7}$ and located at redshift 0.07. The pairs have a flux ratio very close to 1 (1.01 to 1.13), the same color and spectra, and a similar velocity difference error bar (+ 100 km/s). It is clear that this field calls for extremely deep photometry in B up to 29 arc second square (Tyson and Seitzer (1988)). At this detection level one expects about 50 blue distant galaxies at large redshift. Some of them should also trace the string. Cowie and colleagues are investigating this field at the Canada-France-Hawaii Telescope (CFHT).

In the future we expect to have wide field CCD cameras using buttable chips. We will be able to make deep photometric surveys on square degrees within a few months of observational time with a four meter telescope. The density of the background population of blue galaxies is so high that it is predictable that such surveys should allow us to set an upper limit on the number of low curvature strings left in the universe, if we do not see any field with a continuous track of image pairs. Such observations are risky but should be favorably compared to the search of new exotic particles on the biggest colliders. I hope some astronomers will take up the challenge.

4.   Cluster lenses

The luminous arcs discovery

In September 1985, during the first spectroscopic survey of Abell 370 we saw a very strange ringlike structure (Soucail et al. (1987a)) which did not resemble any known extragalactic object (Figure 5). This very thin blue structure was a highly circular arc longer than 100 kpc at the cluster redshift.

Paczynski's paper in Nature (1987) and his guess that the arc in A370 as well as a second one in Cl 2244 (Lynds and Petrosian (1986)) could be gravitational images of a high-redshift object prompted us to place top priority on spectroscopic observations. This was a real challenge with existing telescopes because the brightness of the arcs is about 10 times fainter than the sky background.

After Paczynski's paper, Hammer simulated a simple multimass model of the A370 cluster lens which fairly reproduced the A370 arc (Soucail et al. (1987b)). At the same time Kovner (1987) and Grossman et al. (1988) showed that it is easy to reproduce the observed arcs with single elliptical potential for the clusters.

In October 1987 we succeeded in obtaining a spectrum of the A370 arc with the Puma/EFOSC spectrograph at the ESO 3.6 meter telescope (ESO press release November 7, 1987). The very convincing spectrum showed that the arc was a gravitational image of a star-forming galaxy at a redshift $Z = 0.724$, about twice the cluster distance (Soucail et al. (1988)). This result was also reached approximately at the same time by Lynds and Petrosian (1988) and more marginally (due to bad weather conditions) by Miller and Goodrich (1988). These concordant observations definitively settled down the problem of the origin and nature of the arcs.

Meanwhile Lavery and Henry (1988) published CCD images of the double arc around the cD galaxy located in the center of Abell 963 and Pello et al. (1987) showed several arclets in A2218. Soon after, Giraud (1988) took the opportunity of cluster observations at ESO to search for arcs and found one in Cl0500-24 (Giraud (1988)). Then, Koo (1988)

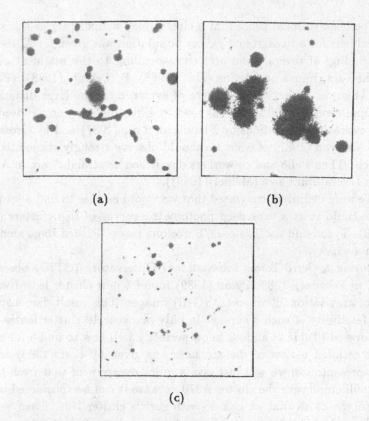

(a)   (b)

(c)

Fig. 5.   (a) The bright arc in A370; (b) an arc in Cl 2244-02; (c) long fragmented arc in Cl 0024+1654.

presented a beautiful case in Cl0024 + 1654 with a large arc segmented in three pieces and its counter-arc: a nearly perfect theoretical case (Fig. 5). Thus by mid-1988, 6 clusters lenses with large arcs were already known.

At approximately the same time Tyson and Seitzer (1988) announced that the background population of very faint galaxies was extremely dense: about 50 galaxies per square arcminute at a detection level of one hundredth of the sky level ($\mu_B \simeq 29$). He showed that their brightness and blue color were fully compatible with a redshift larger than one ($B - R$ from 0 to 1). We immediately understood why the first arcs found had a blue color. All the ingredients were there to open a new observational field: *the study of gravitational lensing of faint background extended objects by rich clusters of galaxies.*

The observational proof that a cluster may actually act as a gravitational lens for a background galaxy population was given for A370 with the finding of several mini arcs corresponding to the brightest object of the background population (Fort (1988); Fort et al. (1988)) (Figure 6). This was in fact the first piece of evidence for the large distance of a population having a still unknown redshift distribution. Indeed the two evocative simulations by Blandford et al. (1987) and by Grossman and Narayan (1987) of what we should observe strongly stimulated our search. Then Pello and co-workers discovered a "straight" arc in A2390 with several mini-arcs (Mellier (1989)).

We were definitely convinced that we would be able to find a lot more if we could start a very deep photometric survey of rich clusters (Fort 1989). Tyson and the Toulouse/Barcelona teams decided to go along for such a survey.

During a Cerro-Tololo International Observatory (CTIO) observing run, in February 1989, Tyson (1989) found 3 new cluster lenses with a lot of gravitationally distorted (GDI) images. This result demonstrates the feasibility of such a survey. In only one year 10 cluster lenses were discovered (Table 2) and we know perfectly well how to find a lot more.

A detailed review of the arc survey is given by Fort (1989) and in this presentation we will just give a quick overview of that fresh topic. We will emphasize the cluster A370 because it can be considered as the prototype of all that we can do with such a cluster lens. Then we will outline what can be learned from this new class of cluster lenses. We consider that Table 2 gives the basic information and the references for all other individual cases already known.

A370

This was the first case in which we obtained the spectroscopic proof for lensing by a cluster as a whole. It has been the most heavily observed cluster and it remains a prototype for all other observations. A370 is bipolar and one of the richest distant Abell clusters. It has a very strong X-ray luminosity and its velocity dispersion ($\sigma = 1350$ km/s) was carefully measured from a large redshift sample (Mellier et al. (1988)). The total mass $M = 10^{14}$ solar masses that we can derive within the large arc radius is fully compatible with the virial mass model and gives a mass to light ratio = 90 (R magnitude).

No deep photometry above $B = 27.5$ per square arcsecond was currently available for this cluster, but this detection level was deep enough

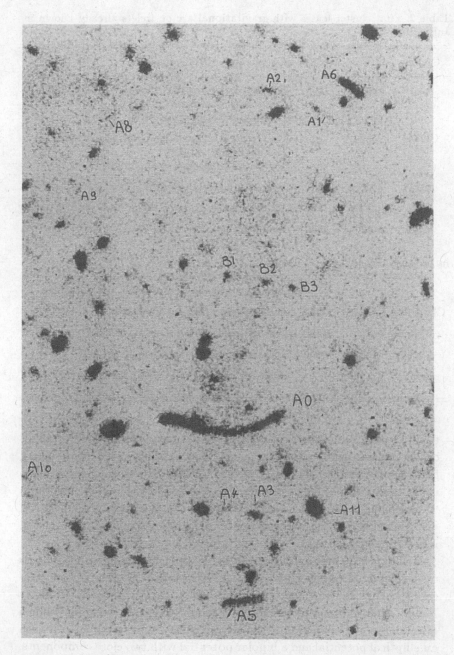

Fig. 6. The bright arc and few arclets in A370. Residuals of the subtraction of the *R* from the *B* image: most galaxies from A370 vanish and several blue arclets A1-A2; A3-A4; A5 appear (from Fort et al. (1989)).

Table 2. First cluster lenses with gravitational arc or arclets already known in July 1989.

Cluster	Z	σ km/s	Comments and references
A370	0.374	1350	One large arc S=0.724 (Soucail et al. (1988); Lynds and Petrosian (1988)); many faint arclets (Fort et al. (1988))
Cl2244	0.32	-	One large arc (Lynds and Petrosian (1988); Hammer et al. (1989); Z arc=2.238 (Mellier et al. (1989))
A2218	0.171	1400	Several arclets (Pello et al. (1987))
A963	0.23	1100	One arc and a counter arc around the CD galaxy
A2390	0.206	1600	One long "straight" arc, Z=0.92 (Lavery and Henry (1988))
0024+1654	0.39	-	One long arc in 3 parts and counter arc (Koo (1988); numerous arclets (Turner and Tyson (1989))
Cl0500-24	0.316	1300	One long arc (Giraud (1988))
3C295	0.461	3x1000	Several arclets (Tyson et al. (1989))
A1689	0.181	1800	Numerous arclets (tyson et al. (1989))
A545	0.154	-	One arc plus many arclets (Tyson (1989))

to find several mini-arcs (Fort et al. (1988)), all tangential and centered on the cluster center. Grossman and Narayan succeeded in modeling these arclets with an elliptical potential (1989). A very interesting point is that using the "Findlens" software developed by Tyson et al. (1989) we were able to map the shape of this potential just using the statistical distortion effect of the background population of blue galaxies, and it indeed appears elliptical.

Mellier (1989b) noticed a small velocity gradient along the major axis of the cluster. Does this mean that A370 is at the end of a merging process and that the dark matter still has a bipolar distribution? Recently Mellier (1989) reconstructed the background shape of the galaxy which gives the large arc and found that its geometry is compatible both with an elliptical potential and a bipolar potential with two close components. An extremely deep photometry followed by the study of the statistical distortion of several tens of background galaxies may give the answer to such an hypothesis.

The large arc in A370 is marginally resolved in thickness. It has also a thicker and tilted western extremity. Several explanations were proposed. But the model which best fits the geometry is obtained with a spiral galaxy with two opposite and symmetric arms as a source (Mellier 1989a). The two arms nicely reproduced the two extremities of the arc. Such image reconstructions were suggested by Shramm et al. (1988) and Kochanek et al. (1988). It is notable that the spectrum of the arc is compatible with a spiral galaxy.

5. What can be learned from cluster lenses?

The study of distant galaxies

Among the first cluster lenses already discovered, 5 already have long arcs. These magnified images are produced when the background galaxy falls close to a caustic cusp. When making spectroscopy or photometry we can sum up all the light coming along the arc in order to get a better signal to noise ratio. Thus we have a natural telescope at our disposal to probe the distant universe. So far we know redshifts for several arcs in different clusters (Table 2). This gives additional proof that we are dealing with gravitational images, but it also provides a valuable spectrum of very distant galaxies that would have been otherwise unobservable. As an example Mellier et al. (1989) give the spectrum of the Cl 2244 arc ($Z = 2.34$). It is, to my knowledge, the spectrum of the most distant galaxy ever obtained. Up to now the question of when the galaxies formed is a fundamental question in observational cosmology. It may well happen that this new class of cluster lenses will help give the answer.

Mapping dark matter

When we try to solve the basic lens equation (Blandford and Kochanek 1987)

$$\vec{i} = \vec{s} - \nabla\phi,$$

it is clear that we just know the image position $\vec{i}$. We do not have an *a priori* idea on the position of the source $\vec{s}$ and of the dimensionless projected potential of the lens ϕ. Thus the modeling of the lens with few arcs does not in general give a unique solution because we have too many free parameters even with the simplest model. As long as we do not have other observations like the mapping of the X-ray gas (Kellog et al. (1989)) (target for the XMM Satellite) to constrain the potential we will be unable to give accurate modeling of the lenses. However it is possible to derive the total mass within the critical line which does depend crucially on the model. Moreover it was pointed out by Tyson (1989) that the population of background galaxies is very dense (50 per arc/minute square) when we reach the limiting detection of about 5/1000 of the sky background, and that we can use it as a grid to map the shape of the potential (Tyson (1989)). In short, all the arclets coming from galaxies at Z between 1 and 2 will appear in a strip which outlines the critical line of the potential and their number is about $N = (\sigma/1000 \, \text{km})^4$ (Fort (1989)). Moreover using a statistical distribution of the length of arclets versus the distance to the cluster center, we can roughly estimate the effective radius $(M(r_{eff}) = M_{total}/2)$ or the cut-off radius for the total mass.

In fact what we can learn about the potential shape, estimate of effective radius and cut-off radius of cluster lenses is already extremely important. At present we have no other tool to map the dark matter on the cluster scale. With the mini-arcs we can hope to see if the visible light always traces the dark matter overdensities.

It is also important to note that crude modeling of the arclets would also give constraints on the distribution of the background population of galaxies, whose distance and color is still a matter of debate (Blandford (1989)). Their number is also a crucial test for the dynamical state of distant clusters (Fort (1989)).

6. Discussion and next observational steps

With the discovery of the first gravitational lens (Walsh et al. (1979)), there were great hopes for the possibility of using an accurate time delay measurement (say within 5% accuracy) to directly derive the Hubble constant H_0. This was probably the main motivation in searching for new multiple QSOs.

Ten years later we finally succeeded for the double QSO 0957 + 561. But meanwhile, theoretical studies of several galaxy lenses, including this QSO, demonstrated how difficult it is to constrain a potential model in order to derive H_0 with an error lower than to about 20%. Indeed for this double QSO, we need to find the third deamplified image and to properly model both the potential of the deflecting galaxy and the nearby cluster. These observations undoubtedly are priority targets for the HST, the radio VLBI and large ground-based telescopes.

More generally, if we try to roughly summarize the ten year scientific output of the observations of galaxy lenses, we can feel a little disappointed if we compare the small number of good cases to the huge observational effort. We did not learn more about the galaxy deflectors than we knew before with classical dynamical studies of galaxies, apart from the fact that in few cases we were surprisingly unable to see the deflector.

It seems that the search for new multiple QSOs will continue to be a long quest. A basic question is to understand why we do not find very close multiple images (< 1 arsecond). Is it connected to the natural deflectors or is it an observational bias coming from atmospheric seeing or VLA resolution? I also bet we will see QSO arcs around galaxies. Speckle interferometry of the brightest QSOs would certainly help, but the HST will be unbeatable to clarify these points.

It appears that theoreticians would appreciate more complete observations of individual cases particularly interesting for modeling (like the radio Einstein rings) rather than an increasing list of dubious candidates. This trend is also clearly visible for the measuring of time delay or microlensing effects. Observers should also try to get light curves for some violently variable objects. Nottale (1986) showed that the three month violent eruption of the BL Lac 0846 + 51 WI could be interpreted in term of microlensing by a 10^{-2} solar mass object. The "missing mass" may well be objects like Jupiter, brown dwarfs or remnants of population III stars which could be detected through their microlensing effect. Such a field would certainly be explored with medium size telescopes (2 meters).

Indeed the claim of an overdensity of galaxies around bright QSOs or radiosources (Webster et al. (1988)) is for the time being an important question. It can bring some evidence about the possible effect of microlensing or macrolensing by smooth haloes of galaxies (or why not of clusters of galaxies!). Such results are a possible key to understanding how much the luminosity function of distant objects is biased by lensing. This is crucial for many cosmological studies.

In this respect the quasi simultaneous discoveries of gravitational arcs and of a huge population of extremely faint and distant blue galaxies (Tyson and Seitzer (1988)) stimulated a big evolution of our views. Observers fully realized that they were looking at distant objects through a very dense and thick clumpy screen of mostly unseen distant galaxies or clusters. This screen could profoundly modify the propagation of light beams, providing both an amplification bias of punctual background sources distant like QSOs or conversely, washing out extended emissions like intrinsic temperature fluctuation of the microwave background and possible large protogalaxies.

We have seen that this screen can be used purposely as an optical distortion grid to map the dark matter or eventually track cosmic strings (if they exist). This is now beautifully used with rich clusters of galaxies acting as gravitational lenses (Tyson (1989)). Such cluster lenses would be found in great number in the sky if we used extremely deep photometry (Fort (1989)). They would be also natural telescopes to probe the distant universe and some hopes still remain of using them to constrain Ω_0 if we succeed in measuring the redshift of several arclets within an almost circular and virialized cluster. Cluster lens observations are a challenge because the sources are extremely faint due to the dimming factor $(1 + Z)^{-4}$ for their surface brightness. But the new generation of Very Large Telescopes will handle such observations with reasonable exposure times. It is also important to note that this topic of cluster lenses will gain a lot from the next generation of X-ray satellites having good imaging capabilities like AXAF and XMM (Kellog et al. (1989)), because the mapping of the gas will provide a good constraint on the potential ϕ of the lens. The discovery of arcs and arclets in clusters of galaxies is the most important discovery in this field during the last years. It will open a large and new avenue to explore during the next decade.

The generation of astronomical instruments in the next century may have a tremendous increase of spatial resolution. They will also have large light collecting power which will allow them to probe the very deep universe. I do not present some of the exotic proposals relevant to gravitational lensing observations with these instruments because nowadays they are too speculative (examples: star light deflection by stars having large proper motion; gravitational glitches in very compact binary systems; microlensing speckles; etc...). But they lead me to conclude that in the near future, observers will have often to consider "Gravitational Optics" as a new optical transfer function in front of our telescopes which forms the images that we are analysing, and as a new tool to probe the Universe.

I thank R. Blandford, I. Kovner, Y. Mellier, R. Pello and G. Soucail for several valuable discussions. I also acknowledge help from all the colleagues in the world who sent me their results or slides prior to communication to the Gravitational Lenses Workshop held in September 13, 1989 in Toulouse, and particularly A. Tyson for the beautiful image of arclets in A1689. Many thanks to E. Davoust and J. Jobard for helpful remarks on an earlier version of this review.

References

Bergmann, A., Petrosian, V. and Lynds, R. (1989), preprint.

Blandford, R. and Kochaneck, C. (1987), Dark Matter in the Universe, *Proceedings of 4th Jerusalem Winter School for Theoretical Physics*, eds. J. Bahcall, T. Piran and S. Weinberg (World Scientific, Singapore), p. 133.

Blandford, R., Phinney, E. S. and Narayan, R. (1987), A Giant Gravitational Lens, *Ap. J.* **313**, 28-36.

Chang, K. and Refsdal, S. (1979), *Nature* **282**, 561.

Cowie, L. and Lilly, S (1989), *Ap. J.* **336**, L41.

Crampton, D., Cowley, A., Hickson, P., Kindl, E., Wagner, R., Tyson, J. and Gullixon, C. (1988), *Ap. J.* **330**, 184.

Fort, B., Prieur, J. L., Mathez, G., Mellier and Y., Soucail, G. (1988), *Astron. Astrophys.* **200**, L17.

Fort, B. (1989a), Astronomy, Cosmology and Fundamental Physics, in *Proceeding of the Third ESO-CERN Symposium, Bologna*, (Kluwer Academic Publishers, Dordrecht), p. 255.

Fort, B. (1989b), Gravitational Lenses, in *Gravitational Lenses Workshop*, Toulouse, September 13-15, preprint.

Fulgmann, W. (1988), *Astron. and Astrophys.* **204**, 73.

Giraud, E. (1988), *Ap. J.* **334**, L69.

Gorenstein, M. V., Bonometti, R. J., Cohen, N. L., Falco, E. E., Shapiro, I. I., Bartel, N., Rogers, A. E. E., Marcaide, J. M. and Clark, T. A. (1989), Center for Astrophysics, Cambridge, Mass. preprint **2526**.

Gott, J. R. III (1985), *Ap. J.* **288**, 422.

Grossman, S. and Narayan, R. (1988a), *Ap. J.* **324**, L37.

Grossmann, S. and Narayan, R. (1988b), Steward Observatory preprint, *Ap. J.*, in press.

Hammer, F. and Le Fèvre, O. (1989), *Ap. J.*, in press.

Hewett, P., Irwin, M., Bunclark, P., Bridgeland, M. and Kibblewhite, E. (1985), *Mon. Not. Roy. Astr. Soc.* **213**, 97.

Hewitt, J. N. (1986), Massachusetts Institute of Technology, Ph.d. Thesis.

Hewitt, J., Burke, B., Turner, E., Schneider, D. P., Lawrence, C., Langston, G. and Brody, J. P. (1988a), Gravitational Lenses, in *Lecture Notes in Physics* **330**, (Springer-Verlag, Berlin), p. 147.

Hewitt, J., Turner, E., Schneider, D. P., Burke, B., Langston, G. and Lawrence, C. (1988b), *Nature* **333**, 537.

Hogan, C. and Rees, M. (1984), *Nature* **311**, 109.

Kayser, R., Surdej, J., Condon, J., Hazard, C., Kellermann, K., Magain, P., Remy, M. and Smette, A. (1989a), *Ap. J.*, in press.

Kayser, R. and Refsdal, S. (1989), *Nature* **338**, 745.

Kellog, E., Falco, E., Jones, C. and Slane, P. (1989), poster paper, this conference.

Kochanek, C. S., Blandford, R. D., Lawrence, C. R. and Narayan, R. (1989), *Mon. Not. Roy. Astr. Soc.*, in press.

Koo, D. (1988), private communication.

Kovner, I. (1987), *Ap. J.* **321**, 686.

Kovner, I. (1989), *Ap. J.* **341**, L1.

Langston, G. I., Schneider, D. P., Conner, S., Carilli, C. R., Lehar, J., Burke, B. F., Turner, E. L., Gunn, J. E., Hewitt, J. N. and Schmidt, M. (1989), *Astron. J.* **97**, 1283.

Lavery, R. and Henry, J. P. (1988), *Ap. J.* **329**, L21.

Lawrence, C. R., Schneider, D. P., Schmidt, M., Bennett, C. L., Hewitt, J. N., Burke, B. F., Turner, E. L. and Gunn, J. E. (1984), *Science* **223**, 46.

Le Fèvre, O., Hammer, F., Nottale, L. and Mathez, G. (1987), *Nature* **326**, 268.

Le Fèvre, O. and Hammer, F. (1988), *Ap. J* **333**, L37.

Lehar, J., Hewitt, J. and Roberts, D. (1988), Gravitational Lenses, in *Lecture Notes in Physics* **330**, (Springer-Verlag, Berlin), p. 147.

Lynds, R. and Petrosian, V. (1986), *Bull. AAS* **18**, 1014.

Lynds, R. and Petrosian, V. (1989), *Ap. J.* **336**, 1.

Magain, P., Surdej, J., Swings, J. P., Borgeest, V., Kaiser, R., Kuhr, H., Refsdal, S. and Remy, M. (1988), *Nature* **334**, 325.

Mathez, G. (1987*b*) *Astron. and Astrophys* **191**, L19.

Mellier, Y. (1989*a*), *Proceeding of the Space Telescope Sciences Institute Workshop, Clusters of Galaxies*, Baltimore 15-17 May, preprint.

Mellier, Y. et al. (1989*a*), *Gravitational Lenses Workshop*, Toulouse, September 13-15.

Mellier, Y. (1989*b*), private communication.

Mellier, Y., Soucail, G., Fort, B. and Mathez, G. (1988), *Astron. and Astrophys.* **199**, 13.

Meylan, G. and Djorgovsky, S. (1989), *Ap. J. Lett.*, in press.

Miller, J. and Goodrich, R. (1988), *Nature* **331**, 685.

Narashima, D. and Chitre, S. (1988), *Ap. J.* **332**, 75.

Narayan, R. (1989), *Ap. J.* **339**, L53.

Nieto, J. L., Roques, S., Llebaria, A., Vanderriest, C., Lelievre, G. di Serego Alighieri, S., Macchetto, F. and Perryman, M. (1988), *Ap. J.* **325**, 644.

Nottale, L. (1986), *Astron. and Astrophys.* **157**, 383.

Paczynski, B. (1987), *Nature* **325**, 572.

Pello, R., Soucail, G., Sanahuaja, B., Mathez, G. and Ojero, E. (1988), *Astron. and Astrophys.* **190**, LII.

Porcas, R., Garrett, M., Quirrenbach, A., Wilkinson, P. N. and Walsh, D. (1988), Gravitational Lenses, in *Lecture Notes in Physics* **330**, (Springer-Verlag, Berlin), p. 82.

Reboul, H., Vanderriest, C., Fringant, A. and Cayrel, R. (1987), *Astron. and Astrophys.* **177**, 337.

Refsdal, S. (1964), *Mon. Not. Roy. Astr. Soc.* **128**, 307.

Schneider, P. (1989), *Astron. and Astrophys* **221**, 221.

Setti, G. and Zamorani, G. (1983), *Astron. and Astrophys* **118**, L1.

Soucail, G., Fort, B., Mellier, Y. and Picat, J. P. (1987*a*), *Astron. and Astrophys.* **172**, L14.

Soucail, G., Mellier, Y., Fort, B., Hammer, F. and Mathez, G. (1987*b*), *Astron. and Astrophys.* **184**, L7.

Soucail, G., Mellier, Y., Fort, B., Mathez, G. and Cailloux, M. (1988), *Astron. and Astrophys.* **191**, L19.

Stoke, J., Schneider, P., Morris, S., Gioia, I., Maccacaro, T. and Schild, R. (1987), *Ap. J.* **315**, L11.

Surdej, J., Magain, P., Swings, J. P., Remy, M., Borgeest, V., Kaiser, R., Refsdal, S. and Kuhr, H. (1988), *Large Scale Structures*, a Fuist DAEC Workshop, eds. C. Balkowski and S. Gordon (Paris).

Surdej, J., Swings, J. P., Magain, P. Borgeest, V., Kaiser, R., Refsdal, S., Courvoisier, T. and Kuhr, H. (1988), *Astron. and Astrophys.* **198**, 49.

Surdej, J. (1989), *Gravitational Lenses Workshop*, Toulouse, September 13-15.

Turner, E. (1980), *Ap. J.* **248**, L89.

Tyson, J. A. (1986), *Ap. J.* **92**, 691.

Tyson, J. A. and Seitzer, P. (1988), *Ap. J.* **335**, 552.

Tyson, J. A. (1989), *Gravitational Lenses Workshop*, Toulouse, September 13-15.

Tyson, J. A., Valdes, F. and Wenk, R. (1989), preprint.

Vanderriest, C. (1989), *Gravitational Lenses Workshop*, Toulouse, September 13-15.

Vanderriest, C., Schneider, J., Herpe, G., Chevreton, M., Moles, M. and Wlerick, G. (1989), *Astron. and Astrophys.*, in press.

Vilenkin, A. (1981), *Phys. Rev. Lett.* **46**, 1169; erratum (1981), *Phys. Rev. Lett.* **46**, 1496.

Walsh, D., Carswell, R. and Weyman, R. (1979), *Nature* **279**, 381.

Webster, R. L. (1989), *Gravitational Lenses Workshop*, Toulouse, September 13-15.

7

Inflation and quantum cosmology

Andrei Linde
CERN, CH-1211,
Geneva 23, Switzerland

We investigate an interplay between elementary particle physics, quantum cosmology and inflation. These results obtained within this approach are compared with the results obtained in the context of Euclidean quantum cosmology. In particular, we discuss relations between the stochastic approach to inflationary cosmology and the approaches based on the investigation of the Hartle-Hawking and tunneling wave functions of the universe. We argue that neither of these wave functions can be used for a complete description of the inflationary universe, but in certain cases they can be used for a description of some particular stages of inflation. It is shown that if the present vacuum energy density ρ_v exceeds some extremely small critical value ρ_c ($\rho_c \sim 10^{-10^7}$ g cm^{-3} for chaotic inflation in the theory $\frac{1}{2}m^2\phi^2$), then the lifetime of mankind in the inflationary universe should be finite, even though the universe as a whole will exist without end. A possible way to justify the anthropic principle in the context of the baby universe theory and to apply it to the evaluation of masses of elementary particles, of their coupling constants and of the vacuum energy density is also discussed.

1. Introduction

With the invention of unified theories of strong, weak, electromagnetic and gravitational interactions, elementary particle physics has entered a very interesting and unusual stage of its development. It appears that in the context of these theories we can predict much more than we can actually verify by the standard experimental methods of high energy physics.

The energy scale at which the unified nature of all four fundamental interactions is expected to become manifest is not very different from the Planck mass M_P, about 10^{-5} grams, where quantum gravity effects become important. (The Planck mass is that mass for which the Compton wavelength l_P, about 10^{-23} cm, equals the Schwarzschild radius.) Its rest energy $M_P c^2$, about 10^{19} GeV, corresponds to the kinetic energy of a small airplane. By contrast, the 80-km-circumference Superconducting Supercollider the Americans hope to build in the near future will accelerate particles up to 10^4 GeV. The largest accelerator ring one could build on Earth, with a circumference of 40,000 km, could not accelerate particles beyond about 10^8 GeV, a center-of-mass energy occasionally to be seen in cosmic-ray collisions. But this still leaves us 12 orders of magnitude short of the energy necessary for a direct test of the unified theories. Of course, there are some indirect tests, such as the searches for proton decay and for supersymmetric partners of ordinary particles. But trying to get a correct theory of all fundamental interactions with only such low-energy experiments is like looking for the correct unified electroweak theory by using nothing but radiotelescopes. (Note that $M_P/E_W \sim E_W/E_\gamma$, where $E_W \sim 10^2$ GeV is the unification energy scale in the theory of weak and electromagnetic interactions, $E_\gamma \sim 10^{-5}$ eV is the typical energy of photons in a radiowave.)

The only accelerator that could ever produce particles energetic enough for a direct testing of the unified theories of all fundamental interactions is our universe itself. The Big Bang scenario as it stood ten years ago, which I will call the hot-universe theory, asserts that the universe was born at some moment $t = 0$ about 15 billion years ago, in a state of infinitely large density ρ. Of course, one cannot really speak of classical space-time for the earliest moments, when kT/c^2 was greater than the Planck mass, and ρ exceeded M_P/l_P^3, making quantum fluctuations of the metric predominant. (We will use a convenient unit system that sets k, c, and $\hbar$ all equal to 1, so that $l_P = 1/M_P$ and the Planck density is M_P^4, roughly 10^{94} g cm^{-3}.) It is just at such times, when the average particle energy exceeded M_P, that the unity of all four fundamental interactions would have been manifest.

With the rapid expansion of the universe, the average energy of particles, given by the temperature, decreases rapidly, and the universe becomes cold. The temperature falls as the reciprocal of $a(t)$, the scale factor, or "radius," of the universe. This means that particle interactions at extremely large energies can have occurred only at the very early stages of the evolution of the universe. One might think it very difficult to extract useful and reliable information from the unique experiment carried out about 10^{10} years ago. Thus, it came as a great surprise to those who study elementary particles that the investigation of physical processes at the very early stages of the universe can rule out most of the existing unified theories.

For example, all the grand unified theories predict the existence of superheavy stable particles carrying magnetic charge: magnetic monopoles. These objects have a typical mass 10^{16} times that of the proton. According to the standard hot-universe theory, monopoles should appear at the very early stages of the universe, and they should now be as abundant as protons. In that case the mean density of matter in the universe would be about 15 orders of magnitude higher than its present value of about 10^{-29} g cm^{-3}.[1]

It was shown that all theories with spontaneous breaking of discrete symmetry (including the simplest $SU(5)$ model, the models of spontaneous CP violation, most of the theories with axions, etc.) lead to the existence of superheavy domain walls, which would be in drastic contradiction with observational cosmology.[2] It was shown that in most of the theories based on $N = 1$ supergravity the primordial abundance of gravitinos contradicts cosmological data by about 10 orders of magnitude,[3] whereas the energy density stored in the so-called Polonyi fields contradicts cosmological data by 15 orders of magnitude for the simplest models,[4] and by 6 orders of magnitude for the no-scale models.[5] Several models based on superstring theory lead to a cosmologically unacceptable type of axion.[6] Most of the Kaluza-Klein theories based on $N = 1$, $d = 11$ supergravity predict the present vacuum energy to be $O(-M_P^4)$, which would contradict the observational data by 125 orders of magnitude. The situation with superstring theories seemed, at first glance, to be somewhat better. To be fair, however, one must say that no sufficiently good cosmological model based on superstrings has been suggested so far, though some interesting ideas have been already proposed, see e.g. Ref. 7.

Despite many efforts, some of the problems listed above still remain unsolved, which probably means that the corresponding theories are actually unrealistic. Fortunately, however, the problems of primordial monopoles, gravitinos, domain walls and some other important cosmological problems such as the flatness problem, the homogeneity and isotropy problem, etc., can be solved simultaneously in the context of a relatively simple scenario.[8-12] This scenario makes it possible also to remove some constraints on the parameters of the elementary particle theory, for example, the constraint on the axion mass $m_A \geq 10^{-5}$ eV ($f_A \leq 10^{12}$ GeV).[13]

The invention of this scenario has modified considerably the standard cosmological paradigm. One of the most important modifications was performed within the last three years. Therefore, we will start with the discussion of the "standard" inflationary universe scenario. Then we will discuss some recent developments in the inflationary cosmology[14,15] and some new trends in the theory of baryogenesis and in the theory of galaxy formation. We will finish with the discussion of exciting possibilities related to baby-universe theory, the Anthropic Principle and the cosmological constant problem.[16-26]

2. The "standard" inflationary scenario

The stage of inflation

Historically, there were several different versions of the inflationary universe scenario. At present, it seems that the simplest, and simultaneously, the most general one is the chaotic inflation scenario.[12] To describe it, let us consider the simplest model based on the theory of a massive non-interacting scalar field ϕ with the Lagrangian

$$L = \frac{M_P^2}{16\pi} R + \frac{1}{2} \partial_\mu \phi \partial^\mu \phi - \frac{m^2}{2} \phi^2 . \tag{1}$$

Here $M_P^{-2} = G$ is the gravitational constant, $M_P \sim 10^{19}$ GeV is the Planck mass, R is the curvature scalar, m is the mass of the scalar field ϕ, $m \ll M_P$. If the classical field ϕ is sufficiently homogeneous in some domain of the universe (see below), then its behavior inside this domain is governed by the equations

$$\ddot{\phi} + 3H\dot{\phi} = -dV/d\phi , \tag{2}$$

$$H^2 + \frac{k}{a^2} = \frac{8\pi}{3M_P^2} (\frac{1}{2}\dot{\phi}^2 + V(\phi)) . \tag{3}$$

Here $V(\phi)$ is the effective potential of the field ϕ (in our case $V(\phi) = \frac{1}{2}m^2\phi^2$), $H = \dot{a}/a$, $a(t)$ is the scale factor of the locally Friedmannian universe (inside the domain under consideration), $k = +1$, -1, or 0 for a closed, open or flat universe, respectively. If the field ϕ initially is sufficiently large ($\phi \geq M_P$), then the functions $\phi(t)$ and $a(t)$ rapidly approach the asymptotic regime

$$\phi(t) = \phi_0 - \frac{mM_P}{2(3\pi)^{1/2}}t, \tag{4}$$

$$a(t) = a_0 \operatorname{Exp}\left(\frac{2\pi}{M_P^2}(\phi_0^2 - \phi^2(t))\right). \tag{5}$$

According to Eqs. (4) and (5), during a time $\tau \sim \phi/mM_P$ the value of the field ϕ remains almost unchanged and the universe expands quasi-exponentially:

$$a(t + \Delta t) \sim a(t) \operatorname{Exp}(H\Delta t) \tag{6}$$

for $\Delta t \leq \tau = \phi/mM_P$. Here

$$H = \frac{2\pi^{1/2}}{\sqrt{3}}\frac{m\phi}{M_P}. \tag{7}$$

Note that $H \gg \tau^{-1}$ for $\phi \gg M_P$.

The regime of quasi-exponential expansion (inflation) occurs for $\phi \geq \frac{1}{5}M_P$. For $\phi \leq \frac{1}{5}M_P$ the field ϕ oscillates rapidly, and if this field interacts with other matter fields (which are not written explicitly in Eq. (1)), its potential energy $V(\phi) \sim m^2\phi^2/2 \sim m^2 M_P^2$ is transformed into heat. The reheating temperature T_R may be of the order $(mM_P)^{1/2}$ or somewhat smaller, depending on the strength of the interaction of the field ϕ with other fields. It is important that T_R does not depend on the initial value ϕ_0 of the field ϕ. The only parameter which depends on ϕ_0 is the scale factor $a(t)$, which grows $\operatorname{Exp}((2\pi/M_P^2)\phi_0^2)$ times during inflation.

If, as is usually assumed, a classical description of the universe becomes possible only when the energy-momentum tensor of matter becomes smaller than M_P^4, then at this moment $\partial_\mu \phi \partial^\mu \phi \leq M_P^4$ and $V(\phi) \leq M_P^4$.

Therefore, the only constraint on the initial amplitude of the field ϕ is given by $\frac{1}{2}m^2/\phi^2 \leq M_P^4$. This gives a typical initial value of the field ϕ:

$$\phi_0 \sim \frac{M_P^2}{m}. \tag{8}$$

Let us consider for definiteness a closed universe of a typical initial size $O(M_P^{-1})$. It can be shown that if initially $\partial_\mu \phi \partial^\mu \phi$ becomes much smaller than $V(\phi)$, the evolution of the universe becomes describable by (2)-(7), and after inflation the total size of the universe becomes larger than

$$l \sim M_P^{-1} \operatorname{Exp}\left(\frac{2\pi}{M_P^2}\phi_0^2\right) \sim M_P^{-1} \operatorname{Exp}\left(\frac{2\pi M_P^2}{m^2}\right). \qquad (9)$$

For $m \sim 10^{-6} M_P$ (which is necessary to produce density perturbations $\delta_\rho/\rho \sim 10^{-5}$, see below)

$$l \sim M_P^{-1} \operatorname{Exp}(2\pi 10^{12}) \geq 10^{10^{12}} \mathrm{cm}, \qquad (10)$$

which is much greater than the size of the observable part of the universe $\sim 10^{28}$ cm.

After such a large inflation the term k/a^2 in Eq. (3) becomes negligibly small compared with H^2, which means that the universe becomes flat and its geometry locally Euclidean. This implies that the total density of the universe ρ becomes almost exactly equal to the critical density $\rho_c = \frac{3M_P^2}{8\pi}H^2$, i.e.

$$\Omega = \frac{\rho}{\rho_c} \sim 1. \qquad (11)$$

For similar reasons the universe becomes locally homogeneous and isotropic. The density of all 'undesirable' objects (monopoles, domain walls, gravitinos) created before or during inflation becomes exponentially small, and they never appear again if the reheating temperature, T_R, is not too large.

We should like to emphasize that for a realisation of this scenario it is sufficient that initially $\partial_\mu \phi \partial^\mu \phi \leq V(\phi) \sim M_P^4$ in a domain of a smallest possible size $l \sim M_P^{-1}$. Since $\partial_\mu \phi \partial^\mu \phi \leq M_P^4$, $V(\phi) \leq M_P^4$ in any classical spacetime, the above-mentioned initial conditions are quite natural. For a more detailed discussion of initial conditions which are necessary for inflation see Ref. 28.

Scalar field fluctuations, perturbations of density and galaxy formation

According to quantum field theory, empty space is not entirely empty. It is filled with quantum fluctuations of all types of physical fields. These fluctuations can be regarded as waves of physical fields with all possible wave-lengths, moving in all possible directions. If the values of these fields, averaged over some macroscopically large time, vanish, then the space filled with these fields seems to us empty and can be called the vacuum.

In the exponentially expanding universe the vacuum structure is much more complicated. The wavelengths of all vacuum fluctuations of the scalar field ϕ grow exponentially in the expanding universe. When the wavelength of any particular fluctuation becomes greater than H^{-1}, this fluctuation stops propagating, and its amplitude freezes at some nonzero value $\delta\phi(x)$ because of the large friction term $3H\dot{\phi}$ in the equation of motion of the field ϕ. The amplitude of this fluctuation then remains almost unchanged for a very long time, whereas its wavelength grows exponentially. Therefore, the appearance of such a frozen fluctuation is equivalent to the appearance of a classical field $\delta\phi(x)$ that does not vanish after averaging over macroscopic intervals of space and time.

Because the vacuum contains fluctuations of all wavelengths, inflation leads to the creation of more and more new perturbations of the classical field with wavelengths greater than H^{-1}. The average amplitude of such perturbations generated during a time interval H^{-1} (in which the universe expands by a factor of e) is given by

$$|\delta\phi(x)| \approx \frac{H}{2\pi}. \tag{12}$$

Perturbations of the field lead to adiabatic perturbations of density $\delta\rho \sim V'(\phi)\delta\phi$, which after inflation grow, and acquire the amplitude[30,29]

$$\frac{\delta\rho}{\rho} = \frac{48}{5}\sqrt{\frac{2\pi}{3}}\frac{V_\phi^{3/2}}{M_P^3 V'(\phi)}, \tag{13}$$

where ϕ is the value of the classical field $\phi(t)$ of Eq. (4), at which the fluctuation we consider has the wavelength $l \sim k^{-1} \sim H^{-1}(\phi)$ and becomes frozen in amplitude. In the theory of the massive scalar field with $V(\phi) = m^2\phi^2/2$,

$$\frac{\delta\rho}{\rho} = \frac{24}{5}\sqrt{\frac{\pi}{3}}\frac{m}{M_P}\left(\frac{\phi}{M_P}\right)^2. \tag{14}$$

Taking account of Eqs. (4), (5) and also of the expansion of the universe by about 10^{30} times after the end of inflation, one can obtain the following result for the density perturbations with the wavelength l (cm) at the moment when these perturbations begin growing and the process of the galaxy formation starts:

$$\frac{\delta\rho}{\rho} \sim 0.8\frac{m}{M_P}\ln l(\text{cm}). \tag{15}$$

At a galaxy scale ($l \sim 10^{21} - 10^{22}$ cm)

$$\frac{\delta\rho}{\rho} \sim 40\frac{m}{M_P},\tag{16}$$

which gives a desirable amplitude $\delta\rho/\rho \sim 10^{-4} - 10^{-5}$ for $m \sim 2 \times 10^{-7} - 2 \times 10^{-6}M_P$. In what follows we will assume that $m \sim 10^{-6}M_P$.

3. Self-reproducing universe and life after inflation

In our previous investigation we only considered local properties of the inflationary universe, which is quite sufficient for a description of the observable part of the universe of the present size $l \sim 10^{28}$ cm. Indeed, in accordance with Eq. (15) our universe remains relatively homogeneous on a scale

$$l \leq l^* \sim \text{Exp}(2\pi\frac{M_P}{m})\text{cm} \sim 10^{3\times10^6}\text{cm}.\tag{17}$$

The corresponding density perturbations were formed at the time, when the scalar field $\phi(t)$ was bigger than ϕ^*, where

$$\phi^* \sim M_P\sqrt{\frac{M_P}{m}} \sim 30M_P,\tag{18}$$

see Eq. (14). (Note that $V(\phi^*) \sim \frac{m^2}{2}(\phi^*)^2 \sim mM_P^3 \ll M_P^4$.) On a scale $l > l^*$ the universe becomes extremely inhomogeneous due to quantum fluctuations produced during inflation. We are coming to a paradoxical conclusion, that the global properties of the inflationary universe are determined not by classical but by quantum effects.

Let us try to understand the origin of such a behavior of the inflationary universe.

A very unusual feature of the inflationary universe is that processes separated by distance l greater than H^{-1} proceed independently of one another. This is so because during exponential expansion any two objects separated by more than H^{-1} are moving away from each other with a velocity v exceeding the speed of light. (This does not contradict special relativity because v is not the speed of any signal; it is just the rate at which the general expansion of the universe separates two distant points.) As a results, any observer in the inflationary universe can see only those processes occurring nearer than H^{-1}.

An important consequence of this general result is that the process of inflation in any spatial domain of radius H^{-1} occurs independently of any events outside it. Any two inflationary domains displaced by more than H^{-1} cannot collide or eat one another, or do each other any damage. Their expansion is due not to the annexation of the territory of their neighbors, but rather to the peaceful (and very rapid) growth in their own volume, as allowed by general relativity. In this sense any inflationary domain of initial size exceeding $2H^{-1}$ can be considered as a separate mini-universe, expanding independently of what occurs outside it.

To investigate the behavior of such a mini-universe, with an account taken of quantum fluctuations, let us consider an inflationary domain of initial size roughly H^{-1} containing a sufficiently homogeneous field whose initial value greatly exceeds M_P. Eq. (4) tells us that during a typical time interval $\Delta t = H^{-1}$ the field inside this domain will be reduced by

$$\Delta \phi = \frac{M_P^2}{4\pi\phi} . \tag{19}$$

By comparison of (12) and (19) one can easily see that if ϕ is much less than $\phi^* \sim M_P\sqrt{M_P/m}$, the decrease of the field ϕ due to its classical motion is much larger than the amplitude of the quantum fluctuations $\delta\phi$ generated during the same time. But for large ϕ (up to the classical limit of $10^4 M_P$), $\delta\phi(x)$ will exceed $\Delta\phi$, i.e. the Brownian motion of the field ϕ becomes more rapid than its classical motion. Because the typical wavelength of the fluctuation field $\delta\phi(x)$ generated during this time is H^{-1}, the whole domain volume after Δt will effectively have become divided into e^3 separate domains (mini-universes) of diameter H^{-1}. In almost half of these domains the field ϕ grows by $|\delta\phi(x)| - \Delta\phi$, which is not very different from $|\delta\phi(x)|$ or $H/2\pi$, rather than decreases. During the next time interval $\Delta t = H^{-1}$, the field grows again in half of these mini-universes. It can be shown that the total physical volume occupied by a permanently growing field ϕ increases with time like $\text{Exp}((3 - \ln 2)Ht)$, and the total volume occupied by a field that does not decrease grows almost as fast as $\frac{1}{2}e^{3Ht}$.

Because the value of the Hubble constant $H(\phi)$ is proportional to ϕ, the main part of the physical volume of the universe is the result of the expansion of domains with nearly the maximal possible field value, M_P^2/m, for which $V(\phi)$ is close to M_P^4. There are also exponentially many domains with smaller values of ϕ. Those domains in which ϕ eventually becomes smaller than about $30M_P$ give rise to the mini-universes of *our* type. In such domains, ϕ eventually rolls down to the minimum of $V(\phi)$, and these mini-universes are subsequently describable by the usual

Big Bang theory. However a considerable part of the physical volume of the entire universe remains forever in the inflationary phase.[14,15]

Thus in our scenario the universe, in which there was initially at least one domain of a size on the order of H^{-1} filled with a sufficiently large and homogeneous field ϕ, unceasingly reproduces itself and becomes immortal. One mini-universe produces many others, and this process goes on without end, even if some of the mini-universes eventually collapse.

But this means that it is vanishingly improbable that our mini-universe would have been the first in the sequence of all mini-universes. Moreover, it no longer seems necessary to assume that there actually was some first mini-universe appearing from nothing or from an initial singularity at some moment $t = 0$ before which there was no space-time at all.

From general topological theorems about singularities in cosmology it does not actually follow that our universe was created *as a whole* at some moment before which the universe did not exist. The usual assumption that the whole universe appears from the unique Big Bang singularity at $t = 0$ is based on the implicit assumption that the universe *as a whole* is sufficiently homogeneous. Indeed, the observable part of our universe is very homogeneous. Observed density fluctuations are less than a part in a thousand and there has been no reason to expect that the universe is inhomogeneous on a larger scale beyond the 10^{10} light-year horizon. In a homogeneous universe one can use the density $\rho(t)$ as a measure of time. In that case it can be shown that the universe appears *as a whole* from a singularity at $t = 0$. The initial density $\rho(0)$ is infinite, and it becomes possible to describe the whole universe in terms of classical space-time after the Planck time M_P^{-1} (about 10^{-43} seconds), when the energy density everywhere simultaneously becomes smaller than the Planck density M_P^4.

With the invention of the inflationary scenario the situation changes drastically. At present only inflation can explain why the observable part of the universe is extremely *in*homogeneous. In some parts of the universe the energy density ρ is now of the order of M_P^4, 125 orders magnitude higher than the 10^{-29} or 10^{-30} g cm^{-3} we can see nearby. In such a scenario there is no reason to assume that the universe was initially homogeneous and that all its causally disconnected parts started their expansions simultaneously.

If the universe is infinitely large (like the Friedmann open or flat universe), then it cannot have had a single beginning; a simultaneous creation of infinitely many causally disconnected regions is totally improbable. Therefore the universe cannot be infinite *ab initio*, or it must exist eternally as a huge self-reproducing entity. Some of its parts appear at different times from singularities, or may die in a singular state. New parts are constantly being created from the space-time foam when $V(\phi)$ exceeds the Planck density, or they may revert to the foamlike state again as a result of large fluctuations in ϕ. But the evolution of the universe as a whole has no end, and it may have had no beginning.

4.   Inflation and the wave function of the universe

One of the most ambitious approaches to cosmology is based on the investigation of the Wheeler-DeWitt equation for the wave function Ψ of the universe.[32] However, this equation has many different solutions, and *a priori* it is not quite clear which of these solutions describes our universe.

A very interesting idea was suggested by Hartle and Hawking.[33] According to their work, the wave function of the ground state of the universe with a scale factor a filled with a scalar field ϕ in the semi-classical approximation is given by

$$\psi_0(a,\phi) \sim \mathrm{Exp}(-S_E(a,\phi)). \tag{20}$$

Here $S_E(a,\phi)$ is the Euclidean action corresponding to the Euclidean solutions of the Lagrange equation for $a(\tau)$ and $\phi(\tau)$ with the boundary conditions $a(0) = a$, $\phi(0) = \phi$. The reason for choosing this particular solution of the Wheeler-DeWitt equation was explained as follows. Let us consider the Green's function of a particle which moves from the point $(0, t')$ to the point $(\mathbf{x}, t)$:

$$< \mathbf{x}, 0 | 0, t' > = \sum_n \psi_n(\mathbf{x})\psi_n(0)\,\mathrm{Exp}(iE_n(t-t'))$$
$$= \int d\mathbf{x}(t)\,\mathrm{Exp}(iS(\mathbf{x}(t))), \tag{21}$$

where ψ_n is a complete set of energy eigenstates corresponding to the energies $E_n \geq 0$.

To obtain an expression for the ground-state wave function $\Psi_0(\mathbf{x})$, one should make a rotation $t \to -i\tau$ and take the limit as $\tau \to -\infty$. In the summation (21) only the term $n = 0$ with $E_0 = 0$ survives, and the integral transforms into $\int dx(\tau)\,\mathrm{Exp}(-S_E(\tau)))$. Hartle and Hawking have argued that the generalization of this result to the case of interest in the semiclassical approximation would yield Eq. (20).

The gravitational action corresponding to the Euclidean section S_4 of de Sitter space dS_4 with $a(\tau) = H^{-1}(\phi) \cos H\tau$ is negative,

$$S_E(a, \phi) = -\frac{1}{2} \int d\eta \left[\left(\frac{da}{d\eta}\right)^2 - a^2 + \frac{\Lambda}{3} a^4 \right] \frac{3\pi M_P^2}{3} = -\frac{3M_P^2}{16V(\phi)}. \quad (22)$$

Here η is the conformal time, $\eta = \int dt/a(t)$, $\Lambda = 8\pi V/M_P^2$. Therefore, according to Ref. 40,

$$\psi_0(a, \phi) \sim \text{Exp}(-S_E(a, \phi)) \sim \text{Exp}(\frac{3M_P^4}{16V(\phi)}). \quad (23)$$

This means that the probability P of finding the universe in the state with $\phi = \text{const}$, $a = H^{-1}(\phi) = 3M_P^2/8\pi V(\phi))^{1/2}$ is given by

$$P(\phi) \sim |\psi_0|^2 \sim \text{Exp}(\frac{3M_P^4}{8V(\phi)}). \quad (24)$$

This expression has a very sharp maximum as $V(\phi) \to 0$. Therefore the probability of finding the universe in a state with a large field ϕ and having a long stage of inflation becomes strongly diminished.

There exists an alternative choice of the wave function of the universe. It can be argued that the analogy between the standard theory (21) and the gravitational theory (22) is incomplete. Indeed, there is an overall minus sign in the expression for $S_E(a, \phi)$ in Eq. (22) which indicates that the gravitational energy associated with the scale factor a is negative. (This is related to the well-known fact that the total energy of a closed universe is zero, being a sum of the positive energy of matter and the negative energy of the scale factor a.) In such a case, to obtain ψ_0 from (22) one should rotate t not to $-i\tau$ but to $+i\tau$ which leads to[34]

$$P \sim |\psi_0(a, \phi)|^2 \sim \text{Exp}(-2|S_E(a, \phi)|) \sim \text{Exp}(-\frac{3M_P^4}{8V(\phi)}). \quad (25)$$

Actually, this result is valid only if the evolution of the field ϕ is very slow, so that this field acts only as a cosmological constant $\Lambda(\phi) = 8\pi V(\phi)/M_P^2$ in Eq. (22). Fortunately, this is indeed the case during inflation. Later the same result was obtained by another method, devised by Zeldovich and Starobinsky,[35] Rubakov,[36] and Vilenkin.[37] This result can be interpreted as the probability of quantum tunnelling of the universe from $a = 0$ (from 'nothing') to $a = H^{-1}(\phi)$. In complete agreement with our previous argument (see Eq. (8)), Eq. (25) shows that a typical initial value of the field ϕ is given by $V(\phi) \sim M_P^4$ (if one does not speculate about the possibility that $V(\phi) \gg M_P^4$), which leads to a very long state of inflation.

It must be said that there is no rigorous derivation of either Eq. (23) or (25), and the physical meaning of creation of everything from 'nothing' is far from clear. Therefore a deeper understanding of the physical processes in the inflationary universe is necessary in order to investigate the wave function of the universe $\Psi_0(a, \phi)$ and to suggest a correct interpretation of this wave function. With this purpose we shall try to investigate the global structure of the inflationary universe, and go beyond the minisuperspace approach used in the derivation of (23) and (25). This can be done with the help of a stochastic approach to inflation,[38,15] which is a more formal way to investigate the Brownian motion of the scalar field ϕ studied in the previous section.

The evolution of the fluctuating field ϕ in any given domain can be described with the help of its distribution function $P(\phi)$, or in terms of its average value $\overline{\phi}$ in this domain and its dispersion $\Delta = (< \delta\phi^2 >)^{1/2}$. However, one will obtain different results depending on the method of averaging: one can consider the distribution $P_c(\phi)$ over the non-growing coordinate volume of the domain (i.e., over its physical volume at some initial moment of inflation), or the distribution $P_p(\phi)$ over its *physical* (proper) volume, which grows exponentially at a different rate in different parts of the domain. It can be shown that the dispersion of the field ϕ in the coordinate volume Δ_c is always much smaller than $\overline{\phi}_c$ for $V(\overline{\phi}_c) \ll M_P^4$. Therefore the evolution of the averaged field $\overline{\phi}_c$, can be described approximately by Eqs. (2)-(5). However, if one wishes to know the resulting spacetime structure and the distribution of the field ϕ after (or during) inflation, it is more appropriate to take an average $\overline{\phi}_P$ over the physical volume, and in some cases the behavior of $\overline{\phi}_P$ and Δ_P differs considerably from the behavior of $\overline{\phi}_c$ and Δ_c.

The Brownian motion of the field ϕ can be described by the diffusion equation

$$\frac{\partial P_c}{\partial t} = \frac{\partial}{\partial \phi} \Big[\frac{\partial(\mathcal{D} P_c)}{\partial \phi} + \frac{P_c}{3H} \frac{\partial V}{\partial \phi} \Big], \tag{26}$$

where the coefficient of diffusion $\mathcal{D} = H^3/8\pi^2$. This equation for the case $H(\phi) = $ const. was first derived by Starobinsky;[38] for a more detailed derivation see Ref. 39. For the special case $\partial V/\partial \phi = 0$ this equation was obtained by Vilenkin.[40]

The stationary solution $(\partial P_c/\partial t = 0)$ would be[38]

$$P_c \sim \mathrm{Exp}(3M_P^4/8V(\phi)), \tag{27}$$

which is equal to the square of the Hartle-Hawking wave function of the universe.[23] At first glance, this result is a direct confirmation of the Hartle-Hawking prescription for the wave function of the universe. (In fact, this is the *only* more or less rigorous confirmation of the Hartle-Hawking prescription which is known to us at present).

However, in all realistic cosmological theories, in which $V(\phi) = 0$ in its minimum the distribution (27) is not normalizable. The source of this difficulty can be easily understood: any stationary distribution may exist only due to a compensation of a classical flow of the field ϕ downwards to the minimum of $V(\phi)$ by the diffusion motion upwards. However, diffusion of the field ϕ discussed above exists only during inflation, i.e. only for $\phi \geq M_P$, $V(\phi) \geq V(M_P) \sim m^2 M_P^2 \sim 10^{-8} M_P^4$. Therefore Eq. (27) would correctly describe the stationary distribution $P_c(\phi)$ in the inflationary universe only if $V(\phi) \geq 10^{-8} M_P^4$ in the absolute minimum of $V(\phi)$, which of course, is unrealistic.[15] Therefore, at present, we do not know how one can use the Hartle-Hawking wave function (20), (23) for the description of the inflationary universe.

Now let us study a general non-stationary solution of Eq. (26) describing the time-evolution of the initial distribution $P_c(\phi, t = 0) = \delta(\phi - \phi_0)$. This solution for the theory $m^2 \phi^2 / 2$ is[15]

$$P_c(\phi, t) = \text{Exp}\left[-\frac{3(\phi - \phi(t))^2 M_P^4}{2m^2(\phi_0^4 - \phi^4(t))}\right], \qquad (28)$$

which shows that at the first stage of the process during the time $\tau \leq \phi_0 2(3\pi)^{1/2}/mM_P$ (see Ref. 4), the maximum of the distribution $P_c(\phi, t)$ almost does not move, whereas the dispersion linearly grows. Then, at $t \gg \tau$, the maximum of $P_c(\phi, t)$ moves to $\phi = 0$, just as the classical field $\phi(t)$.[4] However, during the first period of time the volume of domains filled by the field ϕ increases approximately by $\text{Exp}(3H(\phi)\tau) \sim \text{Exp}(c_1\phi_0\phi/M_P^2)$ where $c_1 = O(1)$. After this time the distribution $P_p(\phi, \tau)$ looks approximately as follows

$$P_p(\phi, \tau) \sim P_c(\phi, \tau) \text{Exp}(3H(\phi)\tau) \sim \text{Exp}\left[-\frac{\phi^2 M_P^4}{c_2 m^2 \phi_0^4} + c_1 \frac{\phi_0 \phi}{M_P^4}\right], \quad (29)$$

where $c_2 = O(1)$. One can easily verify that, if $\phi_0 > \phi^* \sim M_P\sqrt{\frac{M_P}{m}}$, then the maximum of $P_p(\phi, t)$ during the time τ becomes shifted to some field ϕ which is bigger than ϕ_0. This just corresponds to the process of eternal self-reproduction of the inflationary universe studied in the previous section. For completeness we shall mention here another solution of Eq. (26). If the initial value of the field ϕ is very large, $\phi_0 \geq$

M_P^2/m, i.e. if one starts with the spacetime foam with $V(\phi) \geq M_P^4$, then the evolution of the field ϕ in the first stage (rapid diffusion) becomes more complicated (the naively estimated dispersion $\Delta_c^2 \sim H^3 \Delta t$ soon becomes greater than ϕ_0^2). In this case the distribution of the field ϕ is not Gaussian. The solution of Eq. (26) at the stage of diffusion from ϕ_0 to some field ϕ with $V(\phi) \ll M_P^4$ is given by

$$P_c(\phi) \sim \text{Exp}\left(-\frac{3\sqrt{3}\pi M_P^3}{m^3 \phi t}\right). \tag{30}$$

This solution describes quantum creation of domains of a size $l \geq H^{-1}(\phi)$, which occurs due to the diffusion of the field ϕ from $\phi_0 \geq M_P^2/m$ to $\phi \ll \phi_0$. Direct diffusion with formation of a domain filled with the field ϕ is possible only during the time $= c(2\sqrt{3}\pi\phi/mM_P)$, $c = O(1)$. At larger times a more rapid process is a diffusion to some field $\tilde{\phi} > \phi$ and a subsequent classical rolling down from $\tilde{\phi}$ to ϕ. Therefore one may interpret a distribution $P_c(\phi)$ formed after a time $t = c(2\sqrt{3}\pi\phi/mM_P)$ as a probability of a quantum creation of a mini-universe filled with a field ϕ,[15]

$$P_c(\phi) \sim \text{Exp}(-c\frac{3M_P^4}{2m^2\phi^2}) \sim \text{Exp}(-2c\frac{3M_P^4}{8V(\phi)}), \tag{31}$$

which is in agreement with the previous estimate for the probability of quantum creation of the universe, Eq. (26).[34–37]

These results imply that life in the inflationary universe will never disappear. Unfortunately, this conclusion does not automatically mean that one can be very optimistic about the future of mankind. Even though new domains of our type are permanently produced during inflation, later on each particular domain either will become practically empty and unsuitable for life or it will evolve into a huge black hole and collapse.[27] The only possible strategy of survival which we see at the moment is to travel from old domains to the new ones, which are displaced near the self-reproducing inflationary domains and always have large enough density of baryons. In the worst case, if we will be unable to travel to such distant places ourselves, we can try to send some information about us, our life and our knowledge, and maybe even stimulate development of such kinds of life there, which would be able to receive and use this information. In such a case one would have a comforting thought that even though life in our part of the universe will disappear, we will have some inheritors, and in this sense our existence is not entirely meaningless. (At least it would not be worse than what we have here now.)

In order to check whether such a possibility does actually exist, it would be necessary to perform a detailed investigation of the global structure of the universe, and especially of its causal structure, which becomes very nontrivial in the context of inflationary cosmology.[15] One should answer the question whether it is actually possible to travel or to send a signal from our part of the universe to a vicinity of a nearby self-reproducing inflationary domain, and to do it in such a way that after a long time which is necessary for the signal to arrive the density of matter near this domain will be sufficiently large for the existence of life there. At present we do not know any no-go theorems which would imply that it is generally impossible and that one cannot elaborate any successful strategy of survival of life in our part of the universe. One of the main aims of this paper is to show that the answer to this question may depend crucially on the value of the vacuum energy density, i.e. on the absence or presence of a tiny cosmological constant in the Einstein equations.

Namely, according to Refs. 14, 15, the self-reproducing domains appear in the inflationary universe independently of the existence or nonexistence of a small energy density ρ_v of the vacuum state in which we live now (if it is not negative and as big as $-10^{-6} M_P^4$. However, the causal structure of the universe, the possibility of communication between its different domains and the possibility of the existence of life there depend crucially on the value of ρ_v. For example, if ρ_v is negative and its absolute value is much bigger than the present energy density of matter $\rho_0 \sim 10^{-29}$ g cm^{-3}, each domain of the universe after inflation and a sufficiently long subsequent period of expansion behaves as an anti de Sitter space and collapses at a time $t \sim M_P \rho_v^{-1/2} \ll M_P \rho_0^{-1/2} \sim 10^{10}$ years, i.e. much earlier than the time necessary for the development of life of our type. Note, that this process occurs locally, at different times in different places of a self-reproducing universe. Therefore one may escape the local collapse by travelling towards the self-reproducing domains. However, since the typical distance between our place and a nearby self-reproducing domain is extremely large, such a journey is possible only if the lifetime of the anti de Sitter domain $t \sim M_P \rho_v^{-1/2}$ is also extremely large and, consequently the absolute value of ρ_v is extremely small (see below).

If the vacuum energy density is positive and bigger than 10^{-27} g cm^{-3}, formation of galaxies of our type becomes impossible.[41] Smaller values of ρ_v do not affect galaxy formation, but do affect the causal structure of the universe. For example, the universe with $\rho_v > \rho_0$ at the time $t > 10^{10}$ years behaves as de Sitter space with the event horizon of the size $H^{-1} \sim 10^{28}$ cm (i.e., smaller than the particle horizon $\sim t$ now). In such a universe the distance between the galaxies grows as e^{Ht}, and after a time of the order $H^{-1} < 10^{10}$ years there would remain only one (our own) galaxy inside our event horizon. (The galaxy itself is not stretched by the universe expansion since it is a gravitationally bounded system with the energy density much bigger than ρ_0.) It is well known that in the exponentially expanding universe one can travel or send any information only to those regions which were inside the event horizon at the moment when the information was sent. This means that soon after the beginning of the exponential expansion of the universe with $\rho_v > \rho_0$ there will be no possibility to travel or send information to other galaxies, and it is certainly impossible to send a signal to any self-reproducing inflationary domain, since a typical distance from our place to any such domain is much bigger than H^{-1}.

This result can be considerably strengthened in all realistic models of inflation. To do it one should first make an estimate of a typical distance l^* from our place to a nearby inflationary self-reproducing domain. This can be easily done, since there should be many such domains in each region of initial size $H^{-1}(\phi)$ containing field $\phi > \phi^*$ (18), and there should be no such domains inside the regions of initial size H^{-1} containing field $\phi < \phi^*$. For the theory $m^2\phi^2/2$ this gives, according to Eq. (5)

$$l^* \sim H^{-1}(\phi^*) \, \mathrm{Exp}\left(\frac{2\pi\phi^{*2}}{M_P^2}\right)\left(\frac{T_R}{T_0}\right). \tag{32}$$

Here $H^{-1}(\phi^*) = M_P(3/8\pi V(\phi^*))^{1/2}$, T_R is the reheating temperature of the universe after inflation, T_0 is the present temperature of the blackbody radiation in the universe. The last factor appears due to the additional stretching of the universe when the temperature falls down from T_R to T_0. Typically this factor is of the order of 10^{30} and $H^{-1}(\phi^*)$ is of the order of 10^{-30} cm. Therefore from Eqs. (18), (32) it follows that

$$l^* \sim \mathrm{Exp}(2\pi M_P/m)\,\mathrm{cm} \sim \mathrm{Exp}((2\pi)10^6)\,\mathrm{cm}, \tag{33}$$

for $m \sim 10^{-6} M_P$. The universe becomes exponentially expanding and acquires an event horizon $H^{-1}(\rho_v) = M_P(3/8\pi\rho_v)^{1/2}$ at the late stages of its evolution, when the energy density of matter, which decreases as t^{-2}, becomes smaller than the vacuum energy density ρ_v. It occurs at the time which is bigger than the present age of the universe $t_0 \sim 10^{10}$ years by the factor $(\rho_0/\rho_v)^{1/2}$. At that time the distance l^* from us to the nearby inflationary self-reproducing domain grows by a factor $(\rho_0/\rho_v)^{1/3}$, since the scale factor of a cold matter dominated universe grows as $t^{2/3}$. The causal connection between us and this region will be possible only if this distance will remain smaller than $H^{-1}(\rho_v)$,

$$M_P\left(\frac{3}{8\pi\rho_v}\right)^{1/2} > M_P\left(\frac{3}{8\pi V(\phi^*)}\right)^{1/2}\mathrm{Exp}\left(\frac{2\pi\phi^{*2}}{M_P^2}\right)\left(\frac{T_R}{T_0}\right)\left(\frac{\rho_0}{\rho_v}\right)^{1/3}, \quad (34)$$

which implies that the causal connection is possible only if ρ_v is smaller than some critical value ρ_c, where

$$\rho_c \sim \frac{V^3(\phi^*)}{\rho_0^2}\left(\frac{T_0}{T_R}\right)^6\mathrm{Exp}\left(-\frac{12\pi M_P}{m}\right) \sim 10^{-10^7}\mathrm{g\,cm^{-3}}. \quad (35)$$

Here $V(\phi^*) \sim mM_P^3 \sim 10^{-6}M_P^4$, $\rho_0 \sim 10^{-29}\mathrm{g\,cm^{-3}} \sim 10^{-125}M_P^4$, $T_R \sim 10^{10}$ GeV. One can easily see that the constraint $\rho_v < \rho_c$ (Eq. (35)) on the value of the vacuum energy density ρ_v is very stringent and practically does not depend on the particular values of $V(\phi^*)$, ρ_0 and T_R.

Returning to the case of a negative vacuum energy, it can be shown by methods similar to those used above that in this case as well a causal contact between our part of the universe and a nearby self-reproducing inflationary domain is impossible if $|\rho_v| > \rho_c$, since in this case the time which is necessary for the contact is bigger than the time before the local collapse of our part of the universe.

These results are rather unexpected. It was clear *a priori* that the existence of a nonvanishing vacuum energy may slightly change the conditions of our life, but it was difficult to imagine that the development of life in our part of the universe may depend crucially on the existence or nonexistence of a tiny vacuum energy density $|\rho_v| \geq 10^{-10^7}\mathrm{g\,cm^{-3}}$.

Of course, the events we are discussing here will occur in a very distant future (at $t \sim 10^{5\times10^6}$ years), but it is very difficult not to be curious about our common fate as it can be understood at the present level of the development of science. Another possible application of these results would be to use them in order to obtain a strong anthropic bound $|\rho_v| < \rho_c \sim 10^{-10^7}\mathrm{g\,cm^{-3}}$, which would help us to solve the cosmological constant problem. At first glance, such an idea certainly cannot work, since the existence of the vacuum energy $-10^{-29}\mathrm{g\,cm^{-3}} < \rho_v < 10^{-27}\mathrm{g\,cm^{-3}}$ does not prevent formation of galaxies and can hardly have any effect on the appearance of life in the universe.[41] Moreover, until very recently the general attitude of physicists to the Anthropic Principle was rather sceptical, to say the least. It was believed that the weak, strong and electromagnetic interactions are the same in all parts of our universe, that the fundamental constants of Nature are universal and it is meaningless to discuss the possibility to live in a universe of a different type. This attitude was somewhat changed first when it was understood that in accordance with the chaotic inflationary universe scenario the universe should contain an exponentially large (or maybe even infinitely large) number of exponentially large domains (mini-universes), in which all kinds of symmetry breaking and all kinds of compactification which are possible in a theory of a given type are actually realized.[29] In other words, the universe becomes divided into many exponentially large domains with different types of low-energy physics and maybe even with different dimensionality inside each of them. This made it possible to get a justification of a kind of Weak Anthropic Principle: If many exponentially large domains with the low-energy physics of our type do exist in the universe described by a given theory, then it is quite natural that we live in one of such domains rather than in a domain where life of our type is impossible.[29]

To explain it in a more detailed way let us remember that in realistic theories of elementary particles there exist many different types of scalar fields ϕ_i. The potential energy $V(\phi_i)$ often has many different local minima, in which the universe may live for an extremely long time, much greater than the 10^{10} years of our observable domain. For example, in the sypersymmetric $SU(5)$ theory, $V(\phi_i)$ has several different minima of almost equal depth. Because the laws governing the interactions of elementary particles at the low energies at which we do experiments depend on the values of the classical fields ϕ_i, each of these minima corresponds to a different low-energy physics. In one of them the $SU(5)$ symmetry between all types of interactions remains unbroken–that is, the scalar fields ϕ_i remain equal to zero. In other minima various symmetry breaking patterns are realized, and in only one of these minima is the broken symmetry of the weak, strong and electromagnetic interactions that which we in fact observe.

During inflation there are large-scale fluctuations of all the fields ϕ_i. As a result, the inflationary universe becomes divided into an exponentially large number of inflationary mini-universes, with the scalar fields taking all possible values. As inflation ends in some mini-universes, these scalar fields roll down to all possible minima of $V(\phi_i)$. The universe becomes divided into many different exponentially large domains, realizing

all possible types of symmetry breaking between the fundamental interactions. In some of these mini-universes the low-energy physics is quite
different from our own. We cannot now see them because the size of our
own domains is much greater than the size of its 10^{10}-light-year observable portion. We could not live in these domains because our kind of
life requires our kind of low-energy physics.

It is very important that in the inflationary universe there is lots of
room for all possible types of symmetry breaking and for all possible
types of life. There is, therefore, no longer any need to require that
in the true theory the minimum of $V(\phi_i)$ corresponding to our type of
symmetry breaking be the only one or the deepest one. This new cos-
mopolitan viewpoint may greatly simplify the task of building realistic
models of the elementary particles.

The change of the values of the scalar fields ϕ_i–that is to say, the
change of the vacuum state–is the simplest kind of "mutation" that may
occur during inflation. Much more interesting possibilities appear if one
considers chaotic inflation in the higher-dimensional Kaluza-Klein theories. In those domains in which the energy density of the field ϕ grows
to the Planck density, quantum fluctuations of the metric at a length
scale of M_P^{-1} become of order unity. In such domains an inflationary
d-dimensional universe can squeeze locally into a tube of smaller dimensionality $d - n$ (or vice versa). If this tube is also inflationary (in
$d - n$ dimensions) and the initial length of the tube is greater than M_P^{-1}
(which is quite probable near the Planck density), then its further expansion proceeds independently of its prehistory and of the fate of its
mother universe. In an eternally existing universe such processes should
occur even if their probability is very small. In fact the probability of
such processes is small only if $V(\phi)$ is far below M_P^4.

Thus the inflationary universe becomes divided into different mini-
universes in which all possible types of compactification produce all sorts
of dimensionalities.[31] By this argument we find ourselves inside a four-
dimensional domain with our kind of low-energy physics not because
other kinds of mini-universes are impossible or improbable, but simply
because our kind of life cannot exist in other domains.

This may have important implications for the building of realistic
Kaluza-Klein and superstring theories. For example, it is extremely complicated, if not impossible, to construct a theory in which only one type
of compactification can occur, leading precisely to a four-dimensional
inflationary universe with the low-energy particle physics of our experience. But from the point of view discussed here, there is no need to

require that the results compactification and inflation have wrought in our realm be the only possible results, or the best. It is enough to find a theory in which such a compactification is possible. This problem is still difficult, but it is much easier than the one we have been trying to solve.

With the invention of the theory of many universes interacting with each other only globally,[45] and especially with the development of the baby universe theory[19-21] the situation changed even more. As it will be argued below, in the context of the baby universe theory it is possible to justify the Strong Athropic Principle as well.

According to the baby universe theory,[19-21] we may live in different quantum states of the universe $|\alpha_i>$, corresponding to different choices of the "fundamental constants," such as the vacuum energy ρ_v (cosmological constant $\Lambda = 8\pi\rho_v/M_P^2$), gravitational constant $G = M_P^{-2}$, fine structure constant α, etc. By measuring the coupling constants we actually determine in which quantum state $|\alpha_i>$ we live now, and after that we cannot "jump" to another quantum state with other constants. That is why we are used to believing that we live in a classical universe with well-determined and unchanging laws of physics.

It would be very desirable to understand why we do live in a universe in a quantum state $|\alpha_i>$ with an extremely small (or even vanishing) cosmological constant, with $M_P \sim 10^{19}$ GeV, $\alpha = 1/137$, etc. A possible idea is to compute the wave function of the universe and then to prove that it is entirely concentrated near some particular values of the coupling constants.[20,23] However, the validity of the euclidean approach to quantum cosmology used in most of the papers on this problem is far from being clear,[29] and it is quite possible also that the wave function of the universe actually does not have any sharp peak at $\Lambda = 0$.[25,26] Moreover, it is not quite clear whether it is the wave function of the whole universe that is to be computed. It is known that in quantum mechanics only those statements can make sense that can be formulated in a concrete operational way. For example, at the classical level one can speak of the age of the universe t. However, the essence of the Wheeler-DeWitt equation for the wave function of the universe is that this wave function *does not depend on time*, since the total Hamiltonian of the universe, including the Hamiltonian of the gravitational field, vanishes identically.[32] Therefore if one would wish to describe the evolution of the universe with the help of its wave function, one would be in trouble. The resolution of this paradox is rather instructive.[32] The notion of evolution is not applicable to the universe as a whole since there is no external observer with respect to the universe, and there is no external clock as well which would belong to the universe. However, we do not actually ask why the universe *as a whole* is evolving in the way we see it. We are just trying to understand our own experimental data. Thus, a more precisely formulated question is why *we see* the universe evolving in time in a given way. In order to answer this question one should first divide the universe into two main pieces: i) an observer with his clock and other measuring devices; and ii) the rest of the universe. Then it can be shown that the wave function of the rest of the universe does depend on the state of the clock of the observer, i.e., on his "time."[32] This time dependence in some sense is "objective," which means that the results obtained by different (macroscopic) observers living in the same quantum state of the universe and using sufficiently good (macroscopic) measuring apparatus agree with each other.[42]

This example teaches us an important lesson. By an investigation of the wave function of the universe *as a whole* one sometimes gets information which has no direct relevance to the observational data, e.g., that the universe does not evolve in time. (Moreover, it is rather difficult to give any probabilistic interpretation to the wave function of the universe as a whole, since typically it is not normalizable.) In order to describe the universe *as we see it* one should divide the universe into several macroscopic pieces and calculate a conditional probability to observe it in a given state under an obvious condition that the observer and his measuring apparatus do exist. This simple condition, however, is very restrictive. In order to understand its meaning one should remember the famous Einstein-Podolsky-Rosen experiment, where two fermions (or photons) are produced by a decay of a spin zero particle. By a measurement of a spin of one of the particles an observer instantaneously determines the spin of another particle even if this particle and the observer are not in a causal contact at the moment of the measurement. This apparent paradox is resolved if one takes into account that prior to the measurement one should describe both particles by the wave function of the pair of particles rather than by separate wave functions of each of them. The knowledge of this wave function (the fact that it corresponds to a state with zero angular momentum) and the knowledge of the spin of one of these particles makes it possible to determine the spin of another particle without making any causal contact with it, just for the reason that there is a strong correlation between spins of particles in a system with a vanishing total angular momentum. For example, the conditional probability to find the second particle in a state with spin

$s = -1/2$ is equal to unity under the condition that the first particle has a spin $s = +1/2$.

Similarly, any scientific exploration of the properties of the universe starts from the division of the universe into two subsystems (the observer with his measuring devices and the rest of the universe). Due to a strong correlation between the properties of these two subsystems, an investigation of our own properties, the properties of our environment and our measuring devices makes it possible to say a lot about the properties of the rest of the universe such as the vacuum energy, the values of some coupling constants etc.

It is important also, that in the EPR experiment one can choose different apparatus, e.g. the one which measures a linear polarization of the first photon, or the one which measures its circular polarization. After the measurement the second photon will be either linearly or circularly polarized, depending on our choice of measuring device. Similarly, many kinds of life and many types of measuring devices may exist, but what we are studying is the correlation of *our* kind of life, environment and measuring devices with the properties of the rest of the universe. But this means that in the context of the baby universe theory one can actually make sense of the Strong Anthropic Principle and try to explain why the electromagnetic coupling constant, the gravitational coupling constant and some other parameters of the theory have those particular values which we know: With other values of the coupling constants the existence of life *of our type* and of the measuring devices we use would be impossible. (Note, however, that in this interpretation the Strong Anthropic Principle has nothing in common with designing the universe for the benefit of human beings or with changing the whole universe by our own will, just as the determination of the properties of the second particle in the EPR experiment and the dependence of these properties on our choice of measuring device has nothing to do with the acausal action at a distance.)

If this idea is correct, then there is no need to compute the most probable values of the cosmological constant and of the gravitational constant by the investigation of the wave function of the universe as a whole, and it is not surprising that some of the results of the corresponding investigation look as counterintuitive as the statement that the universe as a whole does not depend on time. Rather one should investigate the probability of measuring some particular eigenvalues of the operators corresponding to the "constants" Λ, e, m_e, G in the quantum state of the universe in which we live now under the obvious condition that we can live and perform our measurements here. In some cases (e.g. in the case of the constants e, m_e, G) the corresponding investigation is rather simple and straightforward. For example, according to Ref. 43, an increase of m_e by more than 2.5 times would make impossible the existence of atoms. An increase of e by more than 30% would lead to an instability of protons and nuclei. A change of the strong interaction constant by more than 10% would lead to the absence of heavy nuclei and of hydrogen in the universe. An increase or decrease of the weak interaction constant by an order of magnitude would lead to the absence of complex elements or of hydrogen respectively. An increase of the gravitational constant G by an order of magnitude would make the lifetime of the sun so small that no biological molecules would appear on the Earth. An even bigger increase of G would lead to an extremely efficient nucleosynthesis and to the absence of hydrogen in the universe.[43,44] A smaller value of G would make expansion of the universe more slow. In such a universe the departure from thermal equilibrium during the process of baryosynthesis would be small, this process would be inefficient and the universe now would be practically empty.

This list can be easily continued, but we hope that the results discussed above clearly demonstrate that the correlation between the properties of observers and the properties of the quantum state of the rest of the universe is very strong indeed. Moreover, most of the results discussed above are not so much *anthropic*, they just show that a small deviation from the present values of coupling constants would make impossible the existence of all measuring devices we use. In the context of the baby universe theory this means that the distribution of the conditional probability to make measurements in a quantum state with given values of weak, strong, gravitational and some other "fundamental constants" proves to be sharply peaked near the actual values of these constants we have measured in the quantum state of the universe in which we live now. This peak in the probability distribution should be especially narrow in unified theories of all fundamental interactions where the values of different coupling constant are strongly correlated due to the underlying symmetry of the theory. This indicates that if one wishes to get an order-of-magnitude estimate of the coupling constant, in some cases one can do it without a computation of these constants in the context of the "big fix" paradigm,[23] and if one still wishes to make a computation (which, of course, is a good idea), then one should compute the amplitudes corresponding to the actual experimental environment rather than the amplitudes corresponding to the metaphysical "universe in itself" investigated by an imaginary external observer. (The discussion of the time-independence of the wave function of the universe shows that in the context of quantum cosmology the subtle difference between the amplitudes of these two types can be crucially important.)

Now let us return to the discussion of the cosmological constant problem. The possibility to explain the small value of the vacuum energy density $|\rho_v| < 10^{-29} \mathrm{g\,cm}^{-3}$ with the help of the Anthropic Principle is much less clear, but is not hopeless. From the results of Ref. 41 it follows that formation of galaxies and appearance of life of our type is possible only if $-10^{-29} \mathrm{g\,cm}^{-3} < \rho_v < 10^{-27} \mathrm{g\,cm}^{-3}$. In the context of the baby universe theory this means that the conditional probability to measure the vacuum energy density ρ_v is concentrated in the interval between $-10^{-29} \mathrm{g\,cm}^{-3}$ and $10^{-27} \mathrm{g\,cm}^{-3}$ under the condition that the measurement is made by an observer of our type. This result could be strengthened up to the desirable constraint $|\rho_v| < 10^{-29} \mathrm{g\,cm}^{-3}$ if it would be possible to prove that life on the Earth has an extragalactic origin. (Any communication between galaxies would be hampered by the exponential expansion of the universe with $\rho_v > 10^{-29} \mathrm{g\,cm}^{-3}$.) Unfortunately, at the moment we have no evidence in favor of this extravagant hypothesis. (Of course, one can turn this argument upside down and say the the vanishing of the cosmological constant can be considered as an argument in favor of the extragalactic origin of life. However, this will be true only if no other solution of the cosmological constant problem will be found.)

Another possibility is related to the results discussed above.[46] It may not be unreasonable to assume that the conditional probability to be born in the universe in a quantum state with a given value of ρ_v is proportional to the total volume of those parts of the universe with this value of ρ_v where life of our type can appear. From the investigation performed above it follows that in the quantum states with $|\rho_v| < 10^{-29} \mathrm{g\,cm}^{-3}$ the total volume of the parts of the universe where life can exist is much bigger than in the states with $|\rho_v| > 10^{-29} \mathrm{g\,cm}^{-3}$, and it may be especially big in the states with $|\rho_v| < 10^{-10^7} \mathrm{g\,cm}^{-3}$. If there is some relation between the total volume of the parts of the universe where life can exist and the volume where life of our type can appear, then one can argue that the probability that we measure the vacuum energy density ρ_v in our quantum state of the universe $|\alpha_i>$ is peaked in the region $|\rho_v| < 10^{-10^7} \mathrm{g\,cm}^{-3}$. This would be more than enough to solve the cosmological constant problem.

Unfortunately, this line of reasoning, which may lead to a justification of the so-called Final Anthropic Principle[44] in its strongest form, is still far from being all elaborated. There are many obvious and less obvious objections which can be raised against the arguments suggested above. It is quite possible that the cosmological constant problem eventually will be solved in quite a different way, e.g. by some modification of ideas proposed in Refs. 45, 20, 23. However, the efficiency of the Anthropic Principle for the determination of many other coupling constants in the context of the baby universe theory and the existence of a strong correlation between the value of the cosmological constant and the lifetime of mankind makes it very tempting to continue an investigation of the possibilities discussed above. At the very least, what we have obtained already is a surprising result that the existence of the vacuum energy density as small as $10^{-10^7} \mathrm{g\,cm^{-3}}$ would be fatal for the entire evolution of life in our part of the universe. This is the first indication of the potential practical importance of recent attempts to investigate the cosmological constant problem. Unfortunately (or, maybe fortunately,) it may take as much as $10^{5 \times 10^6}$ years until the significance of the current work on this problem will be fully appreciated.

References

1. Ya. B. Zeldovich and M. Yu. Khlopov (1978), *Phys. Lett.* **79B**, 239; J. P. Preskill (1979) *Phys. Rev. Lett.* **43**, 1365.
2. Ya. B. Zeldovich, M.-Yu Kobzarev and L. B. Okun (1974), *Phys. Lett.* **50B**, 340; S. Parke and S.-Y. Pi (1981), *Phys. Lett.* **121B**, 313; G. Lazarides, Q. Shafi and T. F. Walsh (1982), *Nucl. Phys.* **B195**, 157; P. Sikivie (1982), *Phys. Rev. Lett.* **48**, 1156.
3. J. Ellis, A. D. Linde and D. V. Nanopoulos (1982), *Phys. Lett.* **116B**, 59; M. Yu. Khlopov and A. D. Linde (1984), *Phys. Lett.* **138B**, 265.
4. G. D. Coughlan, W. Fischler, E. W. Kolb, S. Raby and G. G. Ross (1983), *Phys. Lett.* **131B**, 59.
5. A. S. Goncharov, A. D. Linde and M. I Vysotsky (1984), *Phys. Lett.* **147B**, 279.
6. S. M. Barr, K. Choi and J. E. Kim (1987), *Nucl. Phys.* **B283**, 591.
7. I. Antoniadis, C. Bachas, J. Ellis and C. B. Nanopoulos (1988), *Phys. Lett.* **B211**, 393.
8. E. B. Gliner (1965), *Sov. Phys. JETP* **22**, 378; (1970), *Dokl. Akad. Nauk SSSR* **192**, 771; E. B. Gliner and I. G. Dymnikova (1975), *Pis. Astron. Zh.* **1**, 7; I. E. Gurevich (1975), *Astrophys. Space Sci.* **38**, 67.
9. A. A. Starobinsky (1979), *JETP Lett.* **30**, 682; (1980), *Phys. Lett.* **91B**, 99.
10. A. H. Guth (1981), *Phys. Rev.* **D23**, 347.

11. A. D. Linde (1982), *Phys. Lett.* **108B**, 389; (1982), *Phys. Lett.* **114B**, 431; (1982), *Phys. Lett.* **116B**, 335,340; A. Albrecht and P. J. Steinhardt (1982), *Phys. Rev. Lett.* **48**, 1220.
12. A. D. Linde (1983), *Phys. Lett.* **129B**, 177.
13. A. D. Linde (1988), *Phys. Lett.* **201B**, 437.
14. A. D. Linde (1986), *Phys. Lett.* **175B**, 395; (1987), *Physical Scripta* **T15**, 169; (1987), *Physics Today* **40**, 61.
15. A. S. Goncharov, A. D. Linde and V. F. Mukhanov (1987), *Int. J. Mod. Phys.* **A2**, 561.
16. S. W. Hawking (1987), *Phys. Lett.* **B195**, 277; (1988), *Phys. Rev.* **D37**, 904.
17. G. V. Lavrelashvili, V. A. Rubakov and P. G. Tinyakov (1987), *JETP Lett.* **46**, 167; (1988), *Nucl. Phys.* **B299**, 757.
18. S. B. Giddings and A. Strominger (1988), *Nucl. Phys* **B306**, 890.
19. S. Coleman (1988), *Nucl. Phys.* **B307**, 867.
20. T. Banks (1988), *Nucl. Phys.* **B309**, 493.
21. S. Giddings and A. Strominger (1988), *Nucl. Phys.* **B307**, 854.
22. S. W. Hawking and R. Laflamme (1988), *Phys. Lett.* **209B**, 39.
23. S. Coleman (1989), *Nucl. Phys.* **B310**, 643.
24. S. Giddings and A. Strominger (1988), Santa Barbara preprint.
25. J. Polchinski (1989), *Phys. Lett.*, to be published.
26. W. Fischler, I. Klebanov, J. Polchinski and L. Susskind (1989), Stanford University preprint.
27. A. D. Linde (1988), *Phys. Lett.* **B211**, 29.
28. A. D. Linde (1985), *Phys. Lett.* **B162**, 281.
29. A. D. Linde (1990), *Particle Physics and Inflationary Cosmology*, (Gordon and Breach, New York), in preparation.
30. V. F. Mukhanov and G. V. Chibisov (1981), *JETP Lett.* **33**, 523; S. W. Hawking (1982), *Phys. Lett.* **115B**, 339; A. A. Starobinsky (1982), *Phys. Lett.* **117B**, 175; A. H. Guth and S.-Y. Pi (1982), *Phys. Lett.* **49**, 1110; J. Bardeen, P. J. Steinhardt and M. Turner (1983), *Phys. Rev.* **D28**, 679; R. Brandenberger (1985), *Rev. Mod. Phys.* **57**, 1; V. F. Mukhanov (1985), *JETP Lett.* **41**, 493; V. F. Mukhanov, L. A. Kofman and D. Yu. Pogosyan (1987), *Phys. Lett.* **193B**, 427.
31. A. D. Linde and M. I. Zelnikov (1988), *Phys. Lett.* **B215**, 59.
32. J. A. Wheeler (1968), in *Relativity, Groups and Topology*, eds. C. M. DeWitt and J. A. Wheeler (Benjamin, New York); B. S. DeWitt (1967), *Phys. Rev.* **160**, 1113.
33. J. B. Hartle and S. W. Hawking (1983), *Phys. Rev.* **D28**, 2960.
34. A. D. Linde (1984), *JETP* **60**, 211; (1984), *Lett. Nuovo Cim.* **39**, 401.
35. Ya. B. Zeldovich and A. A. Starobinsky (1984), *Sov. Astron. Lett.* **10**, 135.
36. V. A. Rubakov (1984), *Phys. Lett.* **148B**, 280.
37. A. Vilenkin (1984), *Phys. Rev.* **D30**, 549.
38. A. A. Starobinsky (1984), in *Fundamental Interactions*, (MGPI Press, Moscow), p. 55; (1986), in *Current Topics in Field Theory, Quantum Gravity and Strings*, Lecture Notes in Physics No. **206**, eds. H. J. de Vega and N. Sanchez (Springer, Heidelberg), p. 107.
39. A. S. Goncharov and A. D. Linde (1986), *Sov. J. Part. Nucl.* **17**, 369; S. J. Rey (1987), *Nucl. Phys.* **B284**, 706.
40. A. Vilenkin (1983), *Phys. Rev.* **D27**, 2848.

41. S. Weinberg (1987), *Phys. Rev. Lett.* **59**, 2607; (1989), *Rev. Mod. Phys.* **61**, 1.
42. H. Everett (1957), *Rev. Mod. Phys.* **29**, 454; B. S. DeWitt and N. Graham, (1973), *The Many-Worlds Interpretation of Quantum Mechanics* (Princeton University Press, Princeton).
43. I. L. Rosental (1988), *Big Bang, Big Bounce: How Particles and Fields Drive Cosmic Evolution*, (Springer, Berlin).
44. J. D. Barrow and F. J. Tipler (1986), *The Anthropic Cosmological Principle*, (Oxford University Press, Oxford).
45. A. D. Linde (1988), *Phys. Lett.* **200B**, 272.
46. A. D. Linde (1989), *Phys. Lett.* **227B**, 352.

8

Theory and implications of the cosmic background radiation

Mirosław Panek*
Princeton University Observatory,
Princeton, NJ 08544 USA

Anisotropies of the temperature of the Cosmic Background Radiation (CBR) give us unique information on the universe at redshifts of about 1000. By using observational limits on these anisotropies we can constrain the parameters of cosmological models and the spectrum and amplitude of the initial perturbations. Additional information on the thermal history of the universe is provided by distortions of the CBR spectrum. In this paper we give an overview of physical processes leading to anisotropies in popular cosmological models.

1. Introduction

The remarkable isotropy of the CBR in a universe containing so many structures is one of the most fascinating cosmological observations. Strong limits put on possible anisotropies of the CBR let us impose important constraints on models of global structure of the universe, and on models of galaxy and cluster formation.

The difference between the CBR and all the other radiations investigated in astronomy is that the former existed from the beginning of the universe, and the latter were created only after first objects were formed. Also, the CBR has now a $2.74K$ blackbody spectrum, and the radiation of astronomical objects is usually not blackbody.

Following the discovery of the CBR (Penzias and Wilson (1965)), progress in observations has been rapid. Currently, anisotropies on the

---

*On leave from Copernicus Astronomical Center, Warsaw, Poland

level of $10^{-4} - 10^{-5}$ are detectable on a wide range of angular scales and frequencies. (Observations of anisotropy are usually made by comparing intensities of radiation detected by two antennas, separated by an angle θ, the angular scale. Each antenna collects radiation coming from a cone of angle σ, and the ratio σ/θ should be substantially smaller than 1). Observations on angular scales from a few arcseconds to 180° have so far detected only the dipole ($\theta = 180°$) anisotropy and the CBR temperature fluctuations around a few clusters of galaxies (see Section 3). Detection of anisotropy on the angular scale of 8° reported by Davies et al. (1987) has yet to be confirmed.

The dipole anisotropy is interpreted as coming from the motion of the Local Group of galaxies with respect to the uniform radiation background (the amplitude of the dipole gives us the velocity). This is the only anisotropy that can be explained by the observer's motion.

In the early universe the radiation was in thermodynamic equilibrium with the rest of the matter. As the universe expanded and cooled down, more and more reactions providing the thermal contact became ineffective. But the most important reaction, Compton scattering of photons off free electrons ceased to be effective only after the electrons combined with protons to form hydrogen. That happened at a redshift of about a thousand, when free electrons recombined with free protons to form neutral hydrogen. At that epoch photons of the CBR became free streaming and in the course of adiabatic expansion they maintained the blackbody spectrum, with the temperature falling at a rate inversely proportional to the scale factor. Information about inhomogeneities at recombination was imprinted into photons in the form of anisotropies of the CBR temperature. The CBR provides us then with a photograph of the universe at $z \simeq 1000$.

Only after bound structures were formed, the CBR photons were involved in gravitational interactions with them, or scattered off hot electron gas produced in some of them. These interactions disturbed the initial information, but simultaneously they left another information—about structures present at small redshifts.

This review describes different mechanisms that can produce the CBR temperature anisotropies. We discuss also the implications of the anisotropy observations for models of the large scale structure formation in the universe.

Section 2 describes basic methods applied to the standard cosmological model to calculate primary anisotropies originating at hydrogen recombination. Section 3 discusses interaction of photons with matter at small redshifts leading to secondary anisotropies. In Section 4 we mention results for some nonstandard cosmological models (Bianchi models, cosmic strings scenario). Section 5 gives a brief description of spectral properties of the CBR. Finally, Section 6 contains conclusions.

2.   CBR anisotropies in the standard model

Background model

Our choice of cosmological model is determined by its ability to reproduce basic features of the universe and by its relative simplicity. The Friedmann-Lemaitre-Robertson-Walker (FLRW) model is excellent from that point of view. It is homogeneous and isotropic, as the universe is on large scales. We fill the spacetime with perfect fluids representing different forms of matter: baryons, radiation, neutrinos, hypothetical dark matter etc.. Today baryonic matter forms many structures and of course cannot be described any longer as a fluid. However, if we are interested in the early universe, or if we analyze its properties on large scales the fluid model works quite well.

There are many parameters in the FLRW model. Their values are still not well known and some freedom of choice is possible. These parameters, and (in parentheses) typical values are:

- Hubble constant $H_0 = h \cdot 100$ km/s $\cdot$Mpc, $(h = 0.5 - 1)$;
- mean density of the universe today ρ_0, or $\Omega_0 = \rho_0/\rho_{cr}$, where $\rho_{cr} = 3H_0^2/8\pi G$ is the critical density, $(\Omega_0 = 0.1 - 1)$;
- densities of different forms of matter, defined like Ω_0;
- cosmological constant Λ, (0 - 1).

We constrain possible combinations of the parameters by comparing model predictions with observations of structures in the universe, abundances of elements, lower limits on the age of the universe, and with the CBR anisotropy limits.

It should be stressed that observations of the large scale structure of the universe indicate that $\Omega_0 < 1$. However, theorists prefer $\Omega_0 = 1$ because this value is predicted by inflationary models of the early universe, many calculations are simpler, and because it is philosophically attractive. To reconcile aesthetically appealing flat universe with apparent shortage of matter to make universe flat dark matter was introduced, or in models with cosmological constant the choice of $\Omega_0 + \Lambda = 1$ is usually made.

Perturbations

Standard scenarios assume that in the early universe there existed small perturbations to the cosmic fluid. These perturbations grew and led to the formation of galaxies, clusters of galaxies, superclusters, etc..

Perturbations can be classified with respect to their transformation properties as scalar, vector or tensor. Gravitational waves (tensor modes) cannot lead to structure formation, velocity (vector) perturbations are also implausible for this purpose because they decay. We are left with density (scalar) perturbations that grow during the matter-dominated era.

Any of the components of matter can in principle be perturbed separately, but theories of the early universe point out two special cases. The first is adiabatic (isentropic) perturbations, where total density of matter is perturbed initially such that everywhere the relative number density of particles of different kind is constant. The second is isocurvature perturbations. Here at some initial moment perturbations of photons and e.g. baryons and dark matter have opposite signs and are such that the total density (and curvature) is uniform.

Matter inhomogeneities are described as the first order perturbations to components of the energy-momentum tensor. They generate perturbations of components of the metric tensor and evolution of the model is given by the linearized Einstein equations. There were a number of ambiguities related to the problem of evolution of perturbations on scales larger than horizon. It is possible to avoid them by using gauge-invariant formalism (Bardeen (1980)), but there is nothing wrong with using a specific gauge. As long as we are interested in observable physical quantities, like temperature of the CBR, both approaches should give the same results (Bond (1988)). For the gauge-invariant formalism applied to the CBR anisotropies see Panek (1986); Abbott and Schaefer (1986).

Last scattering

As we have mentioned in the Introduction, photons of the CBR became free streaming at the hydrogen recombination. But the last scattering was not simultaneous for all photons as the number of free electrons decreased gradually over some period of time. Therefore positions of the last scattering event of photons are located in a layer of finite width. Dynamics of the recombination gives $z_E \simeq 1000$ ($T \simeq 3000K$) as the mean redshift (temperature) of emission, and the relative thickness of the last scattering zone was $\Delta z/z_E \simeq 0.07$.

We are interested in differences of temperature of photons coming from two directions. The differences come from the fact that photons emitted were affected by the gravitational field of perturbations or had initial temperature fluctuating from the mean value.

Only perturbations on scales larger than the thickness of the last scattering zone can effectively influence the mean temperature of photons coming from a given direction. Smaller scale perturbations of opposite signs cancel out their contributions. The thickness of the last scattering region defines a characteristic angle - θ_c, and we can expect to find no substantial fluctuations at $\theta < \theta_c$ (more precisely, there is an exponential decay of the amplitude of the fluctuations around this angular scale).

On the other hand, for angular scales sufficiently large we can neglect the finite thickness of the last scattering and treat it as a surface.

It is useful to know angular scales subtended by progenitors of observed structures. Typical values are: galaxies — 1′; clusters of galaxies — 20′; and largest suggested structures of size approximately 300 h_{75}^{-1} Mpc — — 3° (Tully (1987)). The thickness of the last scattering is about 8′ and the horizon size at recombination is 2°.

The scale of 2° is used to separate small and large angular scales. For small scales the dynamics of recombination is important and we have to solve the Boltzmann equation for photons, including gravitational and Thomson scattering on density perturbations (see e.g. Peebles and Yu (1970)). Results of calculations made in this case depend strongly on model and parameters used.

For large angular scales it is sufficient to use the concept of the last scattering surface (LSS). The calculations are much simpler. The results are less model-dependent, and therefore more important for our understanding of the early universe. We will concentrate further on this case.

Geodesic propagation

The calculation for large angular scales is performed by investigating photon propagation in the perturbed metric. We use linearized Einstein and geodesic equations to look for first order corrections to photon energy, and for this purpose we assume that they propagate along 0th order geodesics.

The CBR temperature fluctuations come from three effects. The first of them is that at the moment of emission the photons can have temperature higher or lower than the average. The second effect is the Doppler shift of energy of photons emitted by a moving source (density perturbations cause velocity perturbations on corresponding scales). The third, and most interesting effect is the gravitational shift coming from perturbations to the gravitational potential that are caused by total density perturbations.

An important fact is that all described processes provide us with information on the LSS. This is obvious for initial temperature perturbations and Doppler shift, but it is also true for gravitational shifts. Photons remember whether they had to climb gravitational potential or rolled down from it at the LSS. When they encounter e.g. a hole of the potential after they left the LSS then the blueshift from rolling down to it and the redshift from climbing out of it cancel and the initial information is not influenced (however, second order effects associated with evolution of the potential during the photon propagation across it may be important for very bound structures (Rees and Sciama (1968))).

For a single mode of the density perturbations the result for temperature fluctuations today, $(\delta T/T)_R$ contains terms describing effects mentioned above: initial $(\delta T/T)_E$, Doppler shift proportional to first order velocity perturbations at the LSS, and gravitational, or the Sachs-Wolfe term being one-third of the perturbation of gravitational potential (Sachs and Wolfe (1967)). There is also a term that is proportional to the integral of the time derivative of the gravitational potential. Only for the matter-dominated, $\Omega_0 = 1$ universe and linear density perturbations is this derivative equal to 0 to the first order.

At large angular scales the Sachs-Wolfe effect dominates other components, and for an $\Omega_0 = 1$, matter-dominated universe it is

$$\left(\frac{\delta T}{T}\right)_R \simeq \frac{1}{3}\left(\frac{L}{ct_R}\right)^2\left(\frac{\delta\rho}{\rho}\right)_R$$

where L is the scale of a given structure today, t_R is the age of the universe, and $(\delta\rho/\rho)_R$ is the density perturbation at the scale of interest today (assuming that the perturbation is still in the linear regime).

Statistics of density perturbations

We assume that the density perturbations in the early universe were a Gaussian random process, and different modes had random phases. We decompose the initial perturbation field into Fourier modes

$$\delta(\vec{k}) = \int \frac{\delta\rho}{\rho}(\vec{r}) \, \mathrm{Exp}(i\vec{k} \cdot \vec{r}) d^3 r$$

where $k = 2\pi/\lambda$ is the wavenumber. All properties of such chosen random process are given by the power spectrum, usually taken to be a power law

$$P(k) = <|\delta(\vec{k})|^2> \sim k^n.$$

After calculating the temperature fluctuations coming from a single mode we sum up all contributions using their linearity and assumed statistical properties.

Observations of large scale structures in the universe tell us about the amplitude of perturbations on scales of up to a few hundred megaparsecs. In principle we should be able to verify the assumptions about the power spectrum on these scales, or even directly relate present structures to anisotropies caused by their progenitors. On the other hand, for scales larger than corresponding to the biggest observed structures we cannot test our assumptions about the amplitudes of perturbations at early epochs.

At this point we encounter an important theoretical problem that can be illustrated by an example of the quadrupole ($\theta = 90°$) anisotropy. If it exists, this anisotropy would be dominated by contributions coming from perturbations on scales comparable to the size of the present horizon. But no causal processes can generate perturbations on scales larger than the horizon size at a given moment. Therefore perturbations on scales dominating the quadrupole anisotropy could be created only today (if this anisotropy would come from perturbations on smaller scales then it would be impossible to avoid a large amplitude of the anisotropy at appropriately smaller angular scales). The only way out would then be to assume that the perturbations existed on all scales long before they came into the horizon.

In general, causal processes could generate perturbations resulting in primary anisotropies on scales smaller than $\sim 2°$ (horizon size at recombination) or secondary anisotropies on scales $< 10°$ (see Section 3). Any anisotropies on larger scales come in the standard model from products of acausal phenomena, or from processes like inflation, avoiding simple causality constraints. For the discussion of the causality constraints applied to the CBR anisotropy see Traschen and Eardley (1986).

Inflation is now the only theory predicting existence of perturbations on all scales. It predicts a scale-invariant, $n = 1$ spectrum. This initial shape is modified during evolution of the universe by differences in growth factors for perturbations of different scales. For scales larger than the horizon at recombination the initial shape is not changed. More complicated inflationary scenarios can naturally produce some kinds of non scale-invariant spectra (Salopek, Bond and Bardeen (1989) and references therein).

Unfortunately, inflation does not give us any answers about the $\Omega_0 < 1$ universe, because it predicts a flat geometry. The $n = 1$ spectrum is scale-free and cannot be naturally implemented into a non-flat universe, as there is a characteristic scale in such universe. It is associated with the curvature radius, and the characteristic angle is

$$\theta_{CR} = \frac{\Omega_0}{2\sqrt{1 - \Omega_0}} \simeq 6° \quad for \quad \Omega_0 = 0.2 \,.$$

Assuming that the spacetime is approximately flat on scales smaller than the curvature radius we can obtain results for $\theta << \theta_{CR}$ only. For $\theta > \theta_{CR}$ there is no simple generalization of the Fourier analysis, we have no prescription for power spectrum, and no reliable results have been obtained. We have only some indications that in the open universe the large-scale anisotropies are smaller than in the flat case (Peebles (1982a); Wilson (1983); Gorski and Silk (1989)).

Normalization

We do not know what initial amplitude of the density perturbations was, but we can see around us cosmic structures that resulted from them. In principle we should be able to find the initial amplitude by comparing the theoretical predictions with the distribution of objects in the universe. However, none of existing theories gives predictions consistent with all observations. Because of that the normalization is the weakest point of all CBR anisotropy calculations. We investigate distribution of galaxies, but we do not even know whether mass in the universe is distributed like the light of galaxies, or maybe it is less concentrated (biasing hypothesis). We have a number of possible normalization procedures, e.g. counting the rms fluctuations of mass distribution in a sphere of a given radius, integrating the galaxy correlation function etc.. Their results are quite model-dependent and sometimes do not agree well for the same scenario - the discrepancy can be as much as a factor of 3.

Results

Observers are pushing down upper limits for anisotropy at all angular scales (see B. Partridge's contribution to these proceedings). Theoretical predictions for different models can be found in cited literature, and I will not attempt to present them here. In most of the cases observations put constraints on allowed choices of cosmological parameters of models. Only sometimes we can say that a model is ruled out. Models that give excessive anisotropy on at least one angular scale are: baryon-dominated models with adiabatic perturbations, with or without cosmological constant (Wilson and Silk (1981); Vittorio and Silk (1985)), cold dark matter (CDM) dominated, $\Omega_0 = 1$ model with axions and isocurvature perturbations (Efstathiou and Bond (1986)), and CDM, small Ω_0 models with adiabatic perturbations (Vittorio and Silk 1984; Bond and Efstathiou (1984)).

On the safe side are: adiabatic, $\Omega_0 = 1$ CDM models, and baryon-dominated, ionized, $\Omega_0 < 1$, isocurvature perturbations models of Peebles (1987).

Levels of anisotropy for inflationary models should be treated as lower limits, because in these models additional anisotropies are expected from large-scale gravitational waves, inevitably produced during inflation. The waves have a scale-invariant spectrum and their amplitude depends on the details of the inflationary model (Starobinsky (1985)).

For a scale-invariant, $n = 1$ power spectrum of density perturbations the mean square value of the l-th multipole scales as $\sim 1/l(l+1)$ (Peebles (1982b)). If we knew the amplitudes of at least two multipoles, then we would be able to use scaling relations like that to find the slope of the power spectrum.

An important feature of the models discussed so far is the Gaussian nature of the CBR fluctuations map. It results directly from the assumption (or in inflationary models - prediction) about the initially Gaussian field of density perturbations. The statistics lets us calculate the rms values of the temperature fluctuations. It allows us also to answer questions about the statistical significance of observational limits, or to predict e.g. the expected number of hot or cold regions of a given temperature per unit area (Vittorio and Juszkiewicz (1987); Bond and Efstathiou (1987)).

One can also hope to verify Gaussian statistics by examining topological properties of the CBR temperature maps (Gott et al. (1990)).

3. Secondary anisotropies

In the previous section we discussed the primary anisotropies, i.e. those imprinted into the CBR at the epoch of recombination. Now we will concentrate on secondary anisotropies resulting from the interaction of photons of the CBR with cosmological structures formed at small redshifts.

Sunyaev - Zeldovich effect

Clusters of galaxies and galaxies themselves contain hot, nonrelativistic gas of free electrons that were released during the numerous ionizing processes in galaxies. Photons of the CBR passing through clouds of hot electrons interact with them via Compton scattering. Interacting photons gain energy and are moved from the low-energy part of their spectrum to higher energies. No new photons are produced and the result is a distortion of the CBR blackbody spectrum. At the Rayleigh-Jeans (low energy) region of the spectrum, where most of anisotropy observations are made, there is a decrease in the number of photons that looks like a negative temperature fluctuation. The amplitude of this effect is proportional to the number and temperature of electrons in the cloud (Zeldovich and Sunyaev (1969)).

The Sunyaev - Zeldovich effect was observed for some clusters of galaxies at a level of 10^{-4}, and on angular scale of several arcminutes (see Sarazin (1986) and references therein).

This anisotropy is associated with the very existence of clusters and galaxies. We should therefore expect to detect some fluctuations from almost all existing structures, also in earlier stages of their evolution. The detection would provide useful information about the universe at low redshifts.

Secondary ionization

Observations do not exclude the possibility that the cosmic medium was reheated and ionized again at a redshift of about 10 or 100. In such a case photons of the CBR interact again with the free electrons and primary anisotropies are diluted or even erased.

As the universe expands the number density of electrons decreases and even without the hydrogen recombination photons again become free streaming. In the extreme case of early and strong reionization this happens at $z > 15$, but the thickness of the last scattering region is now much bigger: $\Delta z/z \simeq 1$ instead of 0.07. Because of that only perturbations on scales much bigger than at the recombination do not cancel out their contributions to the radiation temperature fluctuations.

Newly generated CBR temperature anisotropies are therefore present only at angular scales larger than approximately $10°$. Their amplitude is, however, the same as those created at $z_E \simeq 1000$. This is because they come from perturbations of gravitational potential (Sachs-Wolfe effect) and in the standard model the potential is constant during the linear stage of evolution of density perturbations. Also the relation between the mass of the perturbation and the associated angular scale is almost the same at $z = 10$ as it was at $z = 1000$.

Reheating was considered to be an effective way to avoid constraints from observations of the CBR anisotropies at small angular scales. However, it was shown (Vishniac (1987)) that at reionization, the second-order couplings of density and velocity perturbations can generate CBR fluctuations on the level of $10^{-6} - 10^{-5}$ at angular scales smaller than $10'$. Detection of the CBR anisotropies on this level at very small ($< 1'$) angular scales would then indicate that the reionization really occurred.

The reionization does not influence large scale anisotropies, and really nothing can be done to get rid of the CBR fluctuations at $\theta > 10°$ (if they were produced at recombination). This fact makes large scale anisotropies especially interesting, because if they are not detected at the level expected in a model, then the model cannot be saved.

Gravitational interaction with large objects

For objects located between us and the LSS the first order effects of their gravitational potential cancel out and give no fluctuations of the CBR. As we mentioned in Section 2, the second order effects can be strong enough if the object is a large, nonlinear density perturbation. Candidates for such objects are clusters of galaxies and voids.

Clusters can be modeled by Swiss cheese solutions: inside FLRW universe there is a spherical region of Schwarzshild solution, and in the center of it there is a spherically symmetric fragment of another (denser) FLRW solution, or a point mass. This is quite a poor model for a cluster, not only because of the assumed spherical symmetry, but also

because it does not reflect realistic density profile of a typical cluster.
The Swiss cheese solution can model only compensated clusters, where
the central overdensity is surrounded by an underdense (empty) region
and the total mass perturbation of a whole structure is zero, so that the
geometry of the outer FLRW universe is not affected. Nevertheless, this
model is useful because it estimates second order effects of the gravitational potential. The amplitude of this effect is comparable to that of
Sunyaev-Zeldovich effect for clusters of a given size (Nottale (1984)).

It is easier to construct a model of a void, as they appear to be
quite empty and almost spherically symmetric. Thompson and Vishniac
(1987) modeled them with a sphere of Minkowski solution separated by
a thin shell from outer Friedman solution. Predicted amplitude of temperature fluctuations for large voids ($\simeq 60h^{-1}Mpc$) is of the order of
10^{-6}.

More realistic, but still spherically symmetric models of cosmic structures could be based on the Tolman-Bondi solutions (Raine and Thomas
(1981)).

Gravitational lensing

Different models predict different formation epoch of bound structures.
It is generally accepted that whatever the scenario, there were some
structures at redshifts of at least 5.

Structures in the universe deflect light rays - this is known as gravitational lensing. The energy of photons is not altered, so no temperature fluctuations can be produced in this process, but previously present
anisotropies can be slightly modified. This comes from differences in deflection angles of two neighbouring photons of the CBR as they cross
the universe containing many gravitational potential perturbations. The
only result of gravitational lensing is slight redistribution of power of
temperature fluctuations on arcminute and smaller scales (see e.g. Cole
and Efstathiou (1988)).

4. Anisotropies in nonstandard models

Some cosmological scenarios do not start from small density perturbations in the FLRW universe. Formation of structures proceeds in different manners, and the CBR temperature fluctuations have nongaussian
properties, distinguishing them from the results of standard scenarios.

Cosmic explosions

In this scenario small scale objects (stars, quasars) form first in the universe which is uniform on large scales. Then at some epoch they explode (e.g. collective supernova explosions) and spherically symmetric shock wave regions are formed. Matter is swept out of spheres and concentrated into massive shells around them. Subsequently, gravitational instability leads to fragmentation of shells and galaxies are formed around empty, spherical regions - voids (Ostriker and Cowie (1981); Ikeuchi (1981)).

There is a lot of ionized gas left inside the spheres after the explosions. This gas produces the CBR fluctuations via the Sunyaev-Zeldovich effect. The microwave sky in this scenario is covered with many overlapping circles of some characteristic size. When observed at low frequencies these circles look like regions of decreased temperature. Statistical fluctuations in number of these regions per area covered by the antenna beam lead to temperature fluctuations. Observational limits for anisotropy at angular scales of $5' - 7'$ give an upper limit of $\sim 10 Mpc$ for the typical size of a void (Hogan (1984)). This is much less than the size of typical observed void - $25 h^{-1} Mpc$. Thus the explosion scenario in its simplest form is ruled out.

Cosmic strings

Cosmic strings are one-dimensional topological defects of a complex scalar field left after a phase transition that had occurred in the very early universe. Their network evolves to a population of closed loops and infinite strings crossing the whole visible universe. In an initially uniform medium (no small density perturbations) they become seeds of structure formation as they interact gravitationally with the rest of the matter. These structures produce the CBR fluctuations by previously described mechanisms, but there is one more mechanism in this scenario which is very specific for cosmic strings (Kaiser and Stebbins (1984); Gott (1985)).

If between us and the LSS there is a long cosmic string moving to the left then the CBR photons coming to us from region on the right hand side of the string will be blueshifted, and those coming from the left hand side - redshifted. This phenomenon is an example of a more general case of scattering off a moving potential well. For example interplanetary spaceships speed up or slow down depending on the side from which they approach a planet. (The same mechanism produces also anisotropy pattern around a moving galaxy cluster; see Birkinshaw and Gull (1983).)

As a result the microwave sky is covered by lines with temperature gradient across them (Bouchet, Bennet and Stebbins (1988)). The anisotropy pattern is highly nongaussian and this property could be used to distinguish between cosmic strings and scenarios using initially small, Gaussian density perturbations. Amplitudes of fluctuations predicted by cosmic strings scenario are lower than present observational limits.

Anisotropic cosmologies

We are accustomed to thinking that we live in a homogeneous and isotropic universe, but small departures from global isotropy were also investigated. We will discuss briefly the results found for two Bianchi cosmologies (see Bajtlik et al. (1986) and references therein).

In Bianchi I cosmology the rate of expansion depends on the direction in space. Photons arriving at different angles were redshifted by different amounts and as the result a strong quadrupole moment of the CBR anisotropy is generated.

Bianchi V models have global anisotropy and negative curvature. Focusing properties of space of negative curvature squeeze the quadrupole pattern into the hot spot. The size of this hot spot is proportional to the density parameter Ω_0 and in principle this fact could be used to measure almost directly the mean density of the universe. But hot spots are also present in models with Gaussian perturbations and to measure Ω_0 this way one should make sure first that the hot spot is of a nonstatistical origin.

5. Spectral properties of the CBR

Energy releases

There are many processes that are able to distort the blackbody spectrum of the CBR. These distortions carry important information about the thermal history of the universe. Energy released by hypothetical processes or objects may not be completely thermalized and can be evidenced in the CBR spectrum. (For a review see Danese and De Zotti (1977).)

If the energy injection (whatever its mechanism) happened at a redshift larger than $10^7 - 10^8$ then any amount of it can be thermalized, because processes coupling matter and radiation and distributing photons all over the spectrum are very effective. Later, the couplings become weaker, reaction rates get smaller and for energy released between $z = 10^7 - 10^8$ and $z \simeq 4 \cdot 10^5$ large inputs may not be completely thermalized. For $4 \cdot 10^5 > z > 4 \cdot 10^4$ any energy injection leaves a distortion, visible mainly in Rayleigh-Jeans part of the spectrum. For energy released at $z < 4 \cdot 10^4$ the distortion is visible over the whole spectrum, and its precise shape can depend on the form of the energy input (injection of high-energy photons, or heating of electron gas).

Observations by Matsumoto et al. (1988) indicate that the CBR spectrum is in fact distorted - there is an excess of energy in the high frequency (Wien) part of the spectrum. Many explanations for the source of the distortion have been proposed: global Sunyaev-Zeldovich effect from reheating, emission from a cosmic dust at $z \simeq 30$, or radiative decays of exotic particles. If confirmed, the distortion would be a very important feature and cosmological scenarios would have to explain it. Let us note that if the distortion comes from cosmic dust (that had to be produced by stars formed earlier) then we have a potential new source for secondary temperature anisotropies (Hogan and Bond (1988)).

Spectral properties of anisotropies

When looking for anisotropies we have to make choices for angular scale and frequency of observations. The second choice is dictated to avoid unwanted sources of fluctuations, like radio galaxies or water vapor in Earth's atmosphere.

If we were able to perform two simultaneous observations at two different frequencies (but at the same angular scale) then we would have a new interesting possibility. The amplitude of fluctuations produced by different mechanisms depends on wavelength. Results of gravitational (Sachs-Wolfe or second order in potential) and Doppler effects do not depend on photon energy. The Sunyaev-Zeldovich effect moves photons from the Rayleigh-Jeans to the Wien region, and negative temperature perturbation at low frequencies is accompanied by positive perturbation at high frequencies. The spectral dependence of anisotropies produced by hypothetical cosmic dust can be even more complex. By comparing amplitudes at different frequencies we should be able to distinguish between possible sources of the CBR anisotropy (Panek and Rudak (1988)).

6. Conclusions

We have described many processes leading to the CBR temperature fluctuations. They can be divided into two groups:

1. Those coming from gravitational interaction of photons with density perturbations; or
2. produced by the interaction of photons with hot electron gas.

Another useful division is based on the epoch of anisotropy production: primary anisotropies are produced at the time of hydrogen recombination, while secondary ones come from interaction of photons with structures present at small redshifts.

Some fluctuations are unavoidable, like Sunyaev-Zeldovich effect from hot gas in clusters of galaxies. Only these fluctuations have been observed so far for several clusters. Most of the processes described in the preceding sections produce small angular scale fluctuations and theoretical predictions in that range are very model-dependent. The detection of small scale anisotropies would give us important clues to understanding structure formation in the universe.

Large scale fluctuations are interesting because of another reason. They are not directly related to the observed structures, but their existence would confirm our theories about origin of perturbations in the very early universe. Detection of fluctuations at large angular scales would support inflationary models of the early universe, because only these models predict perturbations on all scales.

We are looking forward to a successful detection of fluctuations on any angular scale. It would be very useful, although difficult, to map the CBR temperature on some area of the sky. This would enable us to test our statistical assumptions and probe the nature of initial perturbations: Gaussian models of small primordial density perturbations versus nongaussian patterns produced by cosmic strings or Sunyaev-Zeldovich effect.

References

Abbott, L. F. and Schaefer, R. K. (1986), *Ap. J.* **308**, 546.
Bajtlik, S., Juszkiewicz, R., Proszynski, M. and Amsterdamski, P. (1986), *Ap. J.* **300**, 463.
Bardeen, J. M. (1980), *Phys. Rev.* **D22**, 1882.
Birkinshaw, M. and Gull, S. F. (1983), *Nature* **302**, 315.

Bond, J. R. (1988), in *The Early Universe*, eds. W. G. Unruh and G. W. Semenoff (D. Reidel, Dordrecht).

Bond, J. R. and Efstathiou, G. (1984), *Ap. J. Lett.* **285**, L45.

Bond, J. R. and Efstathiou, G. (1987), *Mon. Not. Roy. Astr. Soc.* **226**, 655.

Bouchet, F., Bennet, D. P. and Stebbins, A. (1988), *Nature* **335**, 410.

Cole, S. and Efstathiou, G. (1989), *Mon. Not. Roy. Astr. Soc.* **239**, 195.

Danese, L. and De Zotti, G. (1977), *Rev. Nuovo Cim.* **7**, 277.

Davies, R. L., Lasenby, A. N., Watson, R. A., Daintree, E. J., Hopkins, J., Beckman, J., Sanchez-Almeida, J. and Rebolo, R. (1987), *Nature* **326**, 462.

Efstathiou, G. and Bond, J. R. (1986), *Mon. Not. Roy. Astr. Soc.* **218**, 103.

Gorski, K. M. and Silk, J. (1989), *Ap. J. Lett.* **346**, L1.

Gott, J. R. (1985), *Ap. J.* **288**, 422.

Gott, J. R., Park, C., Juszkiewicz, R., Bies, W. E., Bennett, D. P., Bouchet, F. R. and Stebbins, A. (1990), *Ap. J.*, in press.

Hogan, C. J. (1984), *Ap. J. Lett.* **284**, L1.

Hogan, C. J. and Bond, J. R. (1988), in *NATO ASI on The Post-Recombination Universe*, eds. N. Kaiser and A. N. Lasenby (Kluwer Academic Press, Norwell, MA).

Ikeuchi, S. (1981), *Publ. Astr. Soc. Japan* **33**, 211.

Kaiser, N. and Stebbins, A. (1984), *Nature* **310**, 391.

Matsumoto, T., Hayakawa, S., Matsuo, H., Murakami, H., Sato, S., Lange, A. E. and Richards, P. L. (1988), *Ap. J.* **329**, 567.

Nottale, L. (1984), *Mon. Not. Roy. Astr. Soc.* **206**, 713.

Ostriker, J. P.and Cowie, L. L. (1981), *Ap. J. Lett.* **243**, L127.

Panek, M. (1986), *Phys. Rev.* **D34**, 416.

Panek, M. and Rudak, B. (1988), *Mon. Not. Roy. Astr. Soc.* **235**, 1169.

Peebles, P. J. E. (1982a), *Ap. J.* **259**, 442.

Peebles, P. J. E. (1982b), *Ap. J. Lett.* **263**, L1.

Peebles, P. J. E. (1987), *Ap. J. Lett.* **315**, L73.

Peebles, P. J. E. and Yu, J. T. (1970), *Ap. J.* **162**, 815.

Penzias, A. A. and Wilson, R. W. (1965), *Ap. J.* **142**, 419.

Raine, D. J. and Thomas, E. G. (1981), *Mon. Not. Roy. Astr. Soc.* **195**, 649.

Rees, M. J. and Sciama, D. W. (1968), *Nature* **217**, 511.

Sachs, R. K. and Wolfe, M. A. (1967), *Ap. J.* **147**, 73.

Salopek, D. S., Bond, J. R. and Bardeen, J. M. (1989), *Phys. Rev.* **D40**, 1753.

Sarazin, C. L. (1986), *Rev. Mod. Phys.* **58**, 1.

Starobinsky, A. A. (1985), *Sov. Astr. Lett.* **11**, 133.

Thompson, K. L. and Vishniac, E. T. (1987), *Ap. J.* **313**, 517.

Traschen, J. and Eardley, D. M. (1986), *Phys. Rev.* **D34**, 1665.

Tully, R. B. (1987), *Ap. J.* **323**, 1.

Vishniac, E. T. (1987), *Ap. J.* **322**, 597.

Vittorio, N. and Juszkiewicz, R. (1987), *Ap. J. Lett.* **314**, L29.

Vittorio, N. and Silk, J. (1984), *Ap. J. Lett.* **285**, L39.

Vittorio, N. and Silk, J. (1985), *Ap. J. Lett.* **297**, L1.

Wilson, M. L. (1983), *Ap. J.* **273**, 2.

Wilson, M. L. and Silk, J. (1981), *Ap. J.* **243**, 14.

Zeldovich, Ya. B. and Sunyaev, R. A. (1969), *Astrophys. Space. Sci.* **4**, 301.

9

The cosmic microwave background: present status of observations, and implications for general relativity

R. B. Partridge
Haverford College,
Haverford, PA 19041 USA

Studies of the $\sim 3K$ cosmic microwave background or "relict" radiation, discovered 25 years ago, have substantially improved our understanding of cosmology and of the formation of large-scale structure in the universe. Here, I look at the implications of these studies for gravity theory.

I will begin by reviewing the observations, particularly the spectrum and the large-scale isotropy of the radiation. Measurements of the spectrum, when combined with other astrophysical data like the abundance of light nuclei, establish the temperature and expansion rate of the universe at early times. These values in turn may be used to tell us whether unmodified general relativity adequately describes the dynamics of the early universe. Upper limits on any large-scale anisotropy sharply restrict the range of possible anisotropic cosmological models, and provide supporting evidence for a period of "inflationary" expansion early in the universe. The preceding paper by Dr. Panek explores some of these consequences further.

1. Introduction

It is an honor to have been invited to review the cosmic microwave background radiation (CBR) for this audience. As an observer and experimentalist, I feel particularly privileged since general relativity is sometimes viewed as the province of theoreticians, not those of us with hands dirtied in the laboratory or at the telescope.

As is well known, the CBR was discovered 25 years ago by Penzias and Wilson (1965). It was immediately interpreted (see Dicke et al. (1965))

223

as radiation left over from the hot Big Bang origin of the universe, radiation still present in the universe. It was also quickly recognized that if this interpretation were correct, the CBR would need to meet two observational tests: its spectrum should be close to that of a thermal blackbody, and it should be quite isotropic (and in particular not correlated with foreground sources of radio emission such as the plane of the galaxy). Within 3 years of its discovery, the radiation had passed both tests. Spectral measurements spanning a range from 0.26 - 21 cm were consistent with a blackbody spectrum with $T_0 = 3$ K, including the departures from the Rayleigh-Jeans law expected near the peak of a Planck spectrum at wavelengths below ~ 1 cm. Upper limits on anisotropies on a range of angular scale were set approximately at $\Delta T/T_0 \lesssim 10^{-3}$.

The 3 K CBR is now virtually universally recognized as a relict of the hot, dense early phase in the universe which reaches us from a much earlier epoch in the history of the universe. When observing the CBR, we "see" back to a surface of last scattering where the photons last interacted with matter. That surface is probably at a cosmological redshift, z, of ~ 1000, but conceivably as low as ~ 20. In any case, the epoch of last scattering is very early. Indeed, the discovery of the CBR was the deciding evidence that the universe does have a history; that the Steady State Theory of Bondi, Gold, and Hoyle (see Bondi (1968)) was not the correct cosmological theory. Incidentally, let me now hint at one theme of this paper by noting that the mere existence of the CBR thus served to test fundamental theory. Was Einstein right, or was an additional "creation field," a cornerstone of the Steady State Theory, needed? The answer came down in favor of unmodified general relativity; and, as we shall see, standard general relativity continues to pass all tests set by observations of the CBR.

The discovery of the CBR also vindicated predictions by George Gamow and his colleagues Alpher, Follin, and Herman, of a big Bang origin of the universe (e.g., Alpher and Herman (1948)). Since George Gamow spent the last years of his exciting, eccentric and productive life here at Colorado, it is only appropriate to note that he had predicted the existence of a thermal background filling the universe nearly 20 years before it was discovered. He and his colleagues recognized that such radiation would have played the dominant role in the dynamics of the universe, and that a hot, early phase would synthesize the nuclei of light elements like ^{4}He. We shall have occasion to return to each of these points later.

In the remainder of this paper I will review the present observational situation (see also the paper by Panek, Chapter 8) and then look briefly at some of the consequences for relativity theory.

2.   The present status of the observations of the CBR

Observations of the CBR may be conveniently grouped under seven headings. The seven fundamental parameters I refer to are the thermodynamic temperature of the CBR, written here T_0; the dipole moment of the radiation, T_1; the quadrupole moment, T_2; the large-scale polarization, T_p; and the limits of anisotropy of the CBR on three different angular scales. In this paper, I will review the most recent observational results on all seven of these parameters. Since the anisotropy measurements are closest to my own work, I will emphasize them. The most exciting recent result in CBR studies, however, is a measurement of the submillimeter spectrum of the background and that is where I will begin. Let me add that the material of Section 2 is excerpted almost verbatim from a more technical review of CBR measurements I prepared for a recent Moriond meeting (Partridge (1989)).

The temperature of the CBR, T_0

In general, measurements of the thermodynamic temperature of the CBR at wavelengths longer than a few millimeters are in good agreement with a value of $T_0 = 2.75 \pm 0.03$ K (note the 1% accuracy, rare in cosmology). Many of the recent measurements are the results of an international collaboration between several Italian institutions and the University of California, Berkeley, in which Haverford initially also participated (Smoot et al. (1985); Partridge (1985); Smoot et al. (1988)). Two of the most recent measurements by the Berkeley group are of particular interest: measurements at 3 cm and 3.3 mm have produced values $1 - 2\sigma$ below the general average (the values are $T_0 = 2.61 \pm 0.06$ K and 2.60 ± 0.09 K, respectively; see Kogut et al. (1988); and Bersanelli et al. (1989)). In contrast, the single most precise measurement in the centimeter range is somewhat high at $T_0 = 2.783 \pm 0.025$ K (Johnson and Wilkinson (1987)). As we shall see, the discrepancy, while not overwhelming, may be important.

Supplementing these direct radiometric measurements are determinations of T_0 based on the excitation of interstellar molecules, particularly

CN, by CBR photons. These measurements of T_0 are very precise and have the added advantage of being free of the many possible sources of systematic error which may crop up in observations made beneath all or part of the earth's atmosphere. There are, of course, sources of systematic and statistical error present in the CN measurements as well; but the agreement with other measurements is gratifying. The newest CN measurements (Crane et al. (1989); Meyer et al. (1989)) give $T_0 = 2.79 \pm 0.03$ K and 2.83 ± 0.09 K at 2.64 mm and 1.32 mm, respectively.

The really exciting spectral measurement is in the short wavelength, Wien, region of a 2-3 K spectrum. On the same day as supernova 1987A was seen, a Japanese sounding rocket carried aloft an experiment to measure the submillimeter spectrum of the cosmic background radiation. The instrument was a collaborative effort by Paul Richards and his group at Berkeley, and Toshio Matsumoto and his group at Nagoya (see Matsumoto et al. (1988)). Cooled optics were employed to reduce systematic offsets; only data free of residual atmospheric emission were used in the final analysis. The sensitivity of the bolometric detectors was high enough to ensure $\sim 1\%$ measurements in the few minutes the instrument was above the atmosphere. Figure 1 shows the results. At $\lambda \sim 1$ mm, $T_0 = 2.799 \pm 0.018$ K, in good agreement with both the CN results and the measurement by Johnson and Wilkinson. The shortest wavelength bands employed by Matsumoto and his colleagues were dominated by thermal emission from dust in our own galaxy. It is worth noting that the Berkeley-Nagoya results in these bands are in good agreement with projections based on IRAS measurements of dust emission made at slightly shorter wavelengths (Fig. 1). We thus have some confidence that the shortest and longest wavelength channels were working properly and were correctly calibrated.

The results at two intermediate wavelengths of 0.7 mm and 0.48 mm show an upward trend in T_0 as the wavelength drops–the submillimeter excess seen in Fig. 1. The composite spectrum from 75 cm to 0.48 mm is not well fit by a blackbody at any temperature, or even by a graybody spectrum. Indeed, a small industry has grown up around explanations of this submillimeter excess. The most developed models involve photo-decays of exotic particles at high redshift (see Hayakawa et al. (1987); Fukugita (1988)); inverse Compton scattering of CBR photons by hot electrons (see Hayakawa et al. (1987); Lacey and Field (1988)) and thermal emission by dust in high redshift galaxies (e.g. Negroponte et al. (1981); Bond, Carr and Hogan (1989)). All of these models have

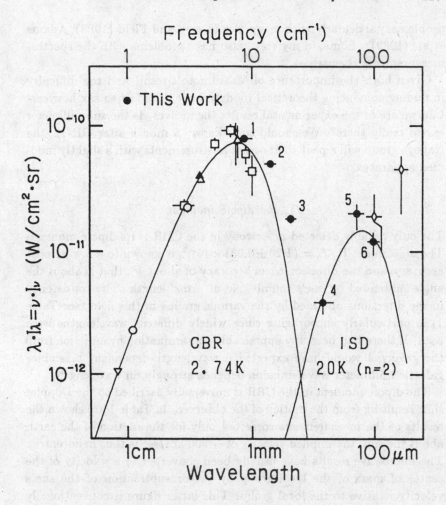

Fig. 1. Taken from Matsumoto et al. (1988): "this work." The open symbols show CN and ground-based measurements of the CBR intensity; the curve is a $T_0 = 2.74$ K blackbody.

problems, particularly with energetics (Lacey and Field (1988); Adams
et al. (1989)). Some, in my view, also have problems with the spectral
measurements themselves.

Given both the importance of Matsumoto's result and the difficulty
in finding convincing theoretical models for it, it is fair to ask how certain we are of the experimental results themselves. Is the submillimeter
excess really there? We should soon know: a month after GR-12, the
Nagoya group will repeat their rocket measurements with a slightly modified apparatus.

T_1, the dipole moment

The only reliably detected anisotropy in the CBR is its dipole moment.
The amplitude, $T_1/T_0 = (1.20 \pm 0.03) \times 10^{-3}$, is known to a few percent
accuracy, and the direction to an accuracy of about 1°, that is about the
angle subtended by one's thumb held at arms' length. The consistency
in the directions obtained by the various groups in this field (see Table
1) is particularly encouraging since widely different wavelengths were
used. If there had been any appreciable contamination by emission from
the galaxy, I would have expected a wavelength-dependent bias since
galactic millimeter wave emission depends strongly on wavelength.

The dipole moment in the CBR is universally ascribed to the Doppler
shift resulting from the motion of the observer. In Table 1 are shown the
results of the measurements corrected only for the motion of the earth
about the sun; the implied velocity of ~ 300 km/sec is thus heliocentric.
The microwave results have usually been converted to a velocity of the
center of mass of the local group by vector subtraction of the sun's
velocity relative to the local group. This latter figure is conventionally
taken to be 300 km/sec towards $\ell = 90°$, $b = 0°$, (or $\alpha = 21^h$, $\delta = +50°$).
In fact the solar motion relative to the center of mass of the local group is
much less precisely known than the CBR dipole velocity. Indeed, Yahil
et al. (1977) find $\ell = 105°$ and $b = -7°$ a better fit. Hence it might be
prudent for theorists to calculate heliocentric velocities directly.

T_2, the quadrupole moment

The best available limits are set by the Soviet space experiment (Table 1). The value I cite is less model-dependent than a slightly more
stringent upper limit given by Klypin et al. (1987): $T_2/T_0 < 2 \times 10^{-5}$.
Improving on this limit will require careful subtraction of galactic emis-

Table 1. Measurements of the dipole moment T_1 and limits on T_2. The results are expressed in thermodynamic temperature.

Group References	Berkeley Lubin et al. (1983)	MIT/UBC Halpern et al. (1988)	Moscow Strukov et al. (1987); Klypin et. al. (1987)	Princeton Fixsen et al. (1983)	Princeton Boughn et al. (1989)
Wavelength, mm	3	~1.7	8	12	15
Vehicle	balloon	balloon	satellite	balloon	balloon
Detector	heterodyne	bolometric	heterodyne	heterodyne	maser amplifier
Dipole amplitude, T_1, mK	3.4 ± 0.2	3.4 ± 0.4	3.16 ± 0.12	3.1 ± 0.2	3.47 ± 0.04
Direction of solar motion: R. A. and Decl.	$11.2^h \pm 0.1^h$ $-6° \pm 1.5°$	$12.1^h \pm 0.24^h$ $-23° \pm 5°$	$11.3^h \pm 0.16^h$ $-7.5° \pm 2.5°$	$11.2^h \pm 0.05^h$ $-8° \pm 0.7°$	$11.1^h \pm 0.1^h$ $-5.9° \pm 1.5°$
Limit on quadrupole moment, T_2, mK	0.4^*	n.a.	$<0.08^†$	<0.19	n.a.

*Calculated from Table I of Lubin et al. (1983) by the present author, by taking the quadrature sum of the measured coefficients $Q_1, \ldots Q_5$.
†Taking the more conservative, model-independent upper limit.

sion (e.g. Boughn et al. (1989)). Even the currently available limits place interesting constraints on anisotropic expansion between the red-shift of last scattering $z \approx 1000$ and now; they also contain some models for the origin of large-scale structure.

T_p, large-scale polarization

As shown by Negroponte and Silk (1980) and Basko and Polnarev (1980), large-scale linear polarization of the CBR will be produced if the expansion of the universe was anisotropic just at the epoch of last scattering. T_p thus measures anisotropy at a specific moment in the past. The best upper limits are still those of Lubin et al. (1983): $T_p/T_0 \lesssim 2 - 4 \times 10^{-5}$ for the dipole and quadrupole components.

Anisotropy on degree scales

If the CBR last scattered at $z \approx 1000$, the present angular scale corresponding to the causal horizon at last scattering is $1° - 2°$. Any anisotropy in the CBR on scales larger than this cannot have been smoothed away by causal processes,[*] and thus carries information about the primordial spectrum of density perturbations in the universe. Recently, Davies and his colleagues (1987) have claimed to detect fluctuations in the CBR corresponding to $\Delta T/T_0 = 3.7 \times 10^{-5}$ at $\lambda = 3$ cm on a scale of $8°$. Some workers in the field are skeptical of this claim and prefer to regard this result as an upper limit; in any case, measurements at degree scales are important and deserve further work. Several efforts are already underway.

Anisotropy on scales of $\sim 10'$

In many (but not all) scenarios for the origin of large-scale structure in the universe, the amplitude of $\Delta T/T$ fluctuations has a maximum at $\sim 10'$ (see, e.g. Bond and Efstathiou (1984); Vittorio and Silk (1984)). This scale is accessible to conventional radio telescopes, and the upper limits on $\Delta T/T$ are consequently very tight: at $\lambda = 1.5$, and for $\theta = 7.5$, $\Delta T/T \leq 1.7 \times 10^{-5}$ (Readhead et al. (1989); see also Uson and Wilkinson (1984) for earlier results). These results are reviewed extensively in

---

[*]Unless an earlier epoch of inflationary expansion occurred; see below.

the literature (e.g. Wilkinson (1986); Partridge (1988)), so I will not consider them further here.

Anisotropy on scales of $< 1'$

In the galaxy formation scenarios referred to above, fluctuations on scales $< 1'$ are smeared out (because the surface of last scattering has a finite thickness). In other models, however, particularly those invoking explosions to produce density perturbations, a higher amplitude of CBR fluctuations on scales $\leq 1'$ may result (Ostriker and Vishniac (1986); Vishniac (1987)). Thus several observers have considered it worthwhile to probe arcsecond scales (Fomalont et al. (1988); Martin and Partridge (1988); Partridge, Nowakowski and Martin (1988)).

To achieve such resolution, we need to employ a different technique,[*] aperture synthesis, in which an array of telescopes, rather than a single antenna, is used to map the sky. This technique produces images of the microwave sky at a resolution set by the size of the array (see Verschuur and Kellermann (1974) for details, or Partridge (1988)). This technique has been used by two groups working at the Very Large Array (VLA) operated by the U.S. National Radio Astronomy Observatory in New Mexico (Fomalont et al. (1984); Fomalont et al. (1988); Knoke et al. (1984); Martin and Partridge (1988); Partridge, Nowakowski and Martin (1988)). Most observations reported to date were made at a wavelength of 6 cm, the wavelength at which the most sensitive receivers have been available. Angular scales of $6''$ to $160''$ were investigated. The sensitivity of this technique is not yet quite comparable to the sensitivity of searches using conventional, filled-aperture radio telescopes: $\Delta T/T \leq 1.3 \times 10^{-4}$ on a scale of $18''$. On the other hand, typically 100-1000 independent sky elements are available in each map.

3.   The CBR and cosmology and gravity theory

The observations listed above have been a bonanza for astrophysicists and cosmologists. Indeed, I think it is fair to say that the discovery of the CBR was second only to Hubble's recognition of the expansion of the universe in establishing modern, scientific cosmology.

---

[*]One exception to this statement is the recent 1 mm observation reported by Kreysa and Chini (1989), made with a single antenna at $\theta = 30''$.

For instance, the mere existence of ~ 3 K blackbody radiation tells us that the dynamics of the early universe were radiation dominated: radiation, not matter, was the main contributor to the stress-energy tensor at early times. This conclusion follows from the different dependences of matter and radiation density on the scale factor of the universe: $\rho_{matter} \propto R^{-3}$ and $\rho_{rad} \propto R^{-4}$.

The observed properties of the CBR have also had a profound impact on theories for the origin of structure in the universe, a topic Dr. Panek treats in more detail. I would say, however, that the limits of $\Delta T/T_0$ combined with other astronomical observations, throw some doubt on all models for galaxy formation (see, for instance, Kaiser and Silk (1986); Bond and Efstathiou (1987); Peebles (1989)).

Now, in keeping with the spirit of this meeting, let us look more specifically at the present and future impact of CBR studies on general relativity and gravity theory.

Observational and experimental tests of general relativity

Allow me to begin with some brief informal remarks on the nature of observational and experimental tests of GR. In its early years, GR was tested in two arenas–the solar system and the cosmos. Beginning with the former, this year marks the 70th anniversary of the eclipse observations of Eddington and others which established the experimental validity of Einstein's theory. In many ways, these eclipse observations are emblematic of all subsequent observational and experimental work on GR:

1. The observations were difficult, and at the limit of contemporary technology.
2. Only the weak field limit of GR was tested.
3. Like most astronomical results, these observations were affected by messy, local astrophysical and atmospheric processes—the system used to test GR was not "clean."

These same features appear in cosmological tests of GR, including those made possible by the discovery of the CBR. In particular, as we shall see, it has often proven difficult to disentangle relativistic effects from astrophysical messiness. With this word of warning, let us now ask how the CBR does (and may in the future) reinforce our trust in GR.

The early universe as a laboratory

One of the crucial predictions of the hot Big Bang theory is that the universe passed through a phase at $t \sim 3$ min after the presumed initial singularity when the conditions of temperature and density were appropriate to synthesize the nuclei of light elements. The predicted abundance of nuclei such as ^{4}He, ^{2}H and ^{7}Li agree astonishingly well with the abundances observed in the oldest stars (see Boesgaard and Steigman (1985), for a review). Those predictions, however, are based on some assumptions and prior knowledge. Among these are: 1) a knowledge of the nuclear physics of the fusion reactions, including relevant crosssections; 2) the temperature of the CBR; 3) the number of lepton families which determines the expansion rate at $t \sim 3$ min; and 4) standard GR. Our knowledge of nuclear physics is good enough to give secure predictions of the abundance of ^{4}He, ^{2}H and ^{7}Li emerging from the Big Bang (see, for instance, Yang et al. (1984); Audouze (1987)), and we know T_0 to a precision of a few percent. That leaves 3) and 4) to be tested by observations. Let us begin by examining the abundance of ^{4}He, since, as Fig. 2 shows, the abundance of this nucleus is essentially independent of the density in the universe. If we *assume* standard general relativity we can use the observed ^{4}He abundance to determine the number of lepton families. The answer, $N \leq 4$, is more precise than any so far provided by elementary particle physics (though checking this result is a major aim of major accelerators coming on line this fall). On the other hand, we may set $N = 3$ as given by standard particle theory. Then the close agreement of predicted and observed ^{4}He abundance becomes a test of gravity theory. GR passes; other theories, such as the scalar-tensor theory of Brans and Dicke, do not necessarily do so. The variants of standard gravity theory which speed up or slow down the expansion rate at the crucial epoch of nucleosynthesis tend to overproduce or underproduce helium.

With our trust in GR thus reinforced, we may turn to additional cosmological parameters derived from observations of the abundances of light elements. In particular, as Fig. 2 shows, the abundance of ^{2}H emerging from the Big Bang has a strong dependence on the baryon density of the universe (physically, deuterons are more likely to encounter each other and fuse to ^{4}He if the universe has a high density). Observations of the ^{2}H abundance in cosmologically "old" environments require a very low value for the baryon density, typically less than a few percent of the closure density. It is this observation (combined with other evi-

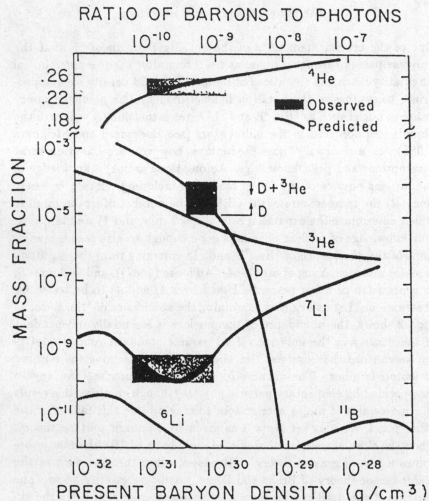

RATIO OF BARYONS TO PHOTONS

Fig. 2. Adapted from *Physics Through the 1990's, Gravitation, Cosmology and Cosmic-Ray Physics* of the Physics Survey Committee (1986). Predictions based on GR are shown; the observations of all light nuclei agree with a present baryon density of $\sim 5 \times 10^{-31}$ gm/cm^3.

dence for additional density in the universe and the strong theoretical argument from inflation theory that the spatial geometry of the universe should be flat) that has prompted so much interest in dark matter (see *IAU Symposium* **117**, (1987)).

Although I believe these conclusions are robust, I should not hide the fact that the astrophysical systems used in these tests are not perfectly "clean." It is well known that stars, including our sun, are busy generating ^{4}He. We must therefore be careful to observe only the *oldest* stars in the search for ^{4}He, and that is technically difficult. Fortunately, perhaps, ^{4}He is very difficult to fission. Thus we can be reasonably confident that the ^{4}He observed in old stars is at least an upper limit on the abundance produced in the Big Bang. The situation with ^{2}H is more complicated, since this flimsy nucleus can easily be broken apart. It is harder to imagine *producing* it in an astrophysical environment, however, so that ^{2}H abundances can reasonably safely be regarded as lower limits; these in turn set upper limits on the baryon density. Finally, it is worth considering the abundance of a trace element produced in the Big Bang, ^{7}Li. As Fig. 2 shows, the ^{7}Li emerging from the Big Bang depends in a complicated way on the baryon density. The observations are consistent only with the minimum value of ^{7}Li shown in the graph, and thus the required baryon density agrees well with that derived from ^{2}H.

To me, the power of these results is quite striking: a knowledge of a few cross sections in nuclear physics and of the temperature of the CBR permit us to check the validity of GR early in the universe, to constrain the number of elementary particle families, and to determine a crucial cosmological parameter.

Anisotropic cosmological models

As is well known, standard GR allows anisotropic as well as isotropic solutions for a dust and/or radiation filled universe (see for instance, MacCallum's review in the Proceedings of the Stockholm meeting). In such models, since the space-time geometry is intrinsically anisotropic, both anisotropic expansion and shear are possible. These in turn introduce large-scale anisotropies into the CBR, since the redshift becomes direction-dependent. Hence limits on the dipole and especially the quadrupole moments of the CBR allow us to constrain both anisotropic expansion and shear. For instance, Hawking (1969) showed two decades ago that the then-available CBR measurements set very stringent limits on rotation in anisotropic cosmological models: in most such models the rotation allowed is $< 1°$ over the 10-20 billion year history of the universe.

The dipole and quadrupole moments of total intensity measure the time-integrated anisotropy from the epoch of last scattering to the present. The large-scale polarization of the CBR, on the other hand, is imprinted by anisotropic expansion just at the surface of last scattering. T_p therefore offers an additional constraint (Basko and Polnarev (1980);

Negroponte and Silk (1980)). Yet another constraint (on anisotropic expansion at $t \sim 3$ min) is provided by the abundance of light elements. All these observations are consistent with entirely isotropic expansion–and hence with Mach's Principle.

Inflation

Although inflation in its modern form (Guth (1981)) arose from particle physics, not relativity, there are two reasons for us to look at it briefly. First, Guth and his colleagues reintroduced and provided a physical explanation for the idea of exponential expansion first considered by Milne and later incorporated in the Steady State Theory. Second, the physical basis of inflation may be linked to supergravity. In addition inflation has had some notable successes in cosmology. It explains why the spatial curvature of the universe is so small and why the universe is in fact so homogeneous on scales larger than the causal horizon at last scattering (causally connected regions are "inflated" to much larger size well before last scattering). In a sense, however, inflation was invented to solve these puzzles lurking in classical cosmology. It is more interesting to ask if there are additional, *testable* predictions of inflation. There is at least one–the spectrum of density perturbations expected to emerge from inflation. Such fluctuations, introduced by quantum processes very early in the universe, are expected to have a scale-invariant (that is, curvature independent of wave number) spectrum, of the type first discussed by Harrison and Zel'dovich twenty years ago. The scale-invariant spectrum implies a specific, power-law relation between the mass of a density perturbation and its amplitude, $\Delta\rho/\rho$. In turn, that establishes a link between the angular scale of CBR temperature fluctuations and their amplitude, $\Delta T/T_0$. Once temperature fluctuations are detected (justified optimism, I believe), we may hope to measure $\Delta T/T_0$ as a function of angular scale, and hence check this prediction of inflation. (Note that the same scale-invariant spectrum is integral to other theories for the origin of large-scale structure, so the confirmation of scale-invariant spectrum is a necessary but not sufficient condition for inflation.)

Cosmic strings

These one-dimensional topological defects, relics of inflation are described elsewhere in this volume. Here, let me note that a moving string will induce a step-like discontinuity in the CBR, whose amplitude proportional to the mass per unit length μ of the string (see Kaiser and Stebbins (1984)). For $G_\mu \sim 10^{-6}$ as predicted by most theories, $\Delta T/T_0 \sim 10^{-5} - 10^{-6}$.

Can we detect such discontinuities in the temperature of the CBR? Not yet, in part because of messy astrophysical details such as the presence of foreground radio sources. Nevertheless, it is important to bear in mind that a network of cosmic strings filling the universe would leave a very distinct signature on the CBR. Because the strings are one-dimensional, the regions of different temperature will be very sharply bounded. Thus the statistics of the angular distribution of CBR temperatures will be very different from Gaussian (e.g., Bouchet et al. (1988)). Observationally, both high angular resolution (to register the discontinuities) and wide sky coverage (to be sure to have at least one string in the field of view) are necessary. The approach most likely to succeed is aperture synthesis, discussed in Section 2 above. We need to improve the sensitivity of these searches by a factor of 5-10.

4. Concluding remarks

It is appropriate to end with a mention of observations not yet made. The power of CBR observations to test general relativity–and much else in cosmology and astrophysics–has not yet been fully realized. But then the field is only one-third as old as GR.

I wish to end by thanking the organizers of this conference, the John Simon Guggenheim Memorial Foundation and the National Science Foundation for funds which enabled me to come to Boulder and to carry out the research described here.

References

Adams, F. C., Freese, K., Levin, J. and McDowell, J. C. (1989), submitted to *Ap. J.*

Alpher, R. A. and Herman, R. C. (1948), *Nature* **162**, 774.

Audouze, J. (1987), in I.A.U. Symposium 124, *Observational Cosmology*, eds. A. Hewitt, G. Burbidge and L. Z. Fang (Reidel, Dordrecht).

Basko, M. M. and Polnarev, A. G. (1980), *Mon. Not. Roy. Astr. Soc.* **191**, 207.

Bersanelli, M., Witebsky, C., Bensadoun, M., De Amici, G., Kogut, A., Levin, S. M. and Smoot, G. F. (1989), *Ap. J.* **339**, 632.

Boesgaard, A. M. and Steigman, G. (1985), *Ann. Rev. Astron. and Astrophys.* **23**, 319.

Bond, J. R., Carr, B. J. and Hogan, C. J. (1989), submitted to *Ap. J.*

Bond, J. R. and Efstathiou, G. (1984), *Ap. J. Lett.* **285**, L45.

Bond, J. R. and Efstathiou, G. (1987), *Mon. Not. Roy. Astr. Soc.* **226**, 655.

Bondi, H. (1968), *Cosmology*, (Cambridge Univ. Press, Cambridge).

Bouchet, F. R., Bennett, D. P. and Stebbins, A. (1988), *Nature* **335**, 410.

Boughn, S. P., Cottingham, D. C., Cheng, E. S. and Fixsen, D. J. (1989), in preparation.

Crane, P., Hegyi, D. J., Kutner, M. L. and Mandolesi, N. (1989), *Ap. J.* in press.

Davies, R. D., Lasenby, A. N., Watson, R. A., Daintree, E. J., Hopkins, J., Beckman, J., Sanchez-Almeida, J. and Rebolo, R. (1987), *Nature* **326**, 462.

Dicke, R. H., Peebles, P. J. E., Roll, P. G. and Wilkinson, D.T. (1965), *Ap. J.* **142**, 414.

Fomalont, E. B., Kellermann, K. I. and Wall, J. V. (1984), *Ap. J. Lett.* **277**, L23.

Fomalont, E. B., Kellermann, K. I., Anderson, M. C., Weistrop, D., Wall, J. V., Windhorst, R. A. and Kristian, J. A. (1988), *A. J.* **96**, 1187.

Fukugita, M. (1988), *Phys. Rev. Lett.* **61**, 1046.

Guth, A. H. (1981), *Phys. Rev.* **D23**, 347.

Hawking, S. W. (1969), *Mon. Not. Roy. Astr. Soc.* **142**, 129.

Hayakawa, S., Matsumoto, T., Matsuo, H., Murakami, H., Sato, S., Lange, A. E. and Richards, P. L. (1987), *Publ. Astron. Soc. Japan* **39**, 941.

Johnson, D. G. and Wilkinson, D. T. (1987), *Ap. J. Lett.* **313**, L1.

Kaiser, N. and Silk, J. (1986), *Nature* **24**, 529.

Kaiser, N. and Stebbins, A. (1984), *Nature* **310**, 391.

Klypin, A. A., Sazhin, M. V., Strukov, I. A. and Skulachev, D.P. (1987), *Soviet Astron. Lett.* **13**, 104.

Knoke, J. E., Partridge, R. B., Ratner, M. I. and Shapiro, I. I. (1984), *Ap. J.* **284**, 479.

Kogut, A., Bersanelli, M., De Amici, G., Friedman, S. D., Griffith, M., Grossan, B., Levin, S., Smoot, G. F. and Witebsky, C. (1988), *Ap. J.* **325**, 1.

Kreysa, E. and Chini, R. (1989), in Third ESO/CERN Symposium, *Astronomy, Cosmology and Fundamental Physics*, eds. Caffo et al. (Kluwer Academic Publishers, Dordrecht).

Lacey, C. G., and Field, G. B. (1988), *Ap. J. Lett.* **330**, L1.

Lubin, P., Melese, P. and Smoot, G. (1983), *Ap. J. Lett.* **273**, L51.

Martin, H. M. and Partridge, R. B. (1988), *Ap. J.* **324**, 794.

Matsumoto, T., Hayakawa, S., Matsuo, H., Murakami, H., Sato, S., Lange, A. E. and Richards, P. L. (1988), *Ap. J.* **329**, 567.

Meyer, D. M., Roth, K. C. and Hawkins, I. (1989), *Ap. J. Lett.* **343**, L1.

Negroponte, J., Rowan-Robinson, M. and Silk, J. (1981), *Ap. J.* **248**, 58.

Negroponte, J. and Silk, J. (1980), *Phys. Rev. Lett.* **44**, 1433.

Ostriker, J. P. and Vishniac, E. T. (1986), *Ap. J. Lett.* **306**, L51.

Partridge, R. B. (1985), in *The Cosmic Background Radiation and Fundamental Physics*, ed. F. Melchiorri (Editrice Compositori, Bologna).

Partridge, R. B. (1988), *Reports on Progress in Physics* **51**, 647.

Partridge, R. B. (1989), Cosmological Parameters Derived from the Cosmic Microwave Background Radiation, in *The 24th Rencontres de Moriond*, ed. J. Audouze (Editions Frontiers, France).

Partridge, R. B., Nowakowski, J. and Martin, H. M. (1988), *Nature* **331**, 146.

Peebles, P. J. E. (1989), in *The Cosmic Microwave Background: 25 Years Later*, eds. N. Mandolesi and N. Vittorio (Kluwer, Dordrecht).

Penzias, A. A. and Wilson, R. W. (1965), *Ap. J.* **142**, 419.

Readhead, A. C. S., Lawrence, C. R., Myers, S. T., Sargent, W. L. W. Hardebeck, H. E. and Moffet, A. T. (1989), submitted to *Ap. J.*

Smoot, G. F., De Amici, G., Friedman, S., Witebsky, C., Sironi, G., Bonelli, G., Mandolesi, N., Cortiglioni, S., Morigi, G., Partridge, R. B., Danese, L. and De Zotti, G. (1985), *Ap. J. Lett.* **291**, L23.

Smoot, G. F., Levin, S. M., Witebsky, C., De Amici, G. and Rephaeli, Y. (1988), *Ap. J.* **331**, 653.

Uson, J. M. and Wilkinson, D. T. (1984), *Ap. J.* **283**, 471.

Verschuur, G. L. and Kellermann, K. I., eds. (1974), *Galactic and Extragalactic Radio Astronomy*, (Springer-Verlag, New York).

Vishniac, E. T. (1987), *Ap. J.* **322**, 597.

Vittorio, N. and Silk, J. (1984), *Ap. J. Lett.* **285**, L39.

Wilkinson, D. T. (1986), *Science* **232**, 1517.

Yahil, A., Tammann, G. A. and Sandage, A. (1977), *Ap. J.* **217**, 903.

Yang, J., Turner, M. S., Steigman, G., Schramm, D. N. and Olive, K. A. (1984), *Ap. J.* **281**, 493.

B1

Mathematical cosmology

John Wainwright
Department of Applied Mathematics,
University of Waterloo,
Waterloo, Ontario, Canada

The workshop on mathematical cosmology was devoted to four topics of
current interest. This report contains a brief discussion of the historical
background of each topic and a concise summary of the content of each
talk.

1. The observational cosmology program

The standard approach for analyzing cosmological observations is to
assume that space-time is isotropic and spatially homogeneous (i.e. that
the cosmological principle holds). It then follows that the universe can be
described by a Friedmann-Lemaitre-Robertson-Walker (FLRW) model,
and the aim is to use the observations to determine the free parameters
that characterize such models. A fundamental question, however, is
whether ideal cosmological observations on our past null cone can be
used to actually determine the geometry of the cosmological space-time,
without introducing *a priori* assumptions about the geometry. This
question provides the rationale for the observational cosmology program,
as described by Ellis et al. (1985). In this paper it was shown that ideal
observations alone do not determine the geometry of spacetime. For
example, even if all observations are isotropic about our position, it
does not follow that the spacetime is spherically symmetric about our
position. However, if the Einstein field equations (EFEs) are assumed

241

to hold, then ideal observations do determine the spacetime geometry off our past null cone.

In the workshop, W. R. Stoeger reported on work in progress with S. D. Nel and G. F. R. Ellis concerning the observational cosmology program. Assuming the Einstein field equations with dust source, they derive the equations that are satisfied by perturbations of an FLRW model, using observational coordinates, (i.e. coordinates that are adapted to our past null cone). The equations can be solved analytically, but the solutions involve hypergeometric functions and are thus difficult to analyze. It is hoped that this procedure will eventually lead to a characterization of "almost FLRW" observational cosmologies.

2.   The cosmological perturbation problem

Following the classical paper of Lifshitz (1946), much work has been done on the cosmological perturbation problem (i.e. the study of perturbations of the FLRW models). Nevertheless, this problem is not fully understood. In order to define perturbations one has to specify a correspondence between the physical universe model and the given background spacetime. A change in this correspondence is called a gauge transformation. The essential point is that if a scalar quantity is nonconstant in the background spacetime, then the values of the perturbation of this quantity will not be invariant under gauge transformations, and hence will not be physically meaningful. This result applies in particular to density perturbations of an FLRW model (see for example Bardeen (1980)).

In a major paper Bardeen (1980) introduced a fully gauge-invariant approach to the cosmological perturbation problem. However, Bardeen's variables, while gauge-invariant, are not directly related to the density perturbations, and lead to a formalism of some complexity.

In the workshop, G. F. R. Ellis reported on recent work with M. Bruni (Ellis and Bruni (1989)), proposing a new approach to the perturbation problem. Their approach can to some extent, be considered as and extension and completion of the approach first used by Hawking (1966). The idea is to use as variables the spatial gradients of the density, pressure and rate of expansion, which are necessarily gauge-invariant. In particular the spatial density gradient leads to a natural and direct gauge-invariant measure of the density perturbation. The EFE's lead to exact evolution equations for these quantities, which can be linearized about the FLRW models. The interpretation of the solutions of these equations is straightforward, since the variables are natural from a cosmological point of view.

3. Isotropic singularities

Observations of isotropy about our position (based primarily on the microwave background radiation) combined with the Copernican principle are customarily used to infer that the recent evolution of the universe can be described by a FLRW model (or perturbation thereof). Whether our universe was more isotropic or less isotropic in the distant past, prior to the decoupling of matter and radiation, is still an open cosmological question, although the current consensus seems to favour the former.

There are two scenarios which lead to an isotropic early universe. In the inflationary scenario, a brief period of exponential expansion leads to a universe which is highly isotropic and close to flatness (e.g. Barrow (1988)). On the other hand, in the so-called quiescent cosmology scenario (Barrow (1978); Goode (1987)) the initial singularity itself is required to be isotropic in some sense. Support for quiescent cosmology comes from the ideas of Penrose, who argues that entropy should be attributed to the gravitational field (e.g. Penrose (1979)).

In the workshop, S. W. Goode gave an introduction to the notion of an "isotropic singularity," which may provide a mathematical framework for studying quiescent cosmology. The essential idea (Goode and Wainwright (1985); Tod (1986)) is that the metric of the physical spacetime is conformally related to a metric which is regular at the singularity, the singularity itself being due to the vanishing of the conformal factor. This notion can be considered as a geometric generalization of the concepts of "quasi-isotropic" singularity and "Friedmann-like" singularity, which were introduced by Lipshitz and Khalatnikov (1963), and Eardley, Liang and Sachs (1972) respectively.

Geometrically, an isotropic singularity can be pictured as a regular hypersurface in the conformally related spacetime. Heuristic considerations suggest that the 3-metric on this hypersurface can be prescribed arbitrarily and that specifying this 3-metric and an equation of state $p = p(\mu)$ determines a unique solution of the Einstein field equations with perfect fluid source. Proving this initial value conjecture is an important unsolved problem in connection with isotropic singularities. J. Ehlers suggested that the conformal formulation of the EFEs due to H. Friedrich (e.g. Friedrich (1986)) could be helpful in solving this problem. A second problem is to investigate, either analytically or numerically, the nature of the density inhomogeneities that arise when a cosmological model evolves from an isotropic singularity.

4.   The evolution of Bianchi cosmologies

The systematic study of Bianchi cosmologies, that is, cosmological models which admit a G_3 of isometries acting on spacelike hypersurfaces, dates back to the paper of Taub (1951). Activity in this area reached a peak in the decade 1966-76, stimulated by the discovery of the cosmic microwave background radiation. For a summary of this work, we refer to MacCallum (1979).

In order to be compatible with the microwave observations (i.e., with the highly isotropic nature of the radiation) a Bianchi cosmology must pass through a highly isotropic epoch during its evolution. It is thus important to describe the asymptotic states of the Bianchi models near the big-bang and at late times, and to describe the principal qualitative features of the evolution between these states. To date, this problem has not been completely solved. Indeed, since the problem is of some complexity, and a variety of different approaches have been used (see for example, Wainwright and Hsu (1989)), it is sometimes difficult to ascertain what has been proved and what has merely been made plausible by heuristic arguments.

Since the EFEs for Bianchi models can be written as an autonomous system of first order differential equations, it is natural to use concepts and techniques from the qualitative theory of DEs (now regarded as part of the theory of dynamical systems, e.g. Anosov and Arnold (1988)) in order to study their evolution. The most comprehensive treatment given to date has appeared in the Russian literature (e.g. Bogoyavlensky and Novikov (1973), and in particular, Bogoyavlensky (1985)). This general approach to the problem has unfortunately not attracted much attention, possibly due to the fact that the variables that are used do not have a direct physical interpretation, and the fact that the description of the asymptotic states near the big-bang is rather complicated. For this reason the present author proposed a formulation (Wainwright (1988)) in which the variables directly describe both the anisotropy in the model and the Bianchi type of its isometry group. This formulation also leads to a simple description of the asymptotic states near the big-bang (Ma (1988)).

In the workshop, K. Rosquist described work in progress with R. Jantzen and C. Uggla, (Rosquist, Uggla and Jantzen (1989)) whose aim is to shed light on the evolution of orthogonal Bianchi models from a different point of view. By using the well-known Misner variables and the so-called Jacobi time gauge, the EFEs for the orthogonal Bianchi models of Hamiltonian type (class A models and class B models with trace $(n) = 0$ in the terminology of Ellis and MacCallum (1969)) were written as the geodesic equations for a conformally flat 3-d Lorentzian metric. This approach suggests the possibility of visualizing the evolution of the models by means of conformal compactification and also leads to a simple description of the known Bianchi exact solutions.

The asymptotic state of Bianchi models of types IX and VIII near the big-bang has attracted considerble attention. In the late 1960's, using heuristic arguments, Belinskii and Khalatnikov (1969) and Misner (1969) independently suggested that in models of type IX, the spacetime geometry has an oscillatory behaviour, and passes through an infinite succession of Kasner-like states as the singularity is approached into the past. Each Kasner solution is characterized by a single real parameter, and it has been shown that the sequence of Kasner states can be described by the iterates of a map of the unit interval onto itself. Furthermore, this map is known to be chaotic (Barrow (1982); Khalatnikov et al. (1985); Ma (1988)). Consequently, it is of interest to calculate the Lyapunov exponents for the system of DEs which govern the evolution of these models, in order to determine whether there is a positive exponent, which is often used as a criterion for chaotic behaviour (e.g. Shimada and Nagashima (1979)). Earlier numerical work by Zardecki (1983) suggested that there was a positive Lyapunov exponent. In the workshop, A. Burd reported on his recent work (Burd, Buric and Ellis (1989)) and on the independent work of (D. Hobill et al. (1989)). It thus appears that although the map which defines the succession of Kasner states is chaotic, the associated system of DEs is not chaotic in the usual sense. The dynamical systems approach will possibly help to clarify this matter.

The author would like to thank the speakers for useful discussions and for providing preprints of their unpublished research.

References

Anosov, D. V. and Arnold, V. I. (1988), *Dynamical Systems I*, (Springer-Verlag), New York).

Bardeen, J. R. (1980), *Phys. Rev.* **D22**, 1882.

Barrow, J. D. (1978), *Nature* **272**, 211.

Barrow, J. D. (1982), *Phys. Rept.* **85**, 1.

Barrow, J. D. (1988), *Q. Journal R. Astr. Soc.* **29**, 101.

Belinskii, V. A. and Khalatnikov, I. M. (1969), *Sov. Phys. JETP* **29**, 911.

Bogoyavlensky, O. I. (1985), Qualitative Theory of Dynamical Systems, in *Astrophysics and Gas Dynamics*, (Springer-Verlag, New York).

Bogoyavlensky, O. I. and Novikov, S. P. (1973), *Sov. Phys. JETP* **37**, 747.

Burd, A., Buric, N. and Ellis, G. F. R. (1989), preprint.

Eardley, D., Liang, E., and Sachs, R. K. (1972), *J. Math. Phys.* **13**, 99-106.

Ellis, G. F. R. and Bruni, M. (1989), preprint.

Ellis, G. F. R. and MacCallum, M. A. H. (1969), *Commun. Math. Phys.* **12**, 108.

Ellis, G. F. R., Nel, S. D., Maartens, R., Stoeger, W. R. and Whitman, A.P. (1965), *Phys. Rept.* **124**, 315.

Francisco, G., and Matsas, G. E. A. (1988), *Gen. Relat. Grav.* **20**, 1047.

Friedrich, H. (1986), *Commun. Math. Phys.* **107**, 587.

Goode, S. W. (1987), *Gen. Relat. Grav.* **19**, 1075.

Goode, S. W., and Wainwright, J. (1985), *Class. Quant. Grav.* **2**, 99-115.

Hawking, S. W. (1966), *Ap. J.* **145**, 544.

Hobill, D., Bernstein, D., Simkins, D. and Welge, M. (1989), *GRG-12 Abstracts.* **B1:16**

Khalatnikov, I. M., Lifshitz, E. M., Khanin, K. M., Shchur, L. N. and Sinai, Ya. G. (1985), *J. Stat. Phys.* **38**, 97.

Lifshitz, E. M. (1946), *J. Phys.* (Moscow) **10**, 116.

Lifshitz, E. M., and Khalatnikov, I. M. (1963), *Adv. Phys.* **12**, 185-249.

Ma, P. K-H. (1988), *A Dynamical Systems Approach to the Oscillatory Singularity in Cosmology*, M. Math. thesis, University of Waterloo, (unpublished).

MacCallum, M. A. H. (1979), in *General Relativity, An Einstein Centenary*, eds. Hawking, S. W. and Israel, W. (Cambridge University Press, Cambridge).

Misner, C. W. (1969), *Phys. Rev. Lett.* **22**, 1071.

Penrose, R. (1979), in *General Relativity, An Einstein Centenary*, eds. Hawking, S. W. and Israel, W. (Cambridge University Press, Cambridge).

Rosquist, K., Uggla, C. and Jantzen, R.T. (1989), preprint.

Shimada, I., and Nagashima, T. (1979), *Prog. Theoret. Phys.* **61**, 1605.

Taub, A. H. (1951), *Ann. Math.* **53**, 472.

Wainwright, J. (1988), in *Relativity Today, Proceedings of the 2nd Hungarian Relativity Workshop 1987*, ed. Z. Perjes (World Scientific, Singapore.)

Wainwright, J. and Hsu, L. (1989), *Class. Quant. Grav.*, to appear.

Zardecki, A. (1983), *Phys. Rev.* **D28**, 1235.

B2

The early universe

Michael S. Turner

NASA/Fermilab Astrophysics Center,
Fermi National Accelerator Laboratory,
Batavia, IL 60510-0500 USA

and

Enrico Fermi Institute,
The University of Chicago,
Chicago, IL 60637-1433 USA

The hot big-bang cosmology is based upon the Friedmann-Robertson-Walker (FRW) solution of general relativity. It is a remarkably successful model, providing a reliable and tested accounting of the history of the universe from about 10^{-2} sec after the bang until today, some 15 Gyr later. It is so successful that it is known as the standard model of cosmology. It accommodates—and in some instances explains—most of the salient features of the observed universe, including the Hubble expansion, the 2.74 K cosmic microwave background radiation (CMBR), the abundance of the light elements D, ^{3}He, ^{4}He, and ^{7}Li, and the existence of structures likes galaxies, clusters of galaxies, etc. In the splendor of its success, it has elevated our thinking about the evolution of the universe, and we have been able to ask a new set of even more profound questions about the universe. These questions include: What is the origin of the baryon number of the universe? What is the origin of the primeval inhomogeneities that gave rise to the structure we see today? Why is the part of the universe we can see (our present Hubble volume) so isotropic and homogeneous—as evidenced by the uniformity of the CMBR temperature—and spatially flat? What is the structure of the universe beyond our Hubble volume? What is the nature of the ubiquitious dark matter? Are there other significant cosmological relics to be discovered? Why is the cosmological constant (equivalently the present vacuum energy density) so small compared to its natural scale: $\Lambda \lesssim 10^{-122} G^{-1}$? What is the origin of the expansion and of space-time itself? Why are there three space-like and one time-like dimensions? How does time emerge from a quantum mechanical treatment of space-time itself? Even more remarkable than the standard model or even our ability to ask this set of questions is the fact that we have been able to begin to address many of these questions.

Recent progress in early universe cosmology has been closely linked to the progress in our understanding of the elementary particles and their fundamental interactions. In the 1960's cosmology ran up against a wall: based upon the exponential increase in the number of hadron states, one could infer that the universe should have a limiting temperature of about 200 MeV (achieved at an age of about 10^{-5} sec). In addition, the known hadrons have finite size and at about this temperature inter-particle spacings would have been comparable to particle sizes, and at earlier times particles would have overlapped. In the 1960's the study of the universe at times earlier than about 10^{-5} sec was clearly nonsensical.

The door to early universe cosmology was opened with the emergence of the realization that the observed hadron states are not the fundamental particles, rather they are comprised of point-like quarks and anti-quarks whose interactions are asymptotically free (i.e., remain perturbatively weak at very high energies and short distances). Today, particle physicists have their own standard model, the $SU(3)_C \otimes SU(2)_L \otimes U(1)_Y$ gauge theory of the strong and electroweak interactions. Their standard model is consistent with a very large body of experimental data that extends to energies as high as 1 TeV (distances as short as 10^{-17} cm). The standard model of particle physics is every bit as successful as the standard model of cosmology. Taking the two standard models together, we can sensibly discuss cosmology back to times as early as about 10^{-12} sec when the temperature was about 300 GeV. At that time the universe was a thermal soup of quarks, leptons, gauge and Higgs bosons.

(There are a couple of pieces in the the standard model still to be filled in: the top quark and the Higgs particle. In that regard, the Higgs sector is crucial, as it provides the mechanism for spontaneous symmetry breaking. Aside from the broken symmetry which we observe with great clarity, there is no evidence for the Higgs mechanism.)

Like its counterpart in cosmology, the standard model of particle physics too allows one to look beyond it and to ask a whole new set of even more profound questions—and even to begin to answer them. These questions include: How are the forces of nature—including gravity—to be unified? How are the quarks and leptons to be unified? Why are there several, if not many, very disparate energy scales in Nature? Why are there three generations of quarks and leptons? (Recent experiments at SLC at SLAC and LEP at CERN exclude the possibility of a fourth generation of quarks and leptons unless the associated neutrino is heavier than about 40 GeV.) Why are there at least four kinds of fields: spin-0, spin-1/2, spin-1, and spin-2? Why is our space-time four dimensional? Why is the vacuum energy density today so very small—less than about 10^{-46} GeV^4?

The mathematical ideas underlying the standard model—the gauge principle and spontaneous symmetry breaking—are so robust that particle physicists have begun to be able to address many of the questions mentioned above. For example, grand unified theories (or GUTs) unify the strong, weak, and electromagnetic interactions, as well as the quarks and leptons. Supersymmetry provides at least a partial explanation for the existence of disparate scales in fundamental physics, and a hint of how the internal symmetries of particle interactions and space-time symmetries might be related. Superstring theories provide the first framework for unifying all the forces and particles of nature in a finite quantum theory. The only rub is that the scale of grand unification–10^{14} GeV or greater–and of string theory–of order the Planck scale (10^{19} GeV)–seem to be hopelessly beyond the capabilities of terrestrial accelerator laboratories.

In the past decade it has become very clear that the exploration of the earliest moments of the universe and the architecture of the fundamental interactions at the shortest distances are inexorably connected: to understand the earliest history of the universe we must understand physics at the very shortest distances (highest energies); conversely, the only "laboratory" in which high enough energies were achieved to study physics at the very shortest distance scales is the early universe itself. Speculations about physics beyond the standard model entail interesting new physics at energy scales greater than about 1 TeV—and in some theories energies less than 1 TeV.

Essentially all unified gauge theories undergo spontaneous symmetry breaking (SSB)—including the standard model. At high temperatures spontaneously broken symmetries are restored, and so one expects a number of cosmological phase transitions associated with SSB. Almost all grand unified theories predict the existence of interactions that do not conserve baryon number and the existence of topological solitons associated with SSB—domain walls, (cosmic) strings, and (magnetic) monopoles. In addition, in some unified gauge theories nontopological solitons can exist and are stablized by dynamics rather than by topology. Many extensions of the standard model predict masses for neutrinos and the existence of new particle states, such as the supersymmetric partners of ordinary particles in supersymmetric theories, the axion, and additional gauge and Higgs particles. Some of these new particles are expected to be stable and should be present today as cosmological relics. Perhaps the most startling prediction of new physics is that there might be additional space-like dimensions that are compactified and very small ($\lesssim 10^{-17}$ cm) today, but could have been "as large as" the familiar three spatial dimensions at very early times. (In superstring theories six additional space-like dimensions may be expected.)

Already many of the theoretical speculations about physics at energies greater than 1 TeV have been applied to the early universe, with the hope of answering some of the cosmological questions posed above and at the same time testing theories that predict them. While we are far from sorting out the earliest history of the universe it has become clear that the answers to the very fundamental questions that we have asked must involve events that took place during the first 10^{-2} sec. Already a number of attractive and very plausible early universe scenarios have been developed. For example, non-equilibrium interactions that violate B and C, CP can explain the origin of the baryon asymmetry of the universe (*Baryogenesis*). The relic abundance of a number of stable, hypothetical particle species is sufficient to account for closure density and thereby explain the nature of the ubiquitous dark matter. Two of the most promising candidates are a 10^{-5} eV axion or a 10-1000 GeV neutralino (sometimes referred to as the photino).

Cosmic inflation provides a very attractive framework for understanding the origin of the large-scale smoothness and flatness of the universe, as well as the origin of the primeval inhomogeneities needed to initiate structure formation. In a phrase, inflation is attractive because it renders the present state of the universe relatively insensitive to the initial state. It could be that inflation may also shed light on the question of why three of the spatial dimensions are large and the others are small (if there are additional spatial dimensions) and perhaps even on the origin of the expansion itself. Together, the relic particle dark matter hypothesis and inflation-produced density inhomogeneities provide the necessary initial data for the structure formation problem, and have led to the most successful and most well motivated paradigm for structure formation: cold dark matter.

A generic prediction of theories that go beyond the standard model is that the universe should have undergone a number of phase transitions during its early history. If it did, then there may be fossil evidence of those transitions—including topological solitons such as monopoles, domain walls, string, and texture, or nontopological such as soliton stars or Q balls. Such relics could provide the primeval inhomogeneity necessary to trigger structure formation, as in the case of cosmic string, domain walls, or texture, or could be the dark matter, as in the case of soliton stars, monopoles, or Q balls. A fossil relic of an early universe phase transition would have tremendous significance for both cosmology and particle physics, even if it did not provide the closure density or the seeds for structure formation.

So much for the past, now for the future. Thanks to a decade or so of fruitful theoretical speculation about the earliest history of the universe we have a number of ideas worthy of being put to the ultimate test: the scrutiny of experiment. Those ideas include the particle-relic dark-matter hypothesis, inflation, cold dark matter as the paradigm for structure formation, and "significant other" relics (a significant other relic is a species whose relic abundance is great enough to detect, but too small to provide closure density). While the testing of early universe cosmology is not an easy task, it is a very important one: without some experimental guidance to narrow and to guide the direction of our theoretical speculations progress in early universe cosmology will not continue.

The majority of contributed papers in Session B2 involved two very interesting and timely early universe topics: inflation and cosmological phase transitions. Inflation has revolutionized the way cosmologists think about the early universe. Its only shortcoming is the lack of *a* compelling model that implements inflation. While numerous models have been constructed to successfully implement inflation, all have a common drawback: the necessity of a very small (or in some cases very large) dimensionless coupling. This has made it very difficult to incorporate inflation into a unified theory of the fundamental interactions in a simple way. A host of different inflationary models were explored in the contributed papers. Phase transitions are intrinsically interesting, and cosmological phase transitions even more so! Especially interesting is the possibility that a relic of some early phase transition was left behind for us to discover. Various aspects of cosmological phase transitions were

addressed in the contributed papers. The papers presented in the Oral
Sessions for Workshop B2 are listed in Table 1 below.

Table 1.

Schedule of oral presentations for workshop B2.

Part I: Inflation

K. I. Maeda, Japan *Cosmology in Generalized Einstein Gravity Theory*

I. M. Khalatnikov, USSR *On the Degree of Generality of inflation in Friedmann Cosmological Models with a Massive Scalar Field*

J. R. Yoon and D. R. Brill, USA *Inflation from Extra Dimensions*

M. T. Ressell and M. S. Turner, USA *Relic Gravitons from Inflation*

L. Amendola and F. Occhionero, Italy *Constraints to the Starobinsky Model from the Quadrupole Moment of the Microwave Background*

Part II: Cosmological Phase Transitions

M. Gleiser, USA *Cosmology of Biased Discrete Symmetry Breaking*

H. Kurki-Suonio, USA *The Quark-Hadron Transition and Nucleosynthesis in the Early Universe*

W. A. Hiscock, P. Anderson and R. Holman, USA *Formation of Domain Walls in the Early Universe*

B3

Relativistic astrophysics

Marek A. Abramowicz*

*Scuola Internazionale Superiore di Studi Avanzati,
Strada Costiera, 11,
34014 Trieste, Italia*

I shall describe here a few subjects which in my opinion were the most interesting among those presented orally or at the poster session during the Workshop.

1) The (still hypothetical) discovery of a half-millisecond pulsar in the Supernova SN 1987A attracted a lot of attention. It could drastically change our understanding of neutron star physics and in particular our understanding of the equation of state at nuclear densities. In this context models of compact stars involving strange (e.g. bosonic) matter are interesting and important.

2) The classical problem of test particle motion in a given gravitational field experienced a surprising new development: it was claimed that the centrifugal force can be attractive to the axis of rotation and some repulsive phenomena may be connected with gravity!

3) Gravitational radiation found a new astrophysical application: it was suggested that energy and angular momentum losses due to gravitational waves can be equivalent to viscous stresses in thick accretion disks around supermassive black holes. This may be relevant for quasars.

The optical variability with frequency 1968.629 Hz, discovered recently by Middleditch et al. (1989) in the supernova SN 1987A, is now generally interpreted as due to rotation of a neutron star. An alternative

---

*Also: International Centre for Theoretical Physics, Trieste, Italy

possibility, that the reported frequency represents a radial oscillation of the neutron star was discussed at the Workshop by J.R. Ipser and L. Lindblom. They made significant progress in the numerical treatment of the normal-mode pulsation equations by re-expressing these equations in terms of a single potential function. It should be stressed, however, that it is still unknown how the oscillations of a neutron star translate into the optical pulses.

The most important and interesting implication of the discovery of this sub-millisecond pulsar (discussed at the Workshop by Lindblom and during a plenary session by J. L. Friedman–see Chapter 2, this volume) is that the fast rotation of the SN 1987A pulsar gives severe constraints on the equation of state at high, nuclear, densities ($\rho > 10^{14}$ g/cm^3). This is because the upper limit set by gravity on the rotation of neutron stars is very sensitive to the equation of state. For a given baryon mass M the maximal possible angular velocity Ω of a rigidly rotating star equals to the Keplerian frequency at the equatorial radius R of the star. The Keplerian frequency is defined as the orbital frequency of a free test particle (moving along a circle) and equals

$$\Omega = \sqrt{\frac{GM}{R^3}}. \tag{1}$$

Since $\Omega = 1.237 \times 10^4$ sec^{-1} for the pulsar in SN 1987A, the mass and radius should satisfy

$$M > 1.15 M_\odot \left(\frac{R}{10\,\text{km}}\right)^3. \tag{2}$$

Neutron star models constructed from equations of state that are *stiff* above nuclear density have substantially larger radii than models constructed from *soft* equations of state. Thus, the limiting frequency decreases (for a given mass) with increasing stiffness of the equation of state.

This is illustrated in Figure 1, where the lines labelled MF (mean field), TI (tensor interaction) BJ (Bethe and Johnson), R (Reid potential) and π (Reid potential with pion condensation) represent computed *non-rotating* neutron star models with increasing softness of equation of state. Too soft equations of state are not consistent with the $1.44 M_\odot$ mass of the non-rotating neutron star observed in the binary pulsar 1913+16. This is indicated in Figure 1 by a horizontal line labelled "$1.44 M_\odot$."

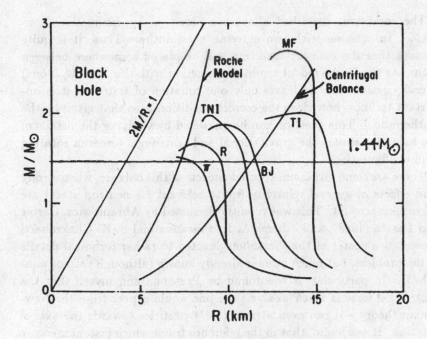

Fig. 1 Constraints for equation of state from the pulsar in SN 1987A (from Nakamura (1989)).

The line labelled "Centrifugal Balance" corresponds to the condition (2) with models which are allowed located to the left of the line. One may expect (from Newtonian theory) that close to this line the models become very oblate due to the strong effect of the centrifugal force. Therefore, the spherical models, assumed in the derivation of the condition (1), are not adequate. One may approximate the shape of the configuration in this case by the relativistic Roche model (Shapiro and Teukolsky (1983)) which assumes that the equipressure surfaces (including the surface of the star) coincide with equipotential surfaces for rigidly rotating test fluid in the Schwarzschild geometry. From this model one finds (Nakamura (1989)) that for the pulsar in SN 1987A it must be

$$M > 3.88 M_\odot \left(\frac{R}{10\,\mathrm{km}} \right)^3. \qquad (3)$$

This condition is marked in Fig. 1 by the line labelled "Roche Model."

The conditions described above are based on approximations connected, in a sense, with two extreme possibilities. Thus, it is quite obvious that the realistic condition will be placed somewhere between them. As the Roche model condition (together with the "$1.44M_\odot$" one) is really restrictive – it leaves only one equation of state ! – it is important to know how close the correct condition is to that given by the Roche model. This question can be answered by studying the details of the balance between the gravitational and centrifugal forces in rotating fluid configurations.

There are some rather unexpected aspects of this balance, when strong field effects of general relativity (quite relevant for neutron stars) are taken into account. This was recently discussed by Abramowicz, Carter and Lasota (1988), and others. A. R. Prasanna and S. K. Chakrabarti presented a poster at the Workshop devoted to rather technical details of the problem, but much more is already known (although yet not published). In particular, it was found by Prasanna and myself that the centrifugal force is *much weaker* than one would expect from the Newtonian theory – it can even attract (*sic!*) particles towards the axis of rotation. It was found, that in the reference frame which instantaneously corotates with a particle, the correct relativistic expression for the centrifugal force **C** is identical with the one wellknown in the Newtonian theory:

$$\mathbf{C} = m\frac{v^2}{\mathcal{R}}\boldsymbol{\lambda}.\tag{4}$$

Here $\boldsymbol{\lambda}$ is the first normal to the particle's trajectory in 3-space defined by projection with respect to the timelike Killing vector,

$$\Omega = \left(\frac{v}{\mathcal{R}}\right)\boldsymbol{\Lambda},\tag{5}$$

is the angular velocity of the instantaneously corotating frame measured with respect to the global rest frame defined by the congruence of trajectories of the timelike Killing vector ($\boldsymbol{\Lambda}$ is the second normal to the particle's spatial trajectory) and **r** is the distance from the instantaneous axis of rotation:

$$\mathbf{r} = \mathcal{R}\boldsymbol{\lambda}.\tag{6}$$

The particle's speed v is measured (along the trajectory) with respect to the global rest frame and the quantity $\mathcal{R}$ is the scalar of curvature of the spatial trajectory of the particle. In the case of circular motion around the Schwarzschild black hole (with the mass M) one has

$$\mathcal{R} = \frac{r^2}{r - 3M}, \quad \lambda_{(r)} = \frac{r - 3M}{|r - 3M|}. \qquad (7)$$

Here r is the proper circumferential radius of the circular orbit, equal to the Schwarzschild radial coordinate, and $\lambda_{(r)}$ is the radial component of the vector normal to the circle. It is clear that the centrifugal force is weaker than it would be according to the Newtonian theory, in which always $\mathcal{R} \equiv r$ and $\lambda_{(R)} \equiv 1$.

One consequence of the weakening of the centrifugal force (recently found by J. C. Miller and myself) is that rapidly rotating compact objects are much less oblate (and thus probably more stable) than the Newtonian theory predicts. This is shown in Figure 2, where the ellipticity of slowly rotating MacLaurin spheroids with fixed total mass and total angular momentum is plotted as a function of the mean radius of the spheroid. Newtonian theory predicts that the ellipticity monotonically increases with decreasing radius, going to infinity when the radius goes to zero (such a limiting case has a shape of an infinitesimally thin rotating disk). However, when the relativistic weakening of the centrifugal force is properly taken into account, one observes that the ellipticity behaves as the Newtonian theory predicts only when the radius is much greater than the gravitational radius: contrary to this theory, the ellipticity reaches a maximum for the value of the mean radius being a few gravitational radii and *decreases* with decreasing radius for still smaller radii. In the range of radii relevant for neutron stars the differences between predictions of the Newtonian theory and general relativity are about 50%.

This obviously is very important for the problem of how fast pulsars can spin, especially because the only detailed stability analysis available to date (Lindblom (1986)) was based in part on Newtonian considerations. According to this analysis the shortest possible period of rotation is 0.6 millisecond.

Some of the jets emerging from quasars and other active galactic nuclei move with super-relativistic velocities (Lorentz factors reaching $\gamma \sim 20$ have been reported). Observations point to a very efficient acceleration mechanism most probably of an electromagnetic nature—e.g. a unipolar induction (Blandford and Znajek (1977); more recently Michel (1987)). The jets are very well collimated. Both the relativistic range of the velocities of jets and their collimation are hard to explain. Some fundamental properties of the test particle motion near the Kerr black hole

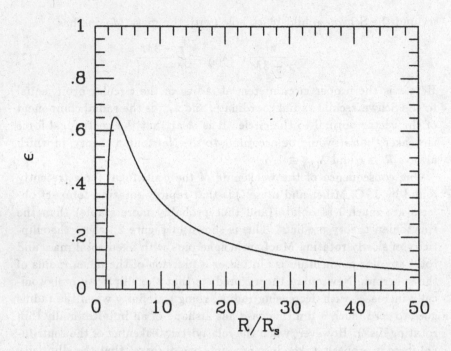

Fig. 2   Ellipticity behaviour for MacLaurin spheroids. R is the mean radius, ϵ the ellipticity and R_S the gravitational radius.

which could be relevant in the context of jets were discussed at the Workshop by J. Bičák who present some new results obtained by him in collaboration with V. Balek, P. Hadrava, O. Semerák, and Z. Stuchlik.

Probably the most interesting part of this presentation was devoted to a possible focusing mechanism. Particles, which are shot out from the vicinity of the hole (or naked singularity) with initial relativistic velocities will be collimated, to some degree, by the gravitational field. No *clear* analytic analysis of this effect can be offered at the moment. Extensive numerical work shows that the collimation effect may be very strong in the case of a naked Kerr singularity but it is always very small for the black holes ($a < m$).

Bičák also discussed the motion of charged test particles in the Kerr-Newman geometry along the axis of symmetry, in the equatorial plane and in the case of a spherical shell free-falling from a large distance. Particle orbits with closed loops for *zero-angular momentum case* were found.

Such a paradoxical situation is possible because of the combined effects of gravitational dragging and the Lorentz force.

M. Gleiser presented very interesting results of his work on boson stars. They are gravitationally bounded states of complex scalar fields. He pointed out, in particular, that the boson stars can be efficient emitters of gravitational radiation both from binary systems of such stars and from excitations of higher modes of the scalar fields. In the latter case, boson stars emit gravitons when they relax to the stable ground state.

Perhaps the most controversial new idea was presented by W.H. Zurek and K. Górski. They suggested that thick accretion disks can lose their angular momentum to gravitational radiation by a rather complicated mechanism which involves the widely discussed Papaloizou-Pringle instability (e.g. Narayan and Goodman (1989)). This instability redistributes angular momentum of (not very big) thick accretion disks on a dynamical time scale and leads to the formation of non-axisymmetric bulges. Zurek and Górski argue that a thick accretion disk has, due to the action of this instability, a large portion of its mass tied up in a single bulge which orbits the central black hole on a very relativistic orbit. They calculated the gravitational radiation emitted by the binary system of the bulge and the hole and concluded that the loss of angular momentum is sufficient to power a typical active galactic nuclei.

There are some points in this very interesting idea which should be examined in a more quantitative way. In particular, the basic assumption that the Papaloizou-Pringle instability always leads to formation of very massive bulges is far from being firmly established. It is more likely that even a small amount of accretion will stabilize the Papaloizou-Pringle instability because accretion advects perturbations inside the sound horizon. This clear idea was fully proved analytically in the case of spherical accretion by Moncrief (1980). The only example of stabilizing effects of accretion in the case of a thick disk was calculated by Blaes (1987).

An even stronger stabilization mechanism is possible. All the detailed studies, both analytic and numerical, of the Papaloizou-Pringle instability show the same general trend: after some critical value of the ratio

$$Q = \frac{\text{(outer radius of the torus)}}{\text{(inner radius of the torus)}} \qquad (8)$$

is reached, increasing Q increases stability (the time scale for growing unstable modes gets longer). For very simple, and therefore probably not very realistic, models there is another critical value of $Q = Q^* < 10^3$ such that for $Q > Q^*$ the models are unconditionally stable. The real torii in astronomical objects have $Q > 10^3$, so this result may indicate that they are Papaloizou-Pringle stable. However, this has not yet been confirmed by numerical analysis of realistic models of torii. For purely numerical reasons it is very difficult to perform such an analysis for models with $Q > 10^3$. Before this is done any prediction of what Papaloizou-Pringle instability does for real astrophysical objects should be treated, in my opinion, with extreme reservation.

Another important, but unrelated, question is how the (hypothetical) loss of angular momentum due to *gravitational* radiation may translate into energy release from the system in the form of *electromagnetic* radiation. This question has not been addressed by Zurek and Górski in their presentation. They have suggested instead that there is a "natural timescale" connected with the loss of angular momentum (which is certainly correct) and when one divides the mass of the bulge by this timescale to get an accretion rate, one may then multiply the accretion rate obtained this way by the efficiency of accretion to get the energy output. This suggestion, in my opinion, does not explain why there should be *any* energy output in the form of electromagnetic radiation. If the angular momentum loss due to gravitational radiation is $\dot{J}$ and energy loss due to the same process is $\dot{E}$ then, for a bulge with angular velocity Ω, one would expect

$$\dot{E} = \Omega \dot{J}. \tag{9}$$

However, exactly the same relation holds for the changes of energy and angular momentum on a nearly Keplerian orbit. If the bulge rotates with nearly Keplerian velocity, there is no energy left which could be possibly radiated in the form of electromagnetic radiation. Although it is true that the orbital speed of the bulge may be far from Keplerian, it is not clear whether this helps to radiate the huge amount of energy in the form of electromagnetic radiation as suggested by Zurek and Górski.

References

Abramowicz, M.A., Carter B. and Lasota, J.-P. (1988), *Gen. Relat. Grav.* **20**, 1173.

Blaes, O. M. (1987), *Mon. Not. Roy. Astr. Soc.* **227**, 975.

Blandford, R. D. and Znajek, R. L. (1977), *Mon. Not. Roy. Astr. Soc.* **179**, 433.

Lindblom, L. (1986), *Ap. J.* **303**, 146.

Michel, F. C. (1987), *Ap. J.* **321**, 714.

Moncrief, V. (1980), *Ap. J.* **235**, 1038.
Middleditch, J., et al. (1989), *IAU Circ.* **4735**.
Nakamura, T. (1989), *Progr. Theor. Phys.* **81**, 1006.
Narayan, R. and Goodman, J. (1989), in *Theory of Accretion Disks*, eds. F. Meyer, W. J. Duschl, J. Frank and E. Meyer-Hofmeister (Kluwer Academic Publishers, Dordrecht).
Shapiro, S. L. and Teukolsky, S. A. (1983), *Ap. J.* **272**, 702.

B4

Astrophysical and
observational cosmology

B. J. Carr

Astronomy Unit, School of Mathematical Sciences,
Queen Mary and Westfield College,
Mile End Road, London E1 4NS, U.K.

Twenty six abstracts were submitted for this workshop, seven of which were selected for oral presentation. The main topics covered were gravitational lensing (9 abstracts), large-scale structure (6 abstracts) and cosmic strings (4 abstracts), all of these topics being represented in the talks. These are not the only areas which have seen important advances recently but they are perhaps the most interesting ones from the perspective of a relativist. In this report I will summarize the contents of the talks, referring to some of the posters where appropriate. Whenever distance scales arise, the Hubble parameter is assumed to be 50 km/s/Mpc.

1. Gravitational lensing

Blandford's plenary contribution (Chapter 5, this volume) illustrates the increasing usefulness of gravitational lensing as a cosmological tool in recent years and this is reflected in the large number of abstracts on the topic. Besides confirming light-bending itself, gravitational lensing can provide evidence of the existence and distribution of dark matter in galaxies, identify the presence of objects on scales from jupiters to supermassive black holes, probe features of large-scale structure, and perhaps even measure the Hubble constant. The posters of Ho Tenlin and Yakimov illustrated how particular instances of gravitational lensing can provide cosmological information. The talks focussed on more general mathematical issues.

The interpretation of observations of distant objects is complicated by the fact that a beam of light may suffer many weak gravitational encounters rather than a single strong one as it propagates through an inhomogeneous background. Thus the Universe itself acts a lens and it is crucial to understand this effect if it is to be disentangled from effects local to the object. Dyer and Harper reported how they have tackled this problem by integrating the optical scalar equations through a Friedmann background containing Swiss Cheese clumps. The optical scalar equations describe the propagation of null geodesics when there are no sources or sinks; the model is characterized by the background Friedmann parameters, the fraction of matter in clumps, the mass spectrum of the clumps and their density profile. Previous work has used statistical versions of the equations (replacing all quantities by their average values) but in this work they integrate the full equations. The most important effect is the amplification and de-amplification of beams relative to the homogeneous case. This is expected because there is less matter in the average beam in the inhomogeneous case, so typical beams are focussed less rapidly and hence de-amplified. Flux conservation then implies that some beams must be amplified. At a redshift of 5, a few percent of beams are highly amplified and this implies that magnitude-limited surveys reach further than expected. A related effect was discussed in the posters of Blanchard and Dyer, and Watanabe.

Gravitational lensing is usually studied in the geometric optics approximation but the picture becomes more complicated if the wavelength exceeds the size of the deflecting object. This might apply, for example, if the lens were a white dwarf, a neutron star or a black hole. As discussed in the talk by Halpern, the problem must then be treated with the full field equations because spin effects associated with polarized waves cannot be described by a scalar wave equation. however, the wave picture is still useful in the Kerr case because the electromagnetic field equations are separable into ordinary differential equations for the radial and angular dependences. The treatment of wave scattering is then as simple as the geometric optics case. If the mass parameter is small, as it would be for a neutron star, the radial equation resembles that for Coulomb scattering. In the black hole case, only approximations can be obtained. one finds that a circularly polarized photon experiences an azimuthal deflection proportional to the ratio of wavelength to deflector size, so the effect of spin becomes important only in the long wavelength limit and it vanishes in the geometric optics approximation. Halpern suggested the usefulness of this effect in discriminating between different types of lenses.

A related point arose in the talk of Ishihara, which described how the polarization vector of a linearly polarized electromagnetic wave propagates in the presence of a rotating gravitational lens. The polarization vector is parallel-propagated along a null geodesic and this leads to a change in its direction. There is also a rotation of the polarization vector about the propagation direction, an effect analogous to Faraday rotation in a magnetized plasma. A linear analysis suggests that this occurs only in the presence of rotating matter (e.g., an accretion disk) but Ishihara showed that it also occurs for a black hole. He finds that the rotation of the polarization plane is proportional to the mass of the hole and the line-of-sight component of its angular momentum. Although the effect is small quantitatively, it is qualitatively important because it gives information about the lens as well as the source.

2.   Large-scale structure

One of the most important observational issues in the last few years has concerned the existence of large-scale streaming motions. The microwave background dipole anisotropy indicates that the Local Group of galaxies has a peculiar velocity of about 600 km/s in a direction 45° away from Virgo. Part of this must come from our infall towards Virgo, which is about 250 km/s, while the rest used to be attributed to Virgo itself having a peculiar velocity towards Hydra-Centaurus, which is at a distance of 30 Mpc. Recently, however, Dressler et al. have concluded from a survey of several hundred elliptical galaxies that our entire neighbourhood within 100 Mpc has a peculiar velocity of 700 km/s. It should be stressed that it is the scale of the velocity field rather than the velocity itself which is surprising. In order to explain it, one may need to invoke a Great Attractor, i.e., a huge concentration of superclusters at a distance of nearly 200 Mpc, although this proposal is highly controversial.

These considerations prompted Lewis to discuss whether such large-scale peculiar velocities could arise from vector perturbations in the early universe. The treatment of vector perturbations is rather sparse in the literature. In the standard Friedmann model it is well-known that peculiar velocities are constant in the radiation era and decay inversely as the scale factor in the matter era but the tricky issue is to know what happens during inflation. Lewis tackles this problem using the variational methods of Brandenberger et al. This is ideal for Lagrangian-based matter formulations and gives a simpler approach than the standard gauge-invariant formalism of Bardeen. It involves varying the Lagrangian with respect to a restricted set of metric coefficients and leads to an evolution equation for a first order vector perturbation. She first uses a synchronous gauge and then a parametric gauge. The linearity of the formalism then allows her to add the equations to give a gauge-invariant result expressed in terms of a metric vector potential. Lewis concludes that the usual scalar-driven inflation cannot give rise to the required peculiar velocities, although vector-driven inflation might do so.

Another important observational development has been the discovery that most galaxies are distributed on walls or filaments with large voids in between, the scale of these structures extending up to 100 Mpc. Some people have argued that the voids are like bubbles with a characteristic radius of 20 mpc. It used to be claimed that this was evidence for the explosion scenario, although the numerical simulations now suggest that it can arise in more general scenarios. There was little discussion of the observational issues themselves but several papers focussed on the evolution of the fluctuations required to explain the observed large-scale structure. For example, Sugiyama's poster used a gauge-invariant formalism to treat the growth of perturbations in the presence of non-baryonic dark matter, while Melott and Shandarin's dealt with numerical simulations of the formation of structure inside pancakes.

One interesting attempt to explain the origin of cosmological fluctuations came in a talk by Zimdahl, who presented a general relativistic theory of hydrodynamic fluctuations. He showed how random fluctuations in the energy-momentum tensor associated with shear viscosity, bulk viscosity and heat conductivity give fluctuations in the Einstein tensor. This generalizes the treatment of Landau and Lifshitz (which does not include gravitational variables) and leads to a general relativistic fluctuation-dissipation theorem. He then applies this to predicting fluctuation spectra. The background is taken to be flat with bulk viscosity and a gauge-invariant formalism for adiabatic hydrodynamic and gravitational perturbations is used. Zimdahl calculates the spectrum of density perturbation on scales larger than the particle horizon when the background equation of state is $p = \rho/3$. For shear viscosity, he finds a k^4 spectrum (as required by causality constraints); for bulk viscosity he gets a k-independent spectrum, a rather puzzling result since such a spectrum is acausal. The amplitude of the fluctuation is the ratio of the relevant mean-free-path to the horizon size divided by number of particles. For galaxy formation one needs fluctuations of order 10^{-4} at horizon epoch and one would only need a small amount of bulk viscosity in the early Universe to explain this.

3. Cosmic strings

One proposal for the origin of large-scale structure is to invoke cosmic strings. The advantage of this proposal is that closed strings can act as seeds for galaxies, so that one does not need to invoke primordial fluctuations. However, the details of this scenario depend upon whether the background dark matter is hot (like neutrinos) or cold (like photinos or axions). In the hot case, any initial galactic-scale fluctuations are erased so that - without strings - galaxies could only form by fragmentation of much larger objects and they would then do so too late. Strings may therefore be required in the hot picture and, although not essential, they could also have important consequences in the cold picture.

In order to assess the relative attractions of the hot and cold scenarios, Perivolaropoulos, Brandenberger and Stebbins reported their studies of the effects of infinite strings. If such strings move at relativistic speeds, they produce wakes (i.e., planar overdensities) with a characteristic length and width of 40 Mpc (the horizon scale at the time when the matter and radiation densities are equal) and a thickness of about 2 Mpc. With cold dark matter, most of the mass in the Universe accretes onto loops and this does not give the low density of galaxies between wakes required by observation. They therefore prefer hot dark matter since loop accretion is then suppressed by free-streaming and the wakes become more important. For the favored parameters, the ratio of wake density to loop density is about 1/4 and a large fraction of galaxies should be concentrated in sheets. The wakes form by today, providing the string parameter $G\mu$ exceeds 5×10^{-7}, which is consistent with the microwave background anisotropy limits. another consequence of straight strings was considered in the poster of Tartaglia, who argued that their effects on the geometry of spacetime could explain the apparent fractal distribution of galaxies.

One development which was not covered in the session was the possible role of superconducting strings. These have recently been invoked as the most plausible source of energy if one wishes to explain large-scale structure through explosions. However, a poster by Helliwell and Konkowski did discuss the effects of such strings on the surrounding geometry. The spacetime is no longer conical and the deficit angle is increased over the zero-current case.

A very different sort of string calculation was described in the talk by Barrabes. He has studied systems of strings by treating them as a fluid and using relativistic kinetic theory. To do this, he introduces a distribution function which specifies the number of strings which intersect a given 2-surface. Previous studies have focussed on thermal equilibrium states in which the distribution function is isotropic and the entropy production vanishes within the 2-surface. In the case, one has a perfect fluid with equation of state $p = \rho/3$ for relativistic strings and $p = -\rho/3$ for nearly static strings (because of string tension). In his most recent work, Barrabes drops the isotropy assumption (so that one no longer has a perfect fluid) but still assumes that there is no entropy production. He focusses on axially symmetric string distributions in a homogeneous background. In particular, he studies an anisotropic Bianchi I model in which the medium consists of a non-interacting mixture of particles and strings, with a string to particle density ratio of 10^{-3}. He finds that, for nearly static strings, the evolution is insensitive to anisotropy and the Universe soon becomes string-dominated. For relativistic strings, the density ratio remains 10^{-3} and the anisotropy disappears rapidly.

4. Microwave background radiation

This topic received rather scant attention in the session. However, there have been several important developments under this heading, so I will include some account of them here for completeness. Firstly, although there is now very good evidence for the black-body nature of the background (the latest CN observations establish the best fit temperature to be 2.796 K at 2.64 mm), a mysterious distortion has been claimed shortward of the peak following a Nagoya-Berkeley rocket experiment. This distortion peaks at 700 microns and corresponds to a submillimetre excess containing about 25% of the energy density in the microwave background itself. If this excess is confirmed, it will be of tremendous cosmological significance. It would probably have to represent either a Compton distortion (due perhaps to a hot intergalactic medium or high energy particles) or some source of pregalactic radiation which has reprocessed by dust. The only reference to this important result in the session was a rather exotic proposal in the poster by Tatewake and Hatsuda which invoked wormholes.

Another important development is the possible detection of small-scale temperature anisotropies. Davies et al. claim a positive detection of 4×10^{-5} at $8°$, although this remains to be confirmed. In any case, the upper limits are now very tight on all scales down to $10''$, as reviewed in Panek's plenary talk (Chapter 8, this volume). This places severe constraints on cosmological models for the density fluctuations. Another interesting development is the controversy over how the amplitude of microwave fluctuations is affected by gravitational lensing. According to the poster of Sasaki, the fluctuations are enhanced on arcsec scales but reduced by a small factor on arcmin scales.

Part C.

Experimental gravitation and gravitational wave detection

10

Experimental tests of the universality of free fall and of the inverse square law

E. G. Adelberger

Department of Physics FM-15,
University of Washington,
Seattle, WA 98195 USA

Recent suggestions of a "fifth force" have stimulated many experiments to search for new macroscopic interactions arising from the exchange of ultra-low mass fundamental bosons. The experiments fall into two categories: searches for violation of the inverse square law, or of the universality of free fall. The principles of both classes of experiments are described and their results are summarized. Because some groups claim positive effects considerably larger than the upper limits established by others, subtle systematic errors that could masquerade as a "fifth force" are briefly discussed. I conclude that there is, at present, no credible evidence for new macroscopic interactions.

1. Introduction

One feature common to essentially all extensions of the standard model is the prediction of additional fundamental scalar or vector bosons. While these particles are ordinarily expected to be very massive ($m_b c^2 > 10^{15}$ eV), the possibility that some have them have such a low mass, $m_b c^2 < 10^{-3}$ eV, that they produce macroscopic forces between unpolarized test bodies, has been considered in a variety of contexts.[2-13] For example, such speculations have been inspired by Kaluza-Klein theories, quantum gravity ideas, scale invariance, CP-violating pseudo-Goldstone bosons, etc. Some of these would have profound astrophysical consequences: ultra-low mass bosons have been invoked to explain the "vanishing" of the cosmological constant,[8] the anomalous rotation curves of galaxies,[12]

273

and the observed "clumpiness" of the universe.[13] One of the most intriguing of these speculations is the suggestion, originally due to Scherk[3] and elaborated upon by Goldman, Hughes and Nieto,[7] that quantum theories of gravity naturally contain scalar and vector partners of the usual tensor graviton that have escaped detection because the gravi-scalar and gravi-vector interations essentially cancel between bodies composed of ordinary matter.

How can one gather experimental information about such bosons? Fortunately the bosons can be detected indirectly via the macroscopic interactions they produce. For example, a vector boson would lead to a macroscopic potential energy between two test bodies which an be written as

$$V_{12}(r) = \alpha_5 (q_5/\mu)_1 (q_5/\mu)_2 e^{-r/\lambda} G \frac{m_1 m_2}{r} ,$$

where α_5 is the square of a dimensionless coupling constant, $\lambda = \hbar/m_b c$ is the interaction range, and G is Newton's constant. The bosons are assumed to couple to a "charge" q_5 which in general will be some linear combination of the apparently conserved quantities B and L (the baryon and lepton numbers). The quantity q_5/μ is the charge per atomic mass unit. The new force, of course, must be distinguished from the (presumably) much stronger gravitational background. This can be done in two ways: by searching for violations of the universality of free fall (weak equivalence principle) or else of the inverse square law (gravitational Gauss' law). I will discuss both of these approaches. Prior to 1986, most of us thought that the very precise equivalence principle tests from Princeton[14] and Moscow[15] did not leave much room for new long-range forces. Furthermore, laboratory and astronomical tests of the $1/r^2$ law showed no evidence for anomalies. There was one hint of something new; Stacey et al.,[16] from a study of the variation of g with depth in mines, claimed evidence for a repulsive interaction $\approx 1\%$ the strength of gravity. However, geophysical data tends to be viewed with suspicion by physicists, especially when it challenges fundamental physical principles.

In 1986 Fischbach et al.[5] made the startling suggestion that experimental evidence for a new macroscopic interaction (a "fifth force") was already at hand. Fischbach et al. reanalyzed the data from the classic von Eötvös experiment, uncovered a striking correlation between the differential acceleration of test body pairs and their baryon/mass ratio (see Figure 1), and suggested a unified explanation of this effect and of Stacey et al.'s claimed anomaly.

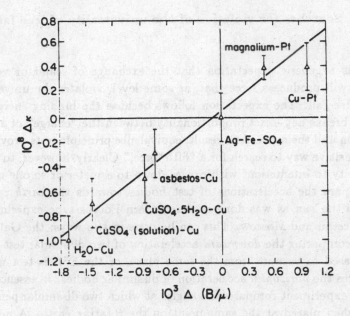

Fig. 1. Fischbach et al.'s reanalysis of the von Eötvös data. The horizontal axis is the difference in B/μ values of a given test body pair, the vertical axis is the corresponding fractional difference in the horizontal accelerations of the test bodies.

Fischbach et al. argued that both results were due to exchange of a new vector boson, which coupled to B, and had a mass in the range 10^{-8} to 10^{-10} eV, producing a force with a range λ between 10 and 1000 m. The range, λ, of this "new force" thus fell in the "geophysical window," $10\,\text{m} < \lambda < 10^4\,\text{m}$, where existing constraints on exotic macroscopic interactions were quite weak. (Ultra-precise satellite ranging data do not provide very stringent constraints on violation of the $1/r^2$ law for ranges much shorter than the radius of satellite orbits, while laboratory $1/r^2$ experiments provide weak constraints for ranges much greater than the $\sim$cm separations of the Cavendish type apparatus.) In this talk I will summarize what has been learned experimentally in the three years since Fischbach et al. launched a renaissance in experimental gravitation by proposing the existence of a new macroscopic interaction with $\alpha_5 \approx 0.01$, $q_5 = B$, and $\lambda \approx 10$ to 1000 m.

2.   Searches for violation of the universality of free fall

There is a generic expectation that the exchange of scalar or vector bosons will produce a force that, at some level, violates the universality of free fall. The expectation follows because the binding energy of matter breaks any exact proportionality between the "charge" of an interaction and the mass. As a result, equivalence principle tests provide a very sensitive way to search for a "fifth force." Clearly, however, to have sensitivity to interactions with ranges down to a meter or so one needs to compare the accelerations of test bodies towards the *earth* rather than to the *sun* as was done in the modern Eötvös type experiments at Princeton and Moscow. This can be done in two ways: the Galilean test of comparing the *downward* acceleration of two dissimilar test bodies released in vacuum from the same place, or the Eötvös test which compares the *horizontal* acceleration of dissimilar bodies. In essence, an Eötvös experiment compares the angles at which two dissimilar pendula hang when placed at the same point on the rotating earth. A pendulum at 45° latitude in the Northern hemisphere of a spherical earth will be deflected by ≈ 1.7 mrad toward the south by inertial effects. If the magnitude of the total external force on two different pendula is not exactly proportional to their mass, this small angle will be slightly different for the two pendula; this can readily be detected by a sensitive torsion balance.

For an Eötvös-type experiment to have good sensitivity for forces with $\lambda \ll r_{earth}$ one needs to take advantage of irregularities in the earth's surface. This occurs because the earth, on the average, has the shape of a fluid in equilibrium under gravitational and centrifugal forces. Hence a pendulum on the open sea will hang perpendicular to the surface; a "fifth force" with $\lambda \ll r_{earth}$ will therefore exert a force exactly along the "vertical." This vertical force cannot change the "angle of the dangle," and thus would be invisible to the experimenter. To see an effect from a short range force the experiment should be performed on sloping terrain (at the edge of a cliff, etc.). For very short range forces ($\lambda \leq 10$ m) one can achieve reasonable sensitivity by searching for differential horizontal acceleration toward a massive laboratory source placed beside the pendulum.

As we shall see, the Eötvös measurements can be done with considerably higher precision than the Galileo tests. However the two types of measurement are to some extent complementary; it is easy to compute the sensitivity of a Galileo result to a given Yukawa interaction of any range whatsoever. This is not true for an Eötvös measurement.

Yukawa interactions with ranges between 100's to 1000's of km produce forces on the test bodies which lie so close to the fiber axis that the torque depends sensitively on horizontal inhomogeneities in the densities deep below the surface of the earth. Fischbach et al.'s paper stimulated many groups to mount experiments capable of detecting the proposed "fifth force." The first two results were reported by Thieberger[17] from Brookhaven and by the Eöt-Wash group[18] in Seattle. Thieberger placed an apparatus containing a hollow Cu sphere floating in H_2O at the edge of a cliff on the Hudson River Palisades. He observed that it moved perpendicular to the cliff edge (see Figure 2) with a magnitude and direction in excellent agreement with the "fifth force" prediction, his result indicating that $\alpha_5 = (1.2 \pm 0.4) \times 10^{-2}$ for $\lambda = 100$ m. On the other hand the Eöt-Wash group, using a continuously rotating torsion balance with

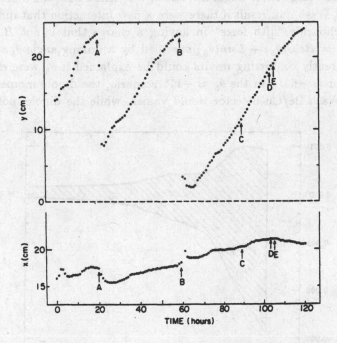

Fig. 2. Horizontal drift velocity of Thieberger's Cu ball (taken from Ref. 17). The y-axis is essentially normal to the cliff face. The ball position was reset at points A and B.

a Be/Cu dipole operated on a hillside on the University of Washington campus, reported a null result, $|\alpha_5| \leq 4 \times 10^{-4}$ for $\lambda = 100$ m, that was in strong disagreement with the "fifth force" prediction as shown in Figure 3.

The apparent strong discrepancy between these results could have been explained in several different ways:

1. an unidentified systematic error in one (or both) of the experiments;

2. differences between the Cu/H_2O and Cu/Be composition dipoles used in the Thiebeger and Eöt-Wash "detectors"; or

3. differences between the cliff and hillside "sources" used in the two experiments. Two factors are potentially important, chemical composition and topography.

The Eöt-Wash group performed three more experiments[19,20,21] designed to distinguish between these three possibilities. In principle, Thieberger's large effect might have been reconciled with the more sensitive Eöt-Wash null result if there were a new interaction that differed from Fischbach's "fifth force" in having a charge that is not B, but rather $q_5 = Bcos(\theta_5) + Lsin(\theta_5)$ specified by a mixing angle θ_5. The two apparently conflicting results could be explained if θ_5 were either $\approx -11°$ or $\approx -63°$. In the $\theta_5 \approx -11°$ scenario, the dipole moment of the Eöt-Wash Be/Cu detector would vanish, while the dipole moment

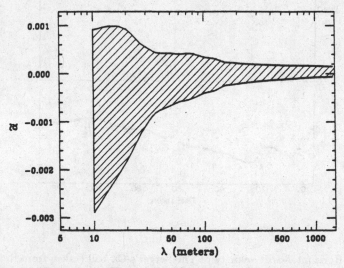

Fig. 3. Constraints from the original Eöt-Wash experiment on α_5 as a function of λ for a fifth force coupled to B (taken from Ref. 18).

of Thieberger's H_2O/Cu detector would be very large. In the $\theta_5 \approx -63°$ scenario, where $q_5 \propto N - Z$, the charges of the terrestrial sources used in both the Thieberger and Eöt-Wash experiments would have been small and would depend sensitively on the details of the chemical composition of the sources. The $\theta_5 \approx -11°$ scenario was ruled out when the Eöt-Wash group repeated their experiment with a Be/Al dipole[19] and obtained a similar null result.

The $\theta_5 \approx -63°$ scenario received support when Boynton et al.[22] operated a Be/Al torsion balance, shown in Figure 4, near a cliff at Index, Washington in the North Cascade mountains.

Boynton reported a 3.5σ result and showed that it, and the previous work of Thieberger and the Eöt-Wash group could all be explained in terms of an interaction whose charge was essentially $N - Z$, the large variation in the reported results being explained by differences in the chemical composition of the various topographic sources (soils *vs.*

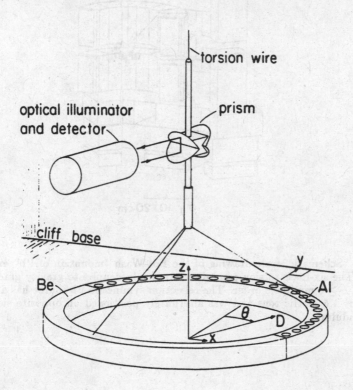

Fig. 4.   Torsion balance used by Boynton et al. (taken from Ref. 22).

E. G. Adelberger

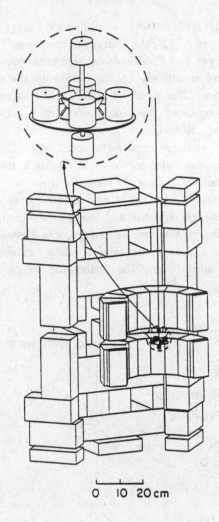

$$0 \quad 10 \quad 20\,cm$$

Fig. 5.   Schematic, scale drawing of the Eöt-Wash "mountain of Pb" experiment. The six-mass torsion pendulum nominally couples to gravity gradients only in $L \geq 5$ multipole order. The reflection symmetric Pb source has a mass of about 1.5 metric tons and was alternately positioned on opposite sides of the pendulum.

rock). Subsequent null results by Fitch et al.,[23] Speake and Quinn,[24] and Cowsik et al.[25] were not inconsistent with Boynton et al.'s proposed interaction.

Meanwhile the Eöt-Wash group had rebuilt their apparatus to improve sensitivity and reduce susceptibility to systematic effects from gravity gradients and thermal effects. With this apparatus, shown in Figure 5, they were able to compare the horizontal acceleration of Be and Al towards an ≈ 1.5 metric ton Pb source[20] with sufficient accuracy to rule out the proposed interaction with $q_5 \propto (B - 2L) \propto (N - Z)$ (see Figure 6). It was possible to obtain such a sensitive result with a "small" laboratory source because Pb has ≈ 120 times the $(N - Z)$ per unit volume of the rock at Mt. Index, and the laboratory source has the nice feature that it can be rotated by 180° while the mountain cannot. Bennett[26] has obtained a null result with somewhat less sensitivity by using the water in a lock on the Snake river as a moveable source.

The $\theta_5 \approx -11°$ and $-63°$ scenarios were basically schemes to reconcile the large result reported by Thieberger with the much smaller upper limits obtained by others. Hence it is very interesting that a group from

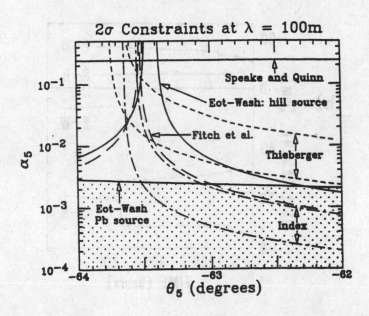

Fig. 6. 2σ constraints on α_5 as a function of θ_5 for an interaction coupled predominantly to $N - Z$ with $\lambda = 100$ m. The dark triangular region shows Boynton et al.'s proposed interaction.

Florence[27] has recently reported the results from a refined version of Thieberger's experiment. Bizzeti et al. floated a plastic sphere (with no projections such as Thieberger needed to keep his ball buoyant) in a stratified fluid whose vertical density gradient was established by dissolved salt. They did not observe any significant motion of the sphere ($v < 10\mu$m/h) which corresponds to a differential acceleration between solid and fluid of 2.4×10^{-9} cm/s^2 (see Figure 7). It is difficult to reconcile this null result with the much larger effect Thieberger saw in his experiment ($\Delta a = (8.5 \pm 1.3) \times 10^{-8}$ cm/s^2).

In the last few months, interesting new null results have been reported that cast even more doubt on positive claims for a new force. At the last Moriond Workshop, an Irvine group reported a very sensitive null result for the differential acceleration of Cu and U towards a Pb source[28] that establishes a 1σ upper limit on vector interactions coupled to $N - Z$ that is ≈ 2 times below the Eöt-Wash value. The Eöt-Wash group, using the improved version of their instrument shown in Figure 8, has recently obtained 1σ upper limits of $\Delta a_\perp = (1.5 \pm 2.3) \times 10^{-11}$ cm/s^2 and $\Delta a_\perp = (0.9 \pm 1.7) \times 10^{-11}$ cm/s^2 for the differential acceleration of Be/Al

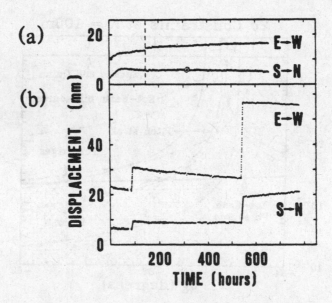

Fig. 7. Data from the floating ball experiment of Bizzetti et al. (taken from Ref. 27).

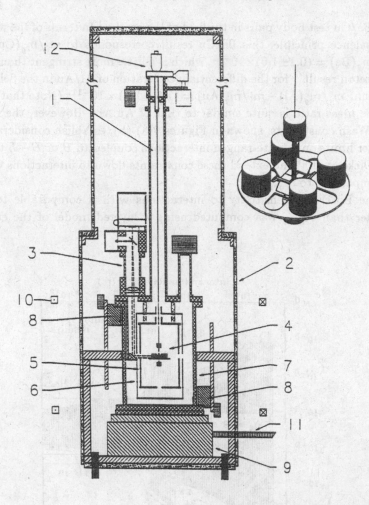

Fig. 8.   Schematic side view of the latest version of the Eöt-Wash differential accelerometer. The major components are labelled: 1. 20µm tungsten fiber; 2. thermal shield; 3. autocollimator; 4. torsion pendulum; 5.-6. magnetic shields; 7. vacuum vessel; 8. gravity gradient compensator; 9. turntable; 10 baseplate vibration isolator; 11. Helmholtz coils; 12. turntable motor.

and Be/Cu test body pairs in the field of the earth.[21] In terms of the weak
equivalence principle, this Be/Cu result corresponds to $m_i/m_g(\text{Cu}) -
m_i/m_g(\text{Be}) = (0.1 \pm 1.0) \times 10^{-11}$, which is a little more stringent than the
Princeton result[14] for the differential acceleration of Al/Au in the field of
the sun: $m_i/m_g(\text{Al}) - m_i/m_g(\text{Au}) = (1.5 \pm 1.5) \times 10^{-11}$. (Note that the
Cu:Be mass ratio is quite similar to that of Au:Al.) However, the new
Eöt-Wash constraints, shown in Figures 9(a)-(b), establish considerably
tighter limits on infinite-ranged interactions coupled to B or $B - L$ than
did Dicke et al.[14] and extend these constraints down to interactions with
λ comparable to r_{earth}.

The Eöt-Wash sensitivity to interactions with λ comparable to or
greater than r_{earth} was computed using a layered model of the earth

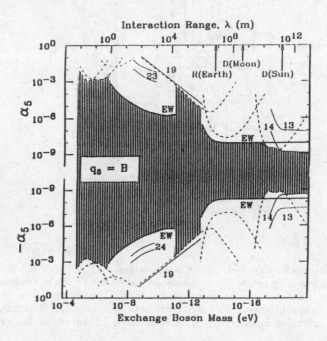

Fig. 9(a). Constraints on ultra-low mass vector bosons coupled to B. Solid
curves are from equivalence principle tests, dashed curves from $1/r^2$ tests
summarized in Ref. 28. The curves labelled EW are from the Eöt-Wash
group, the numbered curves correspond to references in the text. Upper limits
on the absolute value of the dimensionless coupling strength α_5 are shown as
a function of boson mass $m_b c^2$, or range λ. Although a vector interaction
requires a positive α_5, for completeness in interpreting experimental results
that may have either sign, we consider the possibility of negative α_5 as well.

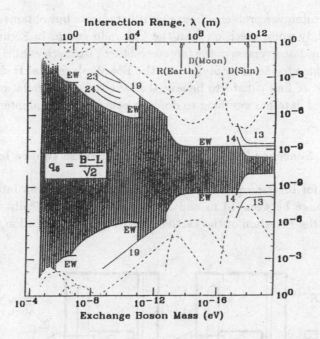

Fig. 9(b).   Constraints on ultra-low mass vector bosons coupled to $(B - L)/\sqrt{2}$. See Fig. 9(a) caption for explanation.

which assumes the gross shape of the earth is that of a fluid in equilibrium under gravitational and centrifugal forces. For $\lambda \leq 20$ km, the sensitivity was computed by detailed integration of the surface and subsurface geology. Constraints were not given in the region from $\lambda = 20$ km to $\lambda = 1000$ km because of sensitivity of the source strength calculation to poorly known details of the subsurface geology. Note that the Eöt-Wash constraints on interactions coupled to B or $B - L$ are considerably tighter than those from $1/r^2$ tests[29] over almost the entire region for $\lambda \geq 1$ m.

In the region between $\lambda = 20$ km and $\lambda = 1000$ km, the tightest constraints on new interactions come from measurements of differential vertical acceleration in Galileo-type experiments. A Boulder group,[30] using the instrument shown in Figure 10, reported that the downward acceleration of Cu and U differed by less than 5×10^{-10} g. This result provides the best constraint in the intermediate regime. Subsequently a Tsukuba group[30] reported null results from a similar experiment with Al/Cu and Al/C test bodies. They incorporated some nice improvements to minimize problems from gravity gradients but obtained results with slightly worse precision than the Colorado group. In Figure 11 we show an updated version of the famous plot by Fischbach et al.;[5] we have simply added to the von Eötvös data the Eöt-Wash results. It should be clear why we find it hard to believe that the striking correlation in the von Eötvös data has anything to do with new fundamental interactions.

3. Searches for violations of the inverse square law

Searches for inverse square law violation due to a Yukawa interaction with a range between 10 m and 10^4 m are most practically done by studying the variation of the earth's gravitational acceleration, g, with

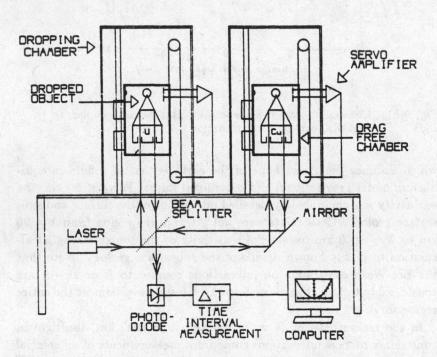

Fig. 10. Interferometric differential accelerometer used by Niebauer et al. The Cu and U test bodies are attached to "drag-free" corner cubes that fall inside shrouds that are driven downward with acceleraion g. The two corner cubes define the arms of a Michelson interferometer. The time variation of the beat frequency is analyzed to find the differential acceleration.

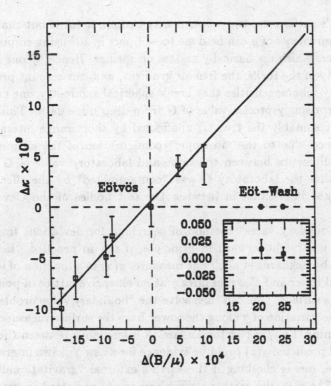

Fig. 11. Updated plot showing the von Eötvös data as re-analyzed by Fischbach et al. and the recent Eöt-Wash results. The vertical scale on the inset has been magnified by a factor of 100 to show the error bars on the Eöt-Wash data.

elevation below or above the surface. Two different methods have been employed: the Airy method and the "boundary value" method. Let us review the principles of these two methods. The Airy method, as practiced for example by Stacey et al.[16] is essentially a geophysical methods for determining G from measurements of g in mines. For simplicity, assume that the earth has spherical symmetry. Then the variation of g with radius r is given by differentiating $g(r) \equiv G/r^2 \int_0^r 4\pi\rho(r_1)r_1^2 dr_1$ with respect to r,

$$\frac{dg}{dr} = 4\pi G\rho(r) - 2\frac{g}{r}.$$

The first term in this expression contains the effect of penetrating spherical shells of matter; the second term, the "free air gradient," contains the effect of getting closer toward the dense core of the earth. Just below the earth's surface the two terms nearly cancel. It turns out that relative measurements of g can be done to ~ 1 part in 10^8 using commercial gravimeters which are basically masses on springs. Hence, *if* one knows the density of the rocks, the free air gradient, *and* can account properly for density inhomogeneities that break spherical symmetry, one can obtain a surprisingly precise value of G from deep mine data. This value of G is presumably the true G, unaffected by short range interactions which cancel due to the "isotropic" configuration of the surrounding rock. A difference between the mine and laboratory value of G could indicate that the laboratory G was "contaminated" by the effect of a short-ranged interaction in between the test bodies of the Cavendish balance.

The "boundary value" method of searching for deviations from the $1/r^2$ law is straight-forward in principle, if not in practice. Here, as practiced by Eckhardt et al.,[32] one measures $g(z)$ as a function of height, z, in a tall tower and also measures g at an extensive lattice of points on the earth's surface. One can then solve the "boundary value problem" to predict the variation of g along the tower from the surface measurements of g assuming the $1/r^2$ Newtonian law. A discrepancy between the measured and predicted $g(z)$ could be evidence for a new Yukawa interaction. In essence, one is checking if the earth's external "gravitational" field satisfies $\nabla^2 \phi = 0$ (the earth's atmosphere is not neglected in practice). The "boundary value" method is superior to the Airy method because it can, in principle, be rendered insensitive to density inhomogeneities within the earth. The first result based on the boundary value method was reported by Eckhardt et al.[32] They took data from a 600 m high TV tower in North Carolina and reported a highly significant discrepancy between the measured tower gradient and the "boundary value" prediction ($\sim 500\,\mu$gal at the top of the tower, 1 gal $= 1$ cm/s^2). Their result, shown in Figure 12, corresponded to a *attractive* force with a strength $\alpha\lambda \approx 5$ m which contrasted with the *repulsive* force claimed by Stacey et al. Eckhardt et al. argued that their results and Stacey's were not inconsistent and indeed were evidence for a pair of Yukawa forces, one attractive and one repulsive. This was particularly intriguing because "quantum gravity inspired" models[7] had predicted attractive and repulsive Yukawa interactions arising from the exchange of hypothetical vector and scalar partners of the usual graviton.

It is worth noting that the Eöt-Wash results tightly constrain the "quantum gravity" two-Yukawa models.[34] Because of the complex topography of the Eöt-Wash source the direction of a Yukawa force is

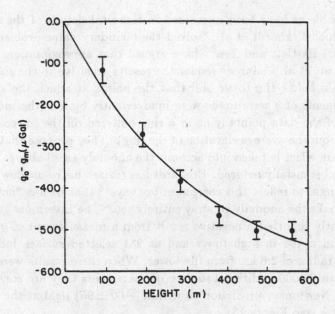

Fig. 12.   Difference between the Newtonian prediction and the observed gravitational acceleration as a function of height in WTVD tower (taken from Ref.
32). The curve corresponds to a scalar interaction with $\alpha \approx 0.02$ and $\lambda = 311$
m.

a strong function of its range. Since two vectors that are not exactly
collinear cannot cancel, the stringent null results require the gravi-vector
and gravi-scalar interactions to have essentially the same range as well
as equal strengths for two different test body pairs. The inherently different nature of vector and scalar charges makes it very unlikely that
this last requirement can be fulfilled.

Because of the potentially "clean" nature of the boundary value methods, Eckhardt et al.'s work attracted much attention. Naturally, questions were raised both about the accuracy of the measurements and the
robustness of the analysis, and groups from Boulder and Livermore undertook independent measurements on similar towers in Colorado and
Nevada, respectively. Eckhardt et al. have taken additional surface
data in the vicinity of their North Carolina tower, and are also studying
the gravity field around another tower. Further work by Eckhardt and
the other groups has confirmed Eckhardt's contention that commercial
gravimeters give reliable readings in a tower environment. On the other

hand, problems have been uncovered in technical details of the analysis by which Eckhardt et al. "solved the boundary value problem." In particular, Bartlett and Tew[33] have argued that approximations made in Eckhardt et al.'s analysis render the results sensitive to the shape of the terrain below the tower and that the points at which the surface measurements of g were made were inadvertently biased (the mean elevation of the data points lying in a ring centered on the tower differs slightly from the average elevation of the ring). They propose that when this terrain effect is taken into account the anomaly ascribable to a new force is substantially reduced. Eckhardt has revised his result downward to $\approx 350\mu$gal to reflect this correction but says[34] that he has "not been able to make the anomaly go away entirely yet." The Livermore group[36] has recently reported a negative result from a measurement of g at 12 heights on a 465 m high tower and at 281 selected surface locations within a radius of 2.6 km from the tower. When these results were combined with 60,000 additional surface measurements they are consistent with the Newtonian prediction to within (-60 ± 95) μgal at the top of their tower (see Figure 13).

Bartlett and Tew[33] have also investigated the topography around Stacey's mine and argue that local terrain corrections (to account for the fact that the mine is effectively located in a valley) can also account for most of the evidence for non-Newtonian effects reported by Stacey et al. Stacey's group are now reanalyzing their data to check for possible terrain effects, and find[37] that these corrections can indeed account for most of the "anomaly."

In summary, there has been considerable progress in understanding subtle systematic biases that can creep into analysis of geophysical data. As the analyses have improved, the anomalies seem to be going away. Within the next year or so we should know whether geophysical measurements really do provide evidence for a violation of the $1/r^2$ law. It is worth noting that if such violations with $\lambda \approx 10^2$ m do occur, unaccompanied by a violation of the universality of free fall, then studies of the earth's gravity field are, at present, the only practical way to detect the phenomenon.

4.   Conclusions

Is there a fifth force? In spite of positive effects claimed in some tests of the universality of free fall and of the inverse square law, one has to

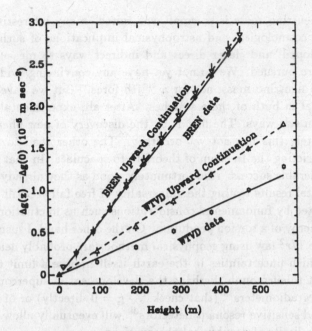

Fig. 13. Observed (solid line) gravitational acceleration and the Newtonian prediction (dashed) as a function of height in the BREN tower (taken from Ref. 36).

say that there is at present no compelling evidence for new fundamental physics. None of the groups that claimed evidence for a new force has reproduced their result, and at least in the composition-dependence arena, considerably more precise null results leave little choice but to discount the positive results as an artifact of some subtle systematic errors. What has been achieved in the last few years of intense activity in this field? First of all, thanks to Fischbach et al., we realized that there was very little hard information that could reveal or rule out the possible existence of new macroscopic interations. This brought forth a considerable improvement in experimental methods, a narrowing of the geophysical window, and greatly improved limits on the interaction of proposed ultra-low mass scalar or vector bosons over an enormous range (10^{11}) of masses. And rumors of new forces have caused a number of people to consider theoretical scenarios in which such forces could arise or play an important role. The Pandora's box, once opened, will be hard to shut. In this respect, the search for new macroscopic interactions is similar to the search for neutrino masses. Once the question of why the neutrino mass is so small was raised, many interesting ideas about the cosmological and astrophysical implications of such masses were developed, and clever direct and indirect ways to measure these masses were pursued. We do not yet have any convincing evidence for a non-zero neutrino mass, nor for a "fifth force." But we have learned a great deal in both of these searches. After all, experimental physics advances in two ways. The first is by the discovery of new phenomena. Unfortunately this has not yet occurred. The other road to progress lies in restricting the freedom of theorists to speculate. In that we have had considerable success. And fortunately, good as they already are, the experimental results testing the universality of free fall are still far from being limited by fundamental considerations such as fluctuations in the thermal energy of a torsion pendulum. On the other hand, I suspect that tests of the $1/r^2$ law using geophysical methods are probably nearing the level at which uncertainties in the earth itself impose a limit that will be difficult to surmount. Perhaps the development of superconducting three-axis gradiometers[39] (that check $\nabla \cdot g = 0$ directly) or of rotating sources and sensitive resonant detectors[38] will eventually allow us to go beyond the limits of geophysical observations.

I would like to acknowledge the enthusiastic and talented efforts of my collaborators in the Eöt-Wash group, and thank Ida Tess for formatting this report. We are supported in part by the National Science Foundation (grant PHY-871939) and the Department of Energy.

References

1. T. D. Lee and C. N. Yang (1955), *Phys.Rev.* **98**, 1501.

2. P. Fayet (1981), *Nucl. Phys.* **B187**, 184; (1986), *Phys. Lett.* **B171**, 261 and (1986), *Phys. Lett.* **B172**, 363.

3. J. Scherk (1979), in *Unification of the Fundamental Particle Interactions*, eds. S. Ferrara, S. Ellis, and P. van Nieuwenhuizen, *Ettore Majorana International Science Series* Vol. 7 (Plenum, New York, 1980); (1979), *Phys. Lett.* **88B**, 265.

4. D. Chang, R. N. Mohapatra and S. Nussinov (1985), *Phys. Rev. Lett.* **55**, 2835.

5. E. Fischbach, D. Sudarsky, A. Szafer, C. Talmadge and S. H. Aronson (1986), *Phys. Rev. Lett.* **56**, 3.

6. A. DeRújula (1986), *Phys. Lett.* **B180**, 213.

7. T. Goldman, R. J. Hughes and M. M. Nieto (1986), *Phys. Lett.* **B171**, 217.

8. R. D. Peccei, J. Sola and C. Wetterich (1987), *Phys. Lett.* **B195**, 183.

9. I. Bars and M. Visser (1987), *Gen. Relat. Grav.* **19**, 219.

10. J. S. Bell (1987), in *Fundamental Symmetries*, eds. P. B. Bloch, P. Pavlopoulos and R. Klapisch (Plenum Press, New York), p. 1.
11. C. T. Hill and G. G. Ross (1988), *Phys. Lett.* **B205**, 125.
12. M. Visser (1988), *Gen. Relat. Grav.* **20**, 77.
13. C. T. Hill, D. N. Schramm and J. N. Fry (1989), *Comments Nucl. Part. Phys.* **19**, 25.
14. P. G. Roll, R. Krotkov and R. H. Dicke (1964), *Ann. Phys. (NY)* **26**, 442.
15. V. B. Braginsky and V. I. Panov (1972), *Sov. Phys. JETP* **34**, 463.
16. F. D. Stacey, G. J. Tuck, G. I. Moore, S. C. Holding, B. D. Goodwin and R. Zhou (1987), *Rev. Mod. Phys.* **59**, 157.
17. P. Thieberger (1987), *Phys. Rev. Lett.* **58**, 1066.
18. C. W. Stubbs, E. G. Adelberger, F. J. Raab, J. H. Gundlach, B. R. Heckel, K. D. McMurry, H. E. Swanson, R. Watanabe (1987), *Phys. Rev. Lett.* **58**, 1070.
19. E. G. Adelberger, C. W. Stubbs, W. F. Rogers, F. J. Raab, B. R. Heckel, J. H. Gundlach, H. E. Swanson and R. Watanabe (1987), *Phys. Rev. Lett.* **59**, 849; (1987), *Phys. Rev. Lett* **59**, 1790 (erratum).
20. C. W. Stubbs, E. G. Adelberger, B. R. Heckel, W. F. Rogers, H. E. Swanson, R. Watanabe, J. H. Gundlach and F. J. Raab (1989), *Phys. Rev. Lett.* **62**, 609. A preliminary account of this work was given by E. G. Adelberger (1988), in *Proc. of the XXIIIrd Rencontre de Moriond*, eds. O. Fackler and J. Tran Thanh Van (Editions Frontiéres, Gif-sur-Yvette), p. 445.
21. B. R. Heckel, E. G. Adelberger, C. W. Stubbs, Y. Su, H. E. Swanson, G. Smith and W. F. Rogers (1989), submitted to *Phys. Rev. Lett.*; a preliminary version of this work was reported by E. G. Adelberger (1989), in *Proc. of the XXIVth Rencontre de Moriond, Tests of Fundamental Laws in Physics*, eds. O. Fackler and J. Tran Thanh Van (Editions Frontiéres, Gif-Sur-Yvette), p. 485.
22. P. E. Boynton, D. Crosby, P. Ekstrom and A. Szumilo (1987), *Phys. Rev. Lett.* **59**, 1385.
23. V. L. Fitch, M. V. Isaila and M. A. Palmer (1988), *Phys. Rev. Lett.* **60**, 1801.
24. C. C. Speake and T. J. Quinn (1988), *Phys. Rev. Lett.* **61**, 1340.
25. R. Cowsik, N. Krishnan, S. N. Tandon, and C. S. Unnikrishnan (1988), *Phys. Rev. Lett.* **61**, 2179.
26. W. R. Bennett Jr. (1989), *Phys. Rev. Lett.* **62**, 365.
27. P.G. Bizzeti, A. M. Bizzeti-sona, T. Fazzini, A. Perego and N. Taccetti (1989), *Phys. Rev. Lett.* **62**, 2901.
28. R. Newman (1989), in *Proc. of the XXIVth Rencontre de Moriond, Tests of Fundamental Laws in Physics*, eds. O. Fackler and J. Tran Thanh Van (Editions Frontiéres, Gif-sur-Yvette), p. 459.
28. C. Talmadge, J.-P. Berthias, R. W. Hellings and E. M. Standish (1988), *Phys. Rev. Lett.* **61**, 1159.
30. T. M. Niebauer, M. P. McHugh and J. R. Faller (1987), *Phys. Rev. Lett.* **59**, 609.
31. K. Kuroda and N. Mio (1989), *Phys. Rev. Lett.* **62**, 1941.
32. D. H. Eckhardt, C. Jekeli, A. R. Lazarewicz, A. J. Romaides, and R. W. Sands (1988), *Phys. Rev. Lett.* **60**, 2567.
33. D. F. Bartlett and W. L. Tew (1989), *Phys. Rev.* **D40**, 673; D. F. Bartlett and W. L. Tew, *J. Geophys. Res.* **B**, to be published.
34. D. H. Eckhardt, private communication.

35. C. W. Stubbs, E. G. Adelberger and E. C. Gregory (1988), *Phys. Rev. Lett.* **61**, 2409.
36. J. Thomas, P. Kasameyer, O. Fackler, D. Felske, B. Harris, J. Kammeraad, M. Millett and M. Mugge (1989), *Phys. Rev. Lett.* **63**, 1902.
37. G. J. Tuck, contribution to this conference.
38. N. Mio, K. Tsubono and H. Hirakawa (1987), *Phys. Rev.* **D36**, 2321.
39. H. J. Paik, Q. Kong, M. V. Moody and J. W. Parke (1988), in *Proc. of the XXIIIrd Rencontre de Moriond*, eds. O. Fackler and J. Tran Thanh Van (Editions Frontiéres, Gif-Sur-Yvette), p. 531.

11

Resonant bar gravitational wave experiments

Guido Pizzella

Physics Department, University of Rome,
Piazzale A. Moro 2,
00185 Rome, Italia

The basic features of resonant antennas for the search for gravitational waves are reviewed. These antennas detect the Fourier component $H(\omega)$ of the metric tensor perturbation $h(t)$ at the antenna resonance. By means of optimum filters it is possible to reach a sensitivity which, for short bursts of gravitational radiation, is basically limited by the noise of the electronic amplifier.

One cryogenic antenna with mass $M = 2300$ kg and temperature $T = 20$ mK should have a sensitivity, for short bursts, of the order of $h_{min} \simeq 3 \times 10^{-21}$. For a monochromatic gravitational wave the sensitivity should be of the order of $h_{min} \sim 3 \times 10^{-26}$ with two months' observation time. The sensitivity values can be improved by employing several resonant detectors.

The bandwidth of the resonant antennas is much larger than the width of the resonance curve; it can reach values up to several tens of Hz if a low noise amplifier is available.

Finally a brief review of the activity going on all over the world by the experimental groups working with resonant antennas is given.

1. Interaction of a gravitational wave with a resonant antenna

Let us first consider two pointlike bodies each with mass m, connected by a spring with elastic constant

$$k = m\omega_R^2 \tag{1}$$

and dissipation per unit mass

$$2\beta_1 = \omega_R/Q, \tag{2}$$

Q being the merit factor. The interaction with a gravitational wave impinging perpendicularly to the oscillator axis is described by the following extension of the geodesic deviation law

$$\ddot{\xi} + 2\beta_1\dot{\xi} + \omega_R^2\xi = -lc^2 R_{030}^3 \tag{3}$$

where l is the distance between the two bodies in the equilibrium configuration along the z coordinate, ξ the change of this distance and R_{030}^3 a component of the Riemann tensor.

As well known if the gravitational wave travels in the oscillator axis direction there is no change of l.

It can be shown that a continuous antenna, say a metallic cylinder with length L and mass M, behaves like many elementary harmonic oscillators of the kind indicated above each one with a resonance frequency equal to one of the various resonance frequencies of the bar. For the longitudinal fundamental mode

$$\omega_0 = \pi v/L, \tag{4}$$

where v is the sound velocity in the bar material ($v = 5400$ m/s in Al at temperature of 4.2 K).

It can be shown (see Amaldi et al. (1979)) that Eq. (3) can be rewritten for the fundamental mode of a cylindrical bar as

$$\ddot{\xi} + 2\beta_1\dot{\xi} + \omega_0^2\xi = \frac{L}{2}\ddot{h}, \tag{5}$$

where ξ indicates the displacement of the bar end with respect to its equilibrium position and the Riemann tensor has been expressed in terms of the metric perturbation h that describes the gravitational wave (we indicate with h, for simplicity, either h_+ or $h_\times$ in the TT gauge).

The solution of Eq. (5) is

$$\xi(t) = \frac{2L}{\pi^2}\omega_0 H(\omega_0)e^{-\beta_1 t}\sin\omega_0 t, \tag{6}$$

where $H(\omega_0)$ is the Fourier component of $h(t)$ at the bar resonance ω_0. For a gravitational wave short burst of duration τ_g we can put, very roughly.

$$H(\omega_0) \approx h(t)\tau_g. \tag{7}$$

In the following we shall conventionally take $\tau_g = 0.001$ s in order to express the antenna sensitivity in terms of $h(t)$ instead of $H(\omega_0)$ as it would be more appropriate.

In the framework of the detection techniques it is also convenient to express the bar vibration in terms of energy instead of displacement as indicated in (6). The maximum total vibration energy of the bar is given by

$$\epsilon = \frac{1}{4} M \omega_0^2 \xi_0^2 ,$$

where $\xi_0 = \frac{2L}{\pi^2} \omega_0 H(\omega_0)$. We obtain

$$\epsilon = \Sigma f(\omega_0) \tag{8}$$

where Σ is the cross section (for the best orientation)

$$\Sigma = \frac{8}{\pi} \left(\frac{v}{c}\right)^2 \frac{G}{c} M \qquad [\text{m}^2 \text{ Hz}] \tag{9}$$

and $f(\omega_0)$ is the spectral energy density of the gravitational wave

$$f(\omega_0) = \frac{c^3}{8\pi G} \omega_0^2 |H(\omega_0)|^2 \qquad \left[\frac{\text{joule}}{\text{m}^2 \text{Hz}}\right]. \tag{10}$$

There have been claims that the cross section may be larger than that indicated by (9). In the following we shall use the above value, which sometimes we shall refer to as the "classical cross section."

Before discussing the gravitational wave antennas it is important to calculate the amplitude of the signals we expect to detect. Without any assumption on the physical mechanism of gravitational wave generation we consider an event taking place at a distance R from the detector, consisting in producing a gravitational wave burst of duration τ_g, isotropically and with total energy $M_{gw} c^2$. Then it is easy to find the corresponding value of h

$$h = \sqrt{\frac{16 G M_{gw} c^2}{c^3 R^2 \omega_0^2 \tau_g}} \tag{11}$$

at the detector location. For $\tau_g = 10^{-3}$ s, $M_{gw} = 10^{-2} M_\odot$, $\nu_0 = 1000$ Hz we find: a) for an event in the center of Galaxy, $R = 8.5$ kpc, $h \simeq 5 \times 10^{-18}$; b) for an event in the Virgo cluster, $R = 19$ Mpc, $h \simeq 2 \times 10^{-21}$.

Thus our goal is to reach a sensitivity $h \leq 2\text{x}10^{-21}$, to be able to observe several events per year.

Finally we consider the antenna response to a monochromatic gravitational wave in resonance with the antenna itself. It is found that

$$\xi(t) = \frac{2L}{\pi^2} Q h_0 sin\omega_0(t) \tag{12}$$

where we have assumed $h(t) = h_0 sin\omega_0 t$.

2. The electromechanical transducer

This is the device which converts the bar mechanical vibrations into electrical signals.

An ideal rough scheme is shown in Figure 1. We imagine the transducer mounted at one end of the bar (it could be mounted also at different places) and the electrical signal coming from it sent to a very low noise electronic amplifier before going into the data processing apparatus.

Most of the present transducers can be classified in three categories: a) inductive transducers; b) capacitive transducers; c) parametric transducers.

A good example of a) is the Stanford transducer (see Boughn et al. (1982)). Basically this type has a small output impedence (ωL) and it produces a current signal.

Examples of b) are the PZT ceramic of Weber and the Rome transducer (see Rapagnani (1982)). They are characterized by having a large output impedance ($1/\omega C$) and by producing a voltage signal.

Finally an example of c) is the Perth transducer (see Veitch et al. (1987)). This type is characterized by the fact that it includes an elec-

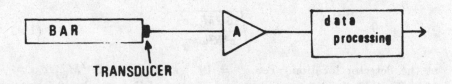

Fig. 1.   Simple scheme of the bar, transducer, low noise amplifier and data processing.

trical pump (that is, it adds energy to the signal), thus magnifying the signal itself.

In the following I shall refer mainly to the Rome transducer, because I wish to treat the problem in a very specific way, but the conclusions are rather general and can be extended to the above three mentioned types.

For a capacitive transducer we have the voltage signal

$$V = \alpha \xi \, ,$$

where α is the transducer constant. An important quantity is the ratio of the electrical energy available in the transducer to the mechanical energy in the bar that has been extracted from the gravitational wave:

$$\beta = \frac{\frac{1}{2} C V^2}{\frac{1}{4} M \omega_0^2 \xi^2} = \frac{2 C \alpha^2}{M \omega_0^2} \, . \tag{13}$$

Typical values are: $C = 4$ nF, $\alpha = 5 \times 10^6$ V/m, $M = 2000$ kg, $\omega_0 = 5000$ rad/s. We get $\beta \simeq 10^{-6}$. Therefore only a very small part of the available energy is, in this case, used for the signal detection (larger β values can be obtained using PZT ceramics but in this case the merit factor Q becomes too small).

Clearly one has to have β as large as possible. This is achieved by making the mechanical part of the electromechanical transducer so as to resonate at the exact bar frequency.

In such a case we have a two coupled oscillator system, as schematically shown in Figure 2, where m_t is the reduced mass of the transducer and $M/2$ the reduced mass of the bar.

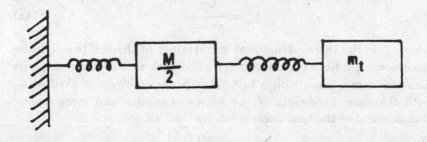

Fig. 2. Two coupled oscillators: the bar with reduced mass $M/2$ and the resonant transducer with reduced mass m_t.

As well known, in this case the mechanical energy transfers completely back and forth between the bar and the transducer. For the β value (electrical energy/mechanical energy) formula (13) still holds where, however, $M/2$ is replaced by m_t which is, typically, 10^4 times smaller, thus providing a β value of the order of 10^{-2}.

Similar considerations can be made for the inductive transducers, for which, using the above resonant method, one can attain β values even larger than 10^{-2}.

As far as the parametric transducers, since they use a particular power source (the electronic pump), they can have β even larger than unity. However, here an additional noise source, due to the pump, is present.

3.   The noise

There can be several sources of noise: thermal, electrical, acoustic, cosmic rays, seismic, etc. Some of these noises can be reduced only by employing several antennas in coincidence. Other noises can be reduced to the minimum by employing proper treatments of the experimental data.

These last noises can be classified in two categories: a) the narrow band noise; b) the wide band noise.

The narrow band noise has a power spectrum of lorentzian type, peaked at the antenna resonance frequency ω_0 with bandwidth $2\beta_1 = \omega_0/Q$, usually very small, as Q is greater than 10^6. This noise has two sources: the brownian motion and the back action due to the electronic amplifier. The brownian noise produces a mean square displacement of the transducer given by

$$\overline{\xi_{br}^2} = \frac{kT}{m_t\omega_0^2}, \tag{14a}$$

where T is the thermodynamical temperature of the bar (we consider from now on a resonant transducer). The back action of the amplifier also produces a noise with a lorentzian power spectrum peaked at ω_0 with the same bandwidth of the brownian motion and mean square displacement of the transducer given by

$$\overline{\xi_b^2} = \overline{\xi_{br}^2} \, \frac{\beta Q T_n}{2\lambda T}. \tag{14b}$$

Here the new parameters T_n and λ are characteristics of the electronic amplifier. In terms of the current and voltage power spectra I_n^2 and V_n^2

of the electronic amplifier we have (for a capacitive transducer)

$$T_n = \frac{\sqrt{I_n^2 V_n^2}}{k},$$

$$\lambda = \sqrt{\frac{V_n^2}{I_n^2}} \omega C.$$

(15)

In total we can consider a narrow band mean square displacement of the transducer

$$\overline{\xi_{nb}^2} = \frac{kT_e}{m\omega_0^2},$$

(16)

with the equivalent temperature

$$T_e = T\left(1 + \frac{\beta Q T_n}{2\lambda T}\right).$$

(17)

The wide band noise is due to the electronic amplifier and has a white spectrum (in the ideal case) that can be expressed by

$$S_0 = \frac{kT_n}{\omega C}(\lambda + 1/\lambda) \quad \left[\frac{\text{Volt}^2}{\text{Hz}}\right].$$

(18)

In the data analysis the following dimensionless quantity plays a fundamental role

$$\Gamma = \frac{S_0}{\alpha^2 \overline{\xi_{nb}^2}/\beta_1} = \frac{T_n(\lambda + 1/\lambda)}{2\beta Q T_e}.$$

(19)

This is the ratio between the wide band and narrow band power noises near the resonance and it turns out to be, in all cases of interest, much less than unity.

4. Optimum filtering and effective temperature

The main objective in the gravitational wave research is to detect small changes in the vibrational status of the antenna, taking into consideration both the amplitude and the phase of the oscillations. These small changes are called "energy innovations" and can be produced both by external forces (say, those due to gravitational wave) or by the brownian and electronic noise.

For a simple visualization of the problem we can consider two successive measurements of the antenna vibrational energy at the times t and

$t + \Delta t$. The energy innovation due to the noise is then roughly given by

$$\Delta\epsilon = KT_e \frac{\Delta t}{Q/\omega_0} + \frac{m\alpha^2}{\omega_0^2} \frac{S_0}{\Delta t} . \tag{20}$$

The first term in the right side takes into account the fact that the narrow band noise mean square displacement changes with time very slowly, that is with a time scale Q/ω_0 (say, one hour). The second term expresses the fact that the wide band noise can be reduced by averaging out over a time Δt.

The optimum filtering techniques provide a more accurate value of the minimum energy innovation that can be observed (see for instance Bonifazi et al. (1978)). With signal to noise ratio (SNR) equal to 1,

$$\Delta\epsilon_{min} = 4kT_e\sqrt{\Gamma} \tag{21}$$

that allows one to introduce the effective temperature

$$T_{eff} = \frac{\Delta\epsilon_{min}}{k} = 4T_e\sqrt{\Gamma} = \sqrt{\frac{8T_e T_n (\lambda + 1/\lambda)}{\beta Q}} . \tag{22}$$

If $\Gamma \ll 1$ then $T_{eff} \ll T_e$. The above formula shows the importance of having a large βQ value. For instance, the largest βQ attainable with a PZT is of the order of 100. With a resonant capacitive transducer one can have instead $\beta Q \sim 10^4 - 10^5$. However with $\beta Q \to \infty$, T_{eff} does not go to zero, because βQ enters also in T_e. It can be immediately seen that the maximum possible value for T_{eff} is, in fact, $2T_n$ (see Giffard (1976)).

The knowledge of T_{eff} immediately gives the sensitivity in terms of $H(\omega)$. Making use of (8) and (9) we find

$$H(\omega) = \frac{L}{v^2}\sqrt{\frac{kT_{eff}}{M}} . \tag{23}$$

Finally it can be shown that the bandwidth $\Delta\nu$ of a resonant antenna is also related to Γ. We have (see Pallottino et al. (1981))

$$\Delta\nu = \frac{\beta_1}{\pi\sqrt{\Gamma}} = \frac{\nu_0}{Q\sqrt{\Gamma}} , \tag{24}$$

which shows that the bandwidth is larger than the lorentzian bandwidth by the factor $1/\sqrt{\Gamma}$. This is due to the fact that the resonant antenna has identical responses both to an external signal and to the brownian noise (the lorentzian curve) and the limit is set by the wide band noise. If the wide band noise were zero, the bandwidth would be infinite.

For completeness we give now the minimum value of $h(t)$ that can be observed (SNR = 1) for a monochromatic gravitational wave within the bandwidth (24). We find easily that

$$h_{min} = 1.9 \times 10^{-26} \left(\frac{T}{0.02 \text{ K}} \frac{2300 \text{ kg}}{M} \frac{10^7}{Q} \frac{50 \text{ days}}{t_m} \right)^{1/2}. \tag{25}$$

5. Present and future sensitivity

We shall now consider the specific case of the Rome experiment. A schematic diagram of the transducer electronics is shown in Figure 3 (see Amaldi et al. (1986)).

The capacitive transducer has capacity C and electrical field E_0. The voltage signal is given by

$$V = E_0 \xi, \tag{26a}$$

with constant electrical charge on the transducer.

This signal needs to go into a superconducting transformer with turn ratio N, in order to match the output impedance $(1/\omega C)$ of the transducer to the input impedance (ωL_{in}) of the SQUID system.

The SQUID measures the magnetic flux ϕ linked to it, thus

$$\phi = \alpha_s \xi. \tag{26b}$$

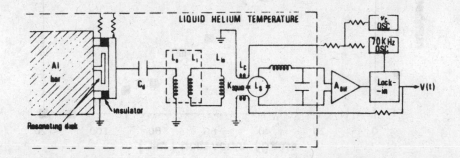

Fig. 3. Scheme of the Rome experiment with the dc SQUID amplifier.

From the characteristic of the Rome apparatus we find

$$\alpha_s = 9.7 \times 10^{10} \frac{E_0}{7 \times 10^6 \, \text{V/m}} \frac{N}{1250} \sqrt{\frac{C}{3.89 \, \text{nF}}} \quad \left[\frac{\phi_0}{\text{m}}\right], \quad (27)$$

where ϕ is expressed in units of the flux quantum $\phi_0 = 2.07 \times 10^{-15}$ weber. The use of a modulating field at 70 kHz allows one to convert linearly the magnetic flux into a voltage signal that, after processing, is recorded on the mass storage.

At present E_0, N and C have the values indicated in (27). The wide band noise is $\phi_N = 2 \times 10^{-6} \phi_0/\sqrt{\text{Hz}}$ and we have $Q = 3 \times 10^6$. With these values we calculate from (22), (23) and (24): $\Delta\nu = 0.8$ Hz, $T_{eff} = 3.4$ mK, $h = 5 \times 10^{-19}$. Measurements have been made with the Rome antenna installed at CERN (the EXPLORER antenna) and operating in superfluid Helium at 2 K. The noise distribution is shown in Figure 4.

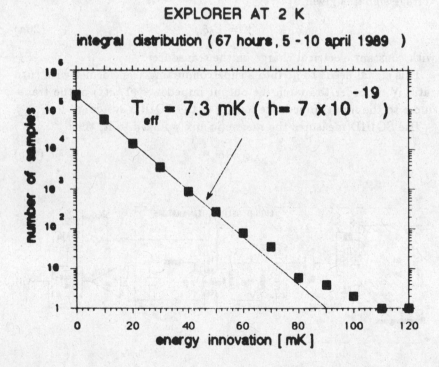

EXPLORER AT 2 K

integral distribution (67 hours , 5 - 10 april 1989)

T_{eff} = 7.3 mK (h= 7 x 10^{-19})

Fig. 4. Integral distribution of the noise for the EXPLORER (Rome) experiment at 2 K during a period of 67 hours (three separate subperiods). There are no energy innovation samples above 120 mK.

We notice that the experimental T_{eff} is a factor of 2 larger than the theoretical value. The reason is due to a poor behaviour of one of the two modes of the coupled oscillator system. However, we also notice a very good noise distribution with very few excess samples. We are planning now to make a few technical improvements in order to recover the factor of 2.

Also we are constructing a new cryostat (NAUTILUS) equipped with a dilution refrigerator for cooling the bar to $T \sim 0.04$ K. If we shall also be able to reach a noise $\phi_N = 10^{-7}\phi_0/\sqrt{\text{Hz}}$ and $Q = 10^7$ we shall get $T_{eff} = 9\mu$K, $h = 2.4 \times 10^{-20}$.

For further improving the sensitivity we should increase E_0, N and C, that is we should achieve a better matching of the system: bar + transducer + SQUID.

The scheme of NAUTILUS is shown in Figure 5.

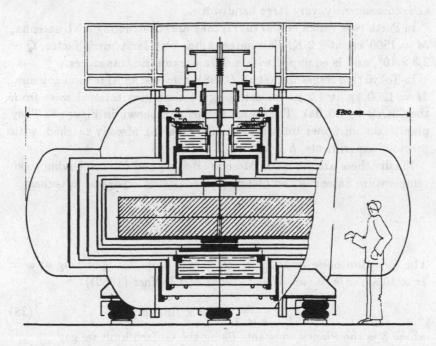

Fig. 5. Scheme of the NAUTILUS (Rome) cryostat equipped with a dilution refrigerator for cooling the Al bar to 40 mK.

6. Brief review of the bar experiments

In Stanford a 4.2 K Al antenna, $M = 4800$ kg, is in operation with an inductive transducer and a microwave SQUID. In the past they have obtained $T_{eff} \simeq 20$ mK (see Boughn et al. (1982)). A new cryostat, Nautilus type, is also in construction. The scheme of this new cryostat is shown in Figure 6.

In Louisiana they are operating an Al bar with $M = 2300$ kg at 4.2 K, whose scheme is shown in Figure 7. We notice a very good mechanical system for filtering out the external disturbances. They have obtained $T_{eff} \simeq 100$ mK because of losses in the inductive transducer and hope to obtain $T_{eff} \sim 5$ mK in the near future.

In Maryland, in addition to room temperature antennas, they are constructing an Al cryogenic antenna, $M = 1400$ kg at $T = 4.2$ K and also planning to go to lower temperature. They use a multimode resonant transducer (see Richard (1986)) in order to reach high β values and, consequently, very large bandwidth.

In Perth (see Veitch et al. (1987)) they are constructing a Nb antenna, $M = 1500$ kg, at 4.2 K. This antenna has very large merit factor, $Q \simeq 2.3 \times 10^8$, and is equipped with a GHz parametric transducer.

In Tokyo (see Nagashima et al. (1988)) they use an Al torsion antenna, $M = 1200$ kg at 4.2 K, for detecting possible gravitational wave from the CRAB (~ 60 Hz). The antenna scheme is shown in Figure 8. They plan to set an upper limit at $h \sim 10^{-22}$ having already reached, with previous experiments, $h \sim 10^{-21}$.

Finally there are groups in Moscow, Beijing and Canton having room temperature antennas and planning to construct cryogenic antennas.

7. Ultimate bar sensitivity

The quantum noise limit for an electronic amplifier operating at $\omega = 2\pi \times 1000$ rad/s, is (see Weber (1954) and Heffner (1962))

$$T_n \simeq \frac{\hbar\omega}{k \ln 2} \simeq 6.8 \times 10^{-8} \text{ K}, \tag{28}$$

where $\hbar$ is the Planck constant. Using the Giffard limit we get

$$T_{eff} = 2T_n \simeq 1.4 \times 10^{-7} \text{ K} \tag{29a}$$

with an ultimate sensitivity for Al and $M = 2300$ kg (see (23))

$$h \simeq 2.9 \times 10^{-21}. \tag{29b}$$

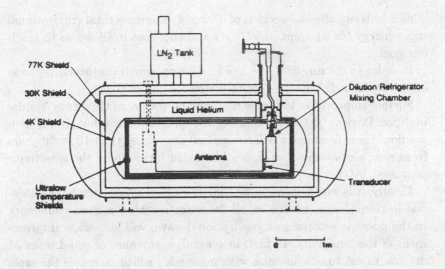

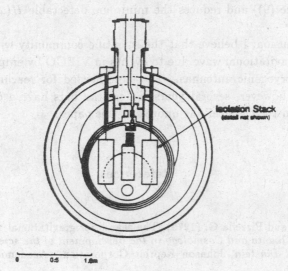

Fig. 6. Scheme of the STANFORD cryostat equipped with a dilution refrigerator for cooling the Al bar to 40 mK.

This sensitivity allows detection of 0.02 of a solar mass total gravitational wave energy for an event in VIRGO and therefore it allows us to reach our goal.

In order to obtain (29a) one has to develop new transducers, perhaps of the parametric type, and/or to construct new SQUIDs.

Further improvements can be obtained by the use of Quantum Nondemolition Devices (QND) for beating the quantum limit, but, in my own opinion, these techniques will be applied not earlier than 10 or 20 years from now, since a lot of work is still needed for reaching the sensitivity expressed by (29).

Finally, it is worth mentioning that the use of several resonant antennas in coincidence, needed in all cases for reaching a good confidence in the possible detection of gravitational wave, will also allow improvement of the sensitivity. In fact, in general, n antennas of equal mass M are equivalent to one antenna with mass nM, which increases the cross section (see (9)) and reduces the minimum detectable $H(\omega)$ value (see (23)).

In conclusion, I believe that the scientific community will be able to observe gravitational wave due to events in VIRGO by employing a few resonant cryogenic antennas. The time needed for reaching this goal could be, however, several years, unless the bars have a cross section larger than the classical one quoted in this paper.

References

Amaldi E. and Pizzella G. (1979), The search for gravitational waves, in *Relativity, Quanta and Cosmology in the development of the scientific thought of Albert Einstein*, Johnson Reprint Corporation, (Academic Press, New York).

Amaldi E., Cosmelli C., Pallottino G. V., Pizzella G., Rapagnani P., Ricci F., Bonifazi P., Castellano M. G., Carelli P., Foglietti V., Cavallari G., Coccia E., Modena I. and Habel R. (1986) *Il Nuovo Cimento* 9C, 829.

Bonifazi P., Ferrari V., Frasca S., Pallottino G. V. and Pizzella G. (1978), Data analysis for gravitational wave antenna, *Il Nuovo Cimento* 1C, 465.

Boughn S. P., Fairbank W. M., Giffard R. P., Hollenhorst J. N., Mapoles E. R., McAshan M. S., Michelson P. F., Paik H. J. and Taber R.C. (1982), Observations with a low temperature, resonant mass, gravitational radiation detector, *Ap. J.* 261, L19.

Giffard R. (1976), Ultimate sensitivity limit of a resonant gravitational wave antenna using a linear motion detector, *Phys. Rev.* D14, 2478.

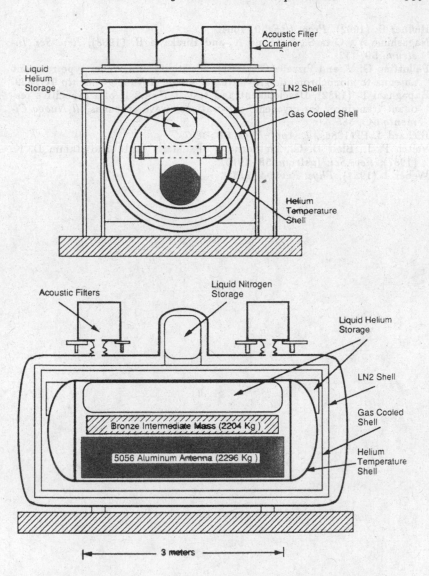

Fig. 7. Scheme of the LSU cryostat. We notice the bronze intermediate mass that is an efficient mechanical filter.

Heffner H. (1962), *Proc. IRE* **50**, 1604.

Nagashima Y., Owa S., Tsubono K. and Hirakawa H. (1988), *Rev. Sci. Instrum.* **59**, 112.

Pallottino G. V. and Pizzella G. (1981), Sensitivity of a Weber type resonant antenna to monochromatic gravitational waves, *Il Nuovo Cimento* **9C**, 829.

Rapagnani P. (1982), Development and test at $T=4.2$ K of a capacitive resonant transducer for cryogenic gravitational wave antennas, *Il Nuovo Cimento* **5C**, 385.

Richard J. P. (1986), *J. Appl. Phys.* **60**, 3807.

Veitch P. J., Blair D. G., Linthorne N. P., Mann L. D. and Ramm D. K. (1987), *Rev. Sci. Instrum.* **58**, 1910.

Weber J. (1954), *Phys. Rev.* **94**, 215.

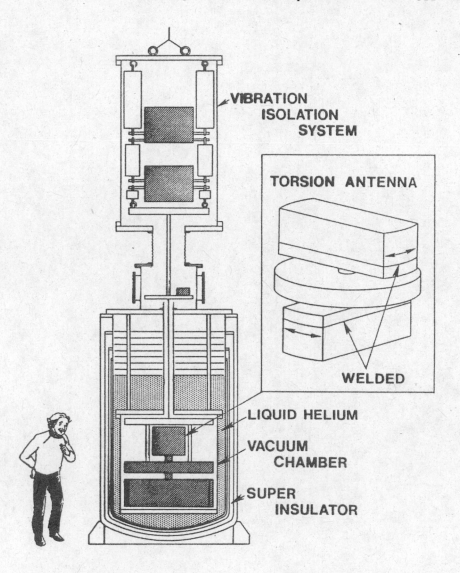

Fig. 8   Scheme of the CRAB IV detector of the Tokyo group.

12

Solar system tests of general relativity: recent results and present plans

Irwin I. Shapiro
Harvard-Smithsonian Center for Astrophysics,
60 Garden Street,
Cambridge, MA 02138 USA

1. Introduction

Experimental tests of general relativity are difficult. Physicists were well aware that pregnant new conceptual insights came mostly from young minds. Now, experimentalists must be cut from that mold, previously the exclusive preserve of theorists. Why? The time elapsed between a good idea for an experimental test and its execution is fast approaching the normal human lifetime. Experimenters must now start young–very young–to live to see the fruits of their ideas realized.

I divide the remainder of this paper into two parts, corresponding, respectively, to the past light cone and the future light cone. Under the former rubric, I discuss, in order, tests of the principle of equivalence, light deflection, signal retardation, perihelia advances, geodetic precession, and the constancy of the gravitational "constant." Under the latter, I mention improved reincarnations of some of these tests, as well as proposed redshift and frame-dragging experiments.

With the proper reference frame established, I move on to the review of recent results and present plans. Because of space and time constraints, much of the treatment is perforce superficial, but, in keeping with the Fourth of July spirit, I shall be democratic and treat all experiments with (almost) equal superficiality.

313

2.   Past light cone

Principle of equivalence

Space tests of the principle of equivalence have been primarily concerned with measuring the equivalence between gravitational and inertial mass in regard to the contribution to each of gravitational self-energy. A violation of the principle in such a case would lead to a non-null Nordtvedt (1968) effect. In particular, the parameter η would be non-vanishing in the relation

$$\frac{M_G}{M_I} = 1 + \eta \left(\frac{U_G}{Mc^2} \right),$$

where M_G, M_I, and U_G represent, respectively, gravitational mass, inertial mass, and gravitational self energy. In the case of observations of a laboratory body, the coefficient of η is quite trivial: about 10^{-23}, for a dense object about 1 meter in diameter. With respect to the orbit of the Moon about the Earth, on the other hand, this coefficient is about 10^{13} times larger. A positive value of η would lead to a displacement of the orbit of the Moon in the direction of the Sun. Lunar laser ranging data now show (Shapiro et al. (1990)): $\eta = 0$ to within the estimated standard error of 0.005. For fully conservative metric theories with preferred frame effects ignored, one can express η in terms of the Parameterized Post-Newtonian (PPN) formalism as

$$\eta = 4\beta - \gamma - 3$$

which implies, using information from light deflection and signal retardation for the value of γ (see the following two sections):

$$\sigma(\beta) \leq 0.002,$$

with β being consistent with unity to within this standard error.

Light deflection

The usual formula given for the deflection of starlight by the Sun is

$$\Delta\Theta \approx 1.75 \left[\frac{1+\gamma}{2} \right] \frac{R_s}{d} \text{ arcsec},$$

where d is the distance of closest approach to the Sun of the path of the detected starlight and R_s is the radius of the Sun.

All experiments to date have been consistent with $\gamma = 1$, to within their respective uncertainties. The most accurate Earth-based experiments to measure the deflection are now done using radio interferometry

(Shapiro (1967)). For these experiments, the theoretical formulation of the deflection is more appropriately given in terms of the difference in arrival times of the wavefront from a distant quasar at the separate antennas of the interferometer. For this situation, we may more accurately represent the effect of the deflection due to the Sun as

$$\Delta\tau \approx \left(\frac{1+\gamma}{2}\right)\frac{2r_0}{r_{se}}\left[\frac{(\mathbf{r}_1 - \mathbf{r}_2)\cdot(\hat{\mathbf{r}}_{se} + \hat{\mathbf{r}}_{sq})}{1+(\hat{\mathbf{r}}_{se}\cdot\hat{\mathbf{r}}_{sq})}\right],$$

where $c = 1$; r_0 is the gravitational radius of the Sun; $\hat{\mathbf{r}}_{se}$ and $\hat{\mathbf{r}}_{sq}$ are unit vectors from the Sun to the Earth and to the quasar, respectively; r_{se} is the Sun-Earth distance; and $\mathbf{r}_i$ ($i = 1, 2$) are the geocentric vectors of the antenna sites.

For connected-element interferometers, the most accurate result has been obtained by Fomalont and Sramek (1976), who found

$$\sigma(\gamma) = 0.02,$$

superseding the result from very-long-baseline interferometry (VLBI) obtained by Counselman et al. (1974). A more recent VLBI result was reported by Carter et al. (1985):

$$\sigma(\gamma) = 0.003.$$

Unfortunately, this result is marred by systematic errors stemming from the use of the then-standard model for the Earth's nutations, which was later shown to be inaccurate. In particular, this result for γ changes by about 2% (Herring and Shapiro (1987)), when the corrected model (Herring et al. (1986)) is used for the nutations of the Earth. Further, the data used by Carter et al. (1985) were obtained primarily from experiments carried out for another purpose, in which virtually no ray paths from the observed extragalactic radio sources passed near the Sun. The lack of sensitivity to the deflection accompanying any one observation was compensated by the use of a very large number of observations ("square root of n" effect). Hence, the estimate obtained for γ was quite sensitive to systematic errors and the discrepancy between the "corrected" result and general relativity cannot be considered significant.

Signal retardation

For metric theories of gravity, the retardation of light signals passing near massive bodies provides a test strongly related to that of the deflection.

For signals transmitted from the Earth to another planet in the solar
system, the contribution to the round-trip delay due to the direct effect
of solar gravity can be represented in isotropic coordinates as

$$\delta\tau \approx \left(\frac{1+\gamma}{2}\right) 4r_0 \, \mathrm{Ln}\left(\frac{r_{se}+r_{sp}+R}{r_{se}+r_{sp}-R}\right),$$

where r_{sp} is the Sun-planet distance and R is the Earth-planet distance.
The result from the Viking experiment, $\gamma = 1$ to within $\sigma(\gamma) = 0.002$,
in the 1970's (Reasenberg et al. (1979)) has not yet been surpassed.

Advances of planetary perihelia

The contribution of the Sun to the secular advances of planetary perihelia is given by the well-known formula,

$$\Delta\omega = \frac{6\pi GM_s}{a(1-e^2)c^2}\left[\frac{2+2\gamma-\beta}{3} + \frac{c^2 R_s^2 J_2}{2GM_s a(1-e^2)}\right] \mathrm{rad/rev},$$

where M_s and R_s are the mass and radius of the Sun, respectively; J_2
is the coefficient of the second zonal harmonic of the Sun's gravitational
field; and a and e are the planet's orbital semimajor axis and eccentricity, respectively. Here, the first term in brackets is the well-known
effect of general relativity, with the modification entailed by the PPN
formalism, and the second term is the Newtonian effect of the gravitational oblateness of the Sun. (The G's and c's have been included to
confuse theoreticians.) A key question: What is the value of J_2? There
are several "levels" of answers to this question; one is based on optical measurements of the brightness contours of the solar surface, and
another on helioseismology. The optical measurements have led some
(e.g., Dicke et al. (1986)) to conclude that the average value of J_2 is far
larger than would be the case were the Sun rotating uniformly throughout at the average angular velocity observed for its surface. The second
level involves inferences on the rotation of the deep interior from the
rotation-induced line-splitting in the acoustic wave power spectrum obtained from observations of oscillations of the solar surface. These latter
measurements (see, e.g., Duvall et al. (1984)) imply that the Sun's gravitational quadrupole moment is very close to the value it would have,
were the Sun rotating uniformly throughout: $\sim 2 \times 10^{-7}$.

Yet a third level of determination, through separation of the two terms
in the above equation, based on comparisons of the perihelia advances of
different planets and asteroids and/or on distinctions from short-period
effects, has yet to yield definitive results.

We consider the determination of the Sun's gravitational quadrupole moment to be most reliably determined at present from helioseismology, implying a value for J_2 of $\sim 2 \times 10^{-7}$ which corresponds to a correction of about 0.05% to the secular advance of the perihelion of Mercury's orbit. From measurement of Mercury's orbit through planetary radar ranging, Shapiro et al. (1989) infer that $\beta = 1$ to within a standard error of

$$\sigma(\beta) = 0.003 .$$

This value for β assumes that γ and its standard error are as determined from the signal retardation experiments (see previous section), and that the value for J_2 inferred from the helioseismological evidence is correct, or nearly so.

Geodetic precession

W. de Sitter (1916) showed that general relativity predicted an "extra" precession of the longitude of the node of the Moon's orbit of nearly two seconds of arc per century. In the PPN formalism, the expression for this precession can be written as

$$\dot{\Omega}_g = \frac{3}{2}\left(\frac{1 + 2\gamma}{3}\right)|\mathbf{r}_{sm} \times \mathbf{v}_{em}|\frac{R_s}{r_{se}^3}$$
$$\approx 2\,\mathrm{arcsec/cy} ,$$

where $\mathbf{r}_{sm}$ represents the Sun-Moon vector and $\mathbf{v}_{em}$ represents the velocity of the Moon with respect to the Earth. The first observational result on this prediction was obtained by Slade (1971) but was only of marginally useful significance:

$$\dot{\Omega}_g \approx (1.5 \pm 0.6)\,\mathrm{arcsec/cy} .$$

More recently, Bertotti et al. (1987) inferred a value of

$$\dot{\Omega}_g \approx (2.0 \pm 0.2)\,\mathrm{arcsec/cy} .$$

Their work did not involve any analysis of data, but rather was based on an inference from other analyses of lunar laser ranging data, which did not disclose any "anomalous" signatures in the residuals. Our direct analysis (Shapiro et al. (1988)), based on almost 20 years of lunar laser ranging data, yielded the following value:

$$\dot{\Omega}_g \approx (2.02 \pm 0.04)\,\mathrm{arcsec/cy} .$$

Use of data that encompassed an interval comparable to the 18.6 year nodal period of the Moon's orbit was necessary in order to break certain degeneracies. This value of nearly 2 arcsec/cy for the geodetic precession, and its estimated standard error, were subsequently confirmed in an independent analysis of virtually the same data set (Dickey et al. (1989)).

Gravitational constant

Over a half-century ago, Dirac (1937) put forward his "large numbers hypothesis" which, in essence, states that the gravitational "constant" would have to be a decreasing function of (atomic) time to maintain the approximate coincidence between the present age of the universe, expressed in atomic units (approximately 10^{40}), and the ratio of the electrostatic to the gravitational forces between two charged elementary particles (also approximately 10^{40}). To test this hypothesis, Shapiro (1964) assumed that for short periods, of the order of a human lifetime, one could represent such a putative time variation by

$$G(t) \approx G_0[1 + (\dot{G}_0/G_0)(t - t_0)],$$

where G_0 and G_0 are the values of G and its rate of change, respectively, at the epoch t_0. Subsequent data from radio and radar ranging observations of the inner planets have yielded two results:

$$(\dot{G}_0/G_0) = (2 \pm 4) \times 10^{-12}\,\mathrm{yr}^{-1}$$

and

$$(\dot{G}_0/G_0) = (-2 \pm 10) \times 10^{-12}\,\mathrm{yr}^{-1}.$$

The former was obtained by Hellings et al. (1983) and the latter by our group, quoted in Anderson et al. (1989). The main difference between the two analyses was in the allowance for uncertainties caused by the gravitational "noise" introduced by the myriad of charted and uncharted asteroids. The difference in the two quoted "standard" errors can thus be considered as a measure of the uncertainty in the uncertainty of this contribution.

Timing data from radio observations of the binary pulsar PSR 1913+16 have also been interpreted in terms of a limit on the possible time variation of the gravitational constant (Damour et al. (1988)):

$$(\dot{G}_0/G_0) = (10 \pm 23) \times 10^{-12}\,\mathrm{yr}^{-1}.$$

This result seems nearly comparable to that from planetary radio and radar ranging. However, there is an important difference. For the timing data from the binary pulsar, the signature of a changing gravitational constant is virtually identical to the signature from gravitational radiation. Thus, one cannot, simultaneously, distinguish a deviation of gravitational radiation from the predictions of general relativity and a deviation from constancy of the gravitational constant. One can obtain only a combined constraint which, at present, shows consistency between the observations and the assumptions that: a) general relativity correctly describes gravitational radiation; and b) the gravitational constant does not vary with (atomic) time.

3. Future light cone

We now discuss present and planned tests.

Redshift Experiments

Distant vs. Terrestrial Clocks. There has been only one experiment with a standard error under 1% that verifies the predicted gravitational redshift. This experiment, which provides verification at nearly the level of one part in 10^4, was carried out by Vessot et al. (1980) using hydrogen maser frequency standards, with one standard launched on a suborbital rocket flight. Another approach, known in principle for at least two decades, involves the annual variation in the rate of terrestrial clocks as the Earth-Sun distance changes. The predicted amplitude is about 1.6 milliseconds. This predicted variation can be observed by comparing a terrestrial clock with one at infinity. In 1967, Jocelyn Bell Burnell discovered that nature had been kind enough to provide us with such clocks. More recently, we were handed a nearly ideal situation when Don Backer and his colleagues discovered a millisecond pulsar for which pulse time of arrivals can be measured with only microseconds or less of uncertainty and for which any slowdown in rate is (almost) too slow to measure.

An accurate redshift experiment thus seems trivial in principle, and should be easy in practice: just compare the readings of the terrestrial and the pulsar clocks as the Earth orbits the Sun. Unfortunately, it is not so simple. Why? Because we do not know *a priori* in which direction the pulsar resides. These very same time-of arrival data would have to be used as well to estimate the pulsar's position in the sky. The annual signature in the pulse time-of-arrival data of a slight difference in position, however, is virtually indistinguishable from the sought-for effect on the terrestrial clock.

Is there a way out? Yes; VLBI, which plays a strong role in the deflection test (see above), may be the savior (Shapiro (1986)). By determining the position of the pulsar by VLBI to milliarcsecond, or better, accuracy, it would be possible to test the "redshift" prediction to about 1 part in 10^3 or better. But for the millisecond pulsars presently known, this determination is distinctly non-trivial, primarily because the spectrum of radio emissions from pulsars is so steep. The flux density decreases rapidly as the radio frequency increases; because the flux density itself, even at the low radio frequencies is so small, it is difficult to correct for ionospheric effects and still obtain a usable signal level. Nonetheless, we—Norbert Bartel, Michael Ratner, other colleagues, and I—made our first test VLBI observations on this pulsar to determine its position relative to a distant quasar nearby on the sky. The experiment was severely hampered by equipment failure at a remote radio telescope, and the standard error in relative position determination was high, about 20 milliarcseconds. We hope to do substantially better in future VLBI observations.

Solar vs. terrestrial clocks. A dramatic improvement in the accuracy of the redshift experiment would be possible were a precise atomic clock to be sent near the Sun. This idea, like many others in this field, is at least a quarter of a century old. It has been discussed seriously since the mid-1970's when it was championed by the late Giuseppe Colombo. Now, again, interest in such a spacecraft is rising, to study the solar wind in situ, as well as to test gravitational theories. A probe, however, would not likely reach the vicinity of the Sun until the first decade of the next century. What would such a probe look like? A sunshield will be needed as well as, likely, a drag-free system, i.e., a freely-falling object inside the spacecraft, largely shielded from non-gravitational forces and having the gravitational forces due to the spacecraft "balanced." The heart of the experiment would likely be a hydrogen maser frequency standard which Robert Vessot would propose to build. He believes it would be relatively straightforward to construct such a maser to operate under conditions near the Sun with a frequency stability of about 1 part in 10^{15}, i.e., with an Allan standard deviation of about 10^{-15} for averaging times of up to several hours.

How would we get the spacecraft and clock near to the Sun, given the daunting energy requirements? The best way seems to be to continue to play celestial billiards and to use a Jupiter swing-by. Of course, the flight would then take a bit more than the eight minute Earth-Sun light travel time—by a factor of nearly 300,000. Of course, once arrived, things happen awfully fast. In less than one day it is all over. But the completion of the analysis may take years.

New light-deflection experiments

Radio interferometry. VLBI still offers the best near-term prospects for improvement of the light-deflection test of theories of gravitation. We—Brian Corey, Dan Lebach, other colleagues, and I—have recently carried out VLBI observations for a new deflection experiment. Preparations for this experiment took more than five years. Our projected (more precisely, hoped-for) accuracy is a determination of γ with a standard error of 0.2%. In this experiment we used the Mark III VLBI system (Rogers et al. (1983)). We also utilized two antennas at each of two sites: the Owens Valley Radio Observatory near Big Pine, California, and the Haystack Observatory in Westford, Massachusetts. The distance between these sites is approximately 3,500 kilometers. Two radio sources were observed—3C273B and 3C279, with the latter being occulted by the sun on or about each October 8th. Our observations, made during the first two weeks of October 1987, utilized three frequency bands: 2.2, 8.4, and 23 GHz. The larger of the two antennas at each site made observations at the highest frequency band. For the majority of the days of the experiment, for the approximately eight hours of daily common visibility of the sources from each site, the four antennas moved back and forth in synchronism from one source to the other, measuring the interferometric group delay (Shapiro (1976)) at the three frequency bands.

The purpose of making observations of two sources was to largely eliminate the effects of differences in drifts of the atomic frequency standards at each site and to decrease the effect of tropospheric effects through use of the difference observations. The magnitude of the corresponding difference in gravitational deflection predicted by general relativity for observations of these two sources is, however, reduced substantially. For example, for observations made three days before the occultation of 3C279, the difference deflection is 100 milliarcseconds in declination and about 175 milliarcseconds in right ascension.

The main sources of error for this experiment are: i) statistical, due to the limited signal-to-noise ratio and spanned bandwidth of the observations at each frequency band; ii) the solar corona which has the most severe effect on the signals, not only because of the high density of the plasma, but mostly because of its turbulent nature. [The delay of the signal caused by refraction in the solar corona is proportional to the inverse square of the frequency whereas the delay introduced by scattering varies as the inverse fourth power of the frequency; were sufficient signal-to-noise ratios obtainable at all three frequency bands simultaneously, we could largely remove the effects of the solar corona on the interferometric signals. The ionosphere, a more benign cousin of the corona, causes us little difficulty because scattering is negligible and sufficiently strong signals were almost always available from two of the frequency bands.]; iii) the Earth's troposphere, which is the largest source of error after the solar corona, is a problem primarily due to the non-hydrostatic nature of the water-vapor distribution. [To try to measure the water-vapor effect, we had both water-vapor radiometers and infrared hygrometers operating at each site.]; iv) source structure. [Neither 3C273B nor 3C279 is a point source at the milliarcsecond resolution of the observations. Such source structure would have no effect on the observations, provided that: a) the observation schedule and its implementation did not vary from day to day; and b) the structure of the sources remained constant over the approximately two-week duration of the experiment. Unfortunately, neither condition held so we must "map" the source brightness distributions.]; v) variations in the Earth's rate of rotation and in its polar motion. [Such Earth-orientation data are available with (almost) sufficient accuracy from other contemporaneous VLBI observations as well as from observations we made that involved a third site at Fort Davis, Texas, on one day of the experiment.]; and vi) delays undergone by the signals from entrance to the receiver until recorded on magnetic tape. [The calibration system used to monitor the variations in this instrumental delay worked well almost all of the time at each site.]

Reduction and analysis of these observations is extraordinarily difficult because we were "pushing" the capabilities of the systems in virtually every respect to try to achieve this factor of ten improvement in accuracy. It remains to be seen whether we will reach our goal.

Optical interferometry. Extension of interferometry to optical wavelengths and the placement of the interferometer in Earth orbit offer the possibility of detecting contributions to the light deflection by solar gravity up to terms of second order in the mass of the Sun. The exterior metric for the Sun can be written as:

$$ds^2 = \left[1 - \frac{2M_s}{r} + \frac{2\beta M_s^2}{r^2} + ...\right]dt^2 - \left[1 + \frac{2\gamma M_s}{r} + \frac{3\epsilon M_s^2}{r^2} + ...\right] dx \cdot dx\,,$$

which yields a post-post-Newtonian contribution to the deflection of

$$\overrightarrow{\Delta\Theta}_{\text{pPN}} = \pi(2 + 2\gamma - \beta + \frac{3}{4}\epsilon)\frac{M_s^2}{r_i^2}\hat{r}_i$$

$$\approx 11\mu\text{as for } r_i = R_s\,,$$

where r_i is the "impact vector." Smaller contributions come, respectively, from the gravitational oblateness of the Sun and from its angular momentum vector. For "typical" geometries

$$\Delta\Theta_{J_2} \sim \frac{2(1 + \gamma)}{r_i^3} M_s R_s^2 J_2$$

$$\sim 0.2\mu\text{as for } r_i = R_s;$$

$$\Delta\Theta_L \sim \frac{2(1 + \gamma)}{r_i^2}L_s$$

$$\sim 0.7\mu\text{as for } r_i = R_s\,,$$

where $J_2 \sim 10^{-7}$ and $L_s \sim 2 \times 10^{48}$ g cm^2 sec^{-1} for an assumed uniform rotation of the Sun (Epstein and Shapiro (1980)).

How could such small effects possibly be measured? One approach (see, for example, Reasenberg et al. (1988), and references therein) involves a precision optical interferometer in space, dubbed POINTS. This instrument consists of two mutually (almost) perpendicular, two-element Michelson interferometers. In the present "straw man" design, each mirror diameter is approximately 0.25 m and each baseline approximately 2 m. A key to obtaining high-accuracy results is an internal laser metrology system, whose errors are at the picometer level. This system is designed to work on the full aperture so as to mimic as closely as possible the lightpath from the observed sources. By having two interferometers at approximately right angles to each other, one can determine the separation of a star whose line of sight passes near to that of the Sun, with respect to the position of a star nearly at right angles to the first. For a given "articulation range" between the two interferometers, this configuration maximizes the probability of finding suitable companion stars, among other benefits.

Overall, the instrument will have an optical efficiency of about 2% and observe stars down to magnitude 10, with the expected standard error in the determination of the arclength between the two stars observed simultaneously being approximately 5 microarcseconds, after 10 minutes of integration. Control of systematic errors is based not only on the metrology system, *per se*, but also on the use of extremely stable materials and active thermal control. To guard against systematic errors, observations will enable use of sky "closure." Each individual observation will involve two stars separated by about 90 degrees on the sky, but subsequent observations will be scheduled to complete arcs that go around the sky. Each appropriate star pair will be observed at least four times per year. All such pairs with one star within four degrees of the Sun will be observed at least once daily. The resultant expected standard error in the determination of the post-post-Newtonian deflection is

$$\sigma(\Delta\Theta_{\mathrm{ppN}}) \sim 3\,\mu\mathrm{as}.$$

When will such an interferometer actually be launched into space? No one knows. However, a reasonable lower bound can be set: the year 2000.

Perihelia advances

Dramatic advances could be made in measuring general relativistic effects on the orbit of Mercury were long-lived landers with radio transponders placed on its surface. Such a hope will most likely not be realized in this century. The next best approach is to continue radar observations of Mercury. Here the major source of error is the poorly-known variation in topography on the planet's surface. Combined with the uncertainties in the radar scattering law, the topography problem has prevented by more than tenfold the realization of the full accuracy inherent in the delay measurements. Modeling the topography to the accuracy of the radar measurements is intrinsically very difficult—we have a rather sophisticated model with hundreds of adjustable parameters and, yet, it is very, very crude. So is there a way out? In principle, yes: the use of "closure" sets of observations, sets in which the (multiple) members of each were obtained with Mercury presenting the same "face" to the Earth, but from different orbital positions. The opportunities for such measurements are not frequent and, when scheduling constraints of the only two systems capable of making these measurements—in Arecibo, PR and Goldstone, CA—are considered, the realizable opportunities drop further to a level of approximately one dozen or less per year. These observations are being pursued cooperatively by Don Campbell, John Chandler, Martin Slade, myself, and others.

The other orbiting entity whose perihelion precession is easily observable is the asteroid Icarus, whose orbital eccentricity is about 0.82. Observations of Icarus have been made during every close approach it has made to the Earth since 1950. The most impressive set of observations–optical and (marginal) radar detections–were made in 1968. A superb opportunity will occur again in 1996 and the Arecibo radar system should be about twentyfold more sensitive on average then than now, and should be able to track Icarus over a relatively long arc. With this thirty-year-long separation between two sets of impressive observations, the relativistic advance should be determined with an uncertainty of only a few percent. Unfortunately, Icarus is not much help in separating out the effects of the Sun's gravitational quadrupole moment since Icarus' orbit has approximately the same semilatus rectum, $a(1 - e^2)$, as does Mercury's.

Frame Dragging

As noted first by Lense and Thirring (1918), the theory of general relativity predicts that rotating massive bodies "drag" inertial frames. Such frame dragging is the analog of the principle of equivalence. All test particles, independent of their mass, are predicted to fall at the same rate when in the same gravitational field, under common circumstances. Correspondingly, all similarly oriented test gyroscopes, independent of their angular momenta, are predicted to precess at the same rate when in free fall about the same rotating mass, under common circumstances. The orbital plane of any test particle in a given orbit about a rotating mass is also predicted to precess at a common rate. These two manifestations of frame dragging can be expressed, respectively, as

$$\frac{d\hat{s}}{dt} = \left(\frac{1+\gamma}{2}\right) \frac{GI\omega}{2c^2a^3(1-e^2)^{3/2}}[\hat{c} \times \hat{s} + 3(\hat{c} \cdot \hat{L})(\hat{L} \times \hat{s})]$$

and

$$\dot{\Omega} = \left(\frac{1+\gamma}{2}\right) \frac{2GI\omega}{c^2a^3(1-e^2)^{3/2}},$$

where $d\hat{s}/dt$ and $\dot{\Omega}$ are the rates, respectively, of change of direction of an orbiting test gyroscope and of the orbital plane of a test particle, due to frame dragging, i.e., due to their orbiting a massive spinning body. The quantities I and ω represent, respectively, the moment of inertia and angular velocity of the massive spinning body; a and e denote, respectively, the semimajor axis and eccentricity of the orbiting body; and $\hat{c}$, $\hat{s}$, and $\hat{L}$ are unit vectors in the directions of the angular momentum vector of the massive spinning body, the spin vector of the gyroscope, and the normal to the orbital plane of the gyroscope, respectively. For a test gyroscope, the effect is maximized for $\hat{s}$ approximately perpendicular to $\hat{c}$ and $\hat{L}$, with the latter two parallel to each other. These predicted effects are, to be sure, not large. For objects in low Earth orbit, the predicted precession rates are of the order of tens of milliarcseconds per year.

Gyroscope precession. The gyroscope experiment was suggested by Pugh (1959) and Schiff (1960), independently, at the start of the space age. The concept for this experiment has been developed primarily at Stanford, with substantial assistance from the Lockheed Corporation and NASA's Marshall Space Flight Center. During the past quarter century of active development of this experiment, awesome technological progress has been made by Francis Everitt and his many colleagues.

The gyroscope system is to consist of four exquisitely precise spherical quartz balls, coated with niobium and cooled to liquid-helium temperatures. These balls are initially spun up via gas pressure and their positions electrostatically controlled. Readout of the gyroscope's spin vectors is accomplished via a SQUID magnetometer which monitors the orientations of the gyroscopes' London moments, induced in their superconducting coating. The ensemble is kept extremely well shielded from external magnetic fields; their effects are attenuated by well over seven orders of magnitude. The spacecraft, which will provide a nearly drag-free environment for the test gyroscopes, is to be placed in a near-circular polar orbit at a mean altitude of about 650 km. The different gyroscopes, in total, will be sensitive not only to the frame dragging, predicted to be about 42 mas/yr for that orbit and appropriate gyroscope orientation, but also to the geodetic precession (see Section 2), which is predicted to be 6600 mas/yr. The projected error budget for the determination of the (free fall) precession of the gyroscopes is about 0.1 mas/yr.

With what are the orientations of the spin axes of the gyroscopes to be compared? In the first step, these orientations are monitored with respect to the quartz housing of the gyroscopes. This housing is to be optically bonded to a quartz telescope that points to a "fixed" star, so the gyroscope's direction can, via this "chain," be monitored with respect to the direction to this reference star. But how "fixed" is the reference star? The star Rigel, originally proposed as a reference, has a proper motion whose uncertainty in the (nearly) inertial reference frame formed by compact, extragalactic objects is at present imprecisely known, the standard error being almost an order of magnitude larger than the uncertainty expected from the performance of the gyroscope system. There are several solutions to this possible degradation of the experiment. One is to place another set of gyroscopes in a different altitude orbit, with use made of the same reference star, so that the star's proper motion could be cancelled in the comparison of the precessions for the two sets of gyroscopes. This possibility, which Francis Everitt reminded me I suggested nearly 15 years ago, is of course not exactly an inexpensive solution. A second possible solution is to await the far more precise measurements of proper motion possible with a space astrometric optical interferometer such as POINTS, described above. A third approach is to seek a suitable reference star that is also sufficiently radio bright to enable its proper motion to be determined with more than sufficient accuracy with VLBI, using techniques similar to, but simpler than, those described above in connection with the light-deflection experiment.

The gyroscopes are expected by the experimenters to be launched in the mid-1990's and to yield standard errors in the precession measurements of well under 1% of the predicted contribution from frame dragging and under 0.01% of the predicted contribution from the geodetic precession.

Orbital plane precession. The possibility of detecting the predicted precession (Lense and Thirring (1918)) of orbital planes was considered for a single Earth satellite by Yilmaz (c. 1959). A clever scheme involving counter-orbiting polar satellites of the Earth was proposed by Van Patten and Everitt (1975). Most recently, Ciufolini (1986) has proposed to determine the frame-dragging induced nodal precession by measuring the sum of the nodal precessions for two satellites in orbits with supplementary orbital inclination angles. The major problem with detecting the contribution of frame dragging to the nodal precession is the uncertainty in our knowledge of the even zonal harmonics of the Earth's (external) gravitational field. These harmonics contribute with equal magnitude, but opposite sign, to the nodal precession of satellites with supplementary orbital inclination angles. Thus the sum of the nodal precessions for the orbits of two such satellites is insensitive to the values of the even zonal harmonics.

In orbit about the Earth now is the LAGEOS I satellite, which has a very low area-to-mass ratio ($\sim 7 \times 10^{-4}$ m^2/kg) to minimize the effects of nongravitational accelerations on the orbit and which is covered with corner-cube retroreflectors to facilitate laser ranging, primarily for studies of the Earth's gravitational field. The orbit of LAGEOS I is nearly circular, with a semimajor axis of about 12,000 km and an orbital inclination $i = 110$ deg. Thus, Ciufolini and his colleagues have proposed the orbiting of an additional satellite, LAGEOS III, with the same orbital parameters as LAGEOS I, save for $i = 70$ deg. For a satellite in such an orbit, the contribution to the nodal precession from frame dragging is predicted to be about 31 mas/yr. [LAGEOS II, soon to be launched, will not be in an orbit suitable for this relativity experiment.]

How accurately will this predicted contribution to the nodal precession be measurable? Two groups undertook extensive analyses to provide an answer: the Italian group used primarily an analytical approach and the University of Texas group performed a covariance analysis, based on a numerical approach. The results from both groups were evaluated by a NASA advisory panel, which concluded that the standard error in a measurement of Ω from these satellites was likely to be about 10%, given three years of data, provided that the deviation of the inclination, due to orbit insertion errors of LAGEOS III, is no more than a few tenths of a degree from the intended supplementary inclination angle.

Launch of LAGEOS III will likely not take place before the mid 1990's.

Epilogue

My predecessor at the Harvard College Observatory, Harlow Shapley, is credited with the claim:

> A hypothesis or theory is clear, decisive and positive but it is believed by no one but the man (sic) who created it. Experimental findings, on the other hand, are messy, inexact things which are believed by everyone except the man who did the work.

This statement is as true today as when it was first enunciated earlier in this century, except when a) the theory is general relativity, or b) the experiment contradicts general relativity. In these latter cases, the roles of theory and experiment are reversed. To follow this prescription, I note that there is not yet one scintilla of believable experimental evidence inconsistent with the predictions of general relativity.

I thank B. E. Corey for showing me the quotation from Harlow Shapley, and the National Aeronautics and Space Administration for partial support of some of the research reported here, through grant NAGW-967. The remarkable similarity between this paper and ones I wrote a decade earlier in connection with Einstein's centenary is a reflection of the agonizingly slow pace of experimental relativity.

References

Anderson, J. D. et al. (1989), in *Advances in Space Research* **9**, (Pergamon, Oxford), p. 71.
Bertotti, B. et al. (1987), *Phys. Rev. Lett.* **58**, 1062.
Carter, W. E. et al. (1985), *J. Geophys. Res.* **90**, 4577.
Ciufolini, I. (1986), *Phys. Rev. Lett.* **56**, 278.
Counselman, C. C. et al. (1974), *Phys. Rev. Lett.* **33**, 1621.
Damour, T. et al. (1988), *Phys. Rev. Lett.* **61**, 1151.
de Sitter, W. (1916), *Mon. Not. Roy. Astr. Soc.* **77**, 155.
Dicke, R. H. et al. (1986), *Ap. J.* **311**, 1025.
Dickey, J. O. et al. (1989), in *Advances in Space Research* **9**, (Pergamon, Oxford), p. 75.
Dirac, P. A. M. (1937), *Nature* **391**, 323.
Duvall, T. L. Jr. et al. (1984), *Nature* **310**, 22.
Epstein, R. and Shapiro, I. I. (1980), *Phys. Rev.* **D22**, 2947.
Fomalont, E. B. and Sramek, R. A. (1976), *Phys. Rev. Lett.* **36**, 1475.
Hellings, R. W. et. al. (1983), *Phys. Rev. Lett.* **51**, 1609.
Herring, T. A. et al. (1986), *J. Geophys. Res.* **91**, 4745.
Herring, T. A. and Shapiro, I. I. (1987), unpublished.
Lense, J. and Thirring, H. (1918), *Phys. Z.* **19**, 156.
Nordtvedt, K. (1968), *Phys. Rev.* **169**, 1017.
Pugh, G. E. (1959), *Res. Memorandum No. 11*, Weapons System Evaluation Group, Department of Defense.
Reasenberg, R. D. et al. (1979), *Ap. J. Lett.* **234**, L219.
Reasenberg, R. D. et al. (1988), *Ap. J.* **96**, 1731.
Rogers, A. E. E. et al. (1983), *Science* **291**, 51.
Schiff, L. I. (1960), *Phys. Rev. Lett.* **4**, 215.
Shapiro, I. I. (1964), *Phys. Rev. Lett.* **13**, 789.
Shapiro, I. I. (1967), *Science* **157**, 806.
Shapiro, I. I. (1976), in *Methods of Experimental Physics* **12C**, (Academic, New York), p. 261.
Shapiro, I. I. (1986), in *Scientific Benefits of an Upgraded Arecibo Telescope*, (National Astronomy and Ionosphere Center, Arecibo), p. 225.
Shapiro, I. I. et al. (1988), *Phys. Rev. Lett.* **61**, 2643.
Shapiro, I. I. et al. (1989), in preparation.
Shapiro, I. I. et al. (1990), in preparation.

Slade, M. A., MIT Thesis (1971).
Van Patten, R. A. and Everitt, C. W. F. (1976), *Cel. Mech.* **13**, 429.
Vessot, R. F. C. et al. (1980), *Phys. Rev. Lett.* **45**, 2081.
Yilmaz, H. (1959), private communication.

13

Interferometric
gravitational wave detectors

Rainer Weiss
Massachusetts Institute of Technology,
Cambridge, Massachusetts 02139 USA

It would all be a lot easier (and more satisfying) if one were reporting on the discoveries being made and the new astrophysical information being observed through the gravitational wave channel but unfortunately this can not yet be done. Instead this talk as many others given by workers in this promising but not yet started field has to dwell on the current technical state and prospects. The prospects are now better than ever and one can only hope that in one of the next GR conferences the groups working in this area will be able to talk about the waveforms they are observing and the physics of the sources that are being uncovered.

The active search for gravitational waves from astrophysical sources has been in process for the past two and one half decades. The search began with J. Weber's initial experiments using resonant acoustic bar detectors. These detectors now much improved by advances in transducer technology, better seismic isolation and operation at cryogenic temperatures have attained rms strain sensitivities of $h \approx 10^{-18}$ in few Hertz wide bands in the 1 kHz region. More important three such detectors (Stanford, Louisiana State University and CERN/Rome) have made triple and paired coincidence measurements, thereby setting new upper limits on the gravitational wave flux incident on the Earth. The prospect for further improvement in sensitivity to possibly $h \approx 10^{-20}$ is realistic and should bring acoustic bar detectors to a point where they could measure the gravitational radiation from such sources as supernova collapses in our and neighboring galaxies. The acoustic bar detectors offer, at present, the best chance for the direct detection of gravitational radiation. A more complete and scholarly review of these systems is given by G. Pizzella in an accompanying paper at this conference (Chapter 11, this volume).

This overview deals primarily with detectors comprised of "free" masses sensed by optical interferometry. Prototype interferometric systems have been constructed in the United States, Germany, Scotland, France and now in Japan. Although these systems are approaching the current sensitivity of the acoustic detectors, there is no strong impetus to make interferometer prototypes into observational instruments. The prime emphasis is to develop them to attain the promise of high sensitivity, $h < 10^{-22}$, over a large detection bandwidth, 10Hz - 10kHz, sufficient to intersect all but the most pessimistic estimates for the astrophysical gravitational wave flux at the Earth. To make good on this promise, the interferometric systems must be scaled from laboratory dimensions of tens of meters to kilometers. The scaling is associated with substantially increased costs and public accountability and requires a major effort which will transform this field, initially populated by independent inventors and scientific entrepreneurs, into integrated groups of engineers and scientists more in the style of large science. At present there are large baseline interferometer efforts being planned by: a Caltech/MIT collaboration in the United States (presentation by R. Vogt at this conference), a Max Planck/Glasgow collaboration in northern Europe (H. Ward), a French/Italian collaboration in southern Europe (A. Giazotto, A. Brillet), and an Australian consortium in western Australia (D. Blair, H. Bachor). All have schedules to be on the "air" by the middle of the decade.

In interferometric detectors an arrangement of "free" masses responds to the metric perturbations of the gravitational wave. In most of the operating and proposed interferometers, the masses are distributed in the shape of an "L" so as to make the interferometer maximally sensitive to one polarization of the tensor gravitational wave and (providing the arms are close to equal) insensitive to frequency fluctuations of the laser light. The light enters the system from the vertex of the "L", is split between the two arms and then after multiple traversals of the arms is interferometrically recombined and detected. The gravitational wave induces phase shifts of opposite sign in the light stored in the two arms of the "L". By proper adjustment of the overall phase difference, the detected intensity can be made proportional to the phase difference of the light having gone the two paths so that the detected intensity becomes proportional to the gravitational wave strain. The conversion of gravitational wave strain to optical phase change is a function of the specific interferometer design.

Gravitational strain sensitivity and photon shot noise

First generation designs use optical delay lines and Fabry - Perot resonators as the optical storage (cavity) elements in the interferometer arms. The optical phase shift, ϕ, per gravitational wave strain, h, in these geometries is roughly proportional to the optical storage time, $\tau_{storage}$, divided by the optical period, τ_{opt}, up to storage times that are comparable to the gravitational wave period, τ_{gwave}. With longer storage times the response saturates but does not attenuate. The ability to measure the optical phase at the interferometer output is limited by photon counting statistics (shot noise) to roughly the reciprocal of the square root of the number of photons in a signal integration time, n_{int}. The allowed integration time depends on the nature of the source signal. If it is a pulse the integration cannot be longer than some fraction of the pulse length; on the other hand if it is a periodic source or a stochastic background source the integration time, using suitable signal processing, can become as long as the observation time.

$$h_{rms} \approx (\Delta h/\Delta\phi)\Delta\phi_{rms} \approx (\tau_{opt}/\tau_{storage})(1/\sqrt{n_{int}}) .$$

Second generation designs, some already tested in prototypes while others are being planned for test, are interferometers which use additional mirrors to enhance the light power circulating in the storage arms or to alter the bandwidth of the interferometer to gain optical phase sensitivity in a narrower band. By adding another mirror to the interferometer at the input to the system the entire interferometer can be made into a resonant cavity ("broadband recycling" due to R.Drever). The additional mirror does not change the bandwidth of the system but increases the circulating power in the arms by approximately the reciprocal of the total interferometer loss, thereby increasing the phase sensitivity against shot noise by the square root of the ratio of the cavity storage time to the total interferometer storage time, τ_{interf}.

$$h_{rms} \approx (\tau_{opt}/\sqrt{\tau_{storage}\tau_{interf}})(1/\sqrt{n_{int}}) .$$

The value of this procedure depends on the magnitude of the mirror and other optical losses in the system. It is always best to achieve a cavity storage time comparable to the gravitational wave period for optimum sensitivity. However if the mirror losses dominate and are small enough so that the interferometer storage time can be made longer than the cavity storage time, broadband recycling offers a net gain. This will become important in long baseline systems where the equivalent number of beam "hits" per mirror for a given storage time is smaller than in the present short interferometer prototypes.

Another second generation variant is to incorporate still another mirror in the interferometer at the output (dual resonant recycling due to B. Meers, a derivative of resonant recycling due to R. Drever). With these additional mirrors at the input and output of the interferometer; the system, now consisting of a set of coupled cavities, can be made resonant at the gravitational wave period with a bandwidth, Δf_{gwave}, roughly equal to the reciprocal of the interferometer storage time. The resonant buildup of the optical phase in the band embracing the gravitational wave frequency increases the optical phase sensitivity to gravitational strain linearly in the ratio of the interferometer to cavity storage time.

$$h_{rms} \approx (\tau_{opt}/\tau_{interf})(1/\sqrt{n_{int}}),$$

$$\tau_{storage} \approx \tau_{gwave},$$

and

$$\Delta f_{gwave} \approx 1/\tau_{interf}.$$

This appears to be a promising system for the detection of periodic sources and a stochastic background of gravitational waves, but much as broadband recycling, is only applicable in a search for gravitational waves below 10 kHz when used in a large baseline system.

Sensitivity limits due to random forces

An exquisite sensitivity to changes in the optical phase is not sufficient to detect gravitational waves, it is just as critical to make sure that the fluctuations in the optical phase are really due to gravitational waves rather than other perturbing influences (a trap some have fallen into when proposing new and extremely sensitive phase measuring concepts.) A substantial effort has been made by the research groups working with gravitational wave interferometers to reduce the random motions of the mirrors (test masses) that define the optical storage cavities. There is a litany of known noise sources that must be controlled, understood or measured. The most important of these are thermal and seismic noise in the terrestrially detectable gravitational wave band, gravitational gradient noise which defines this band and finally quantum noise which, in the absence of new ideas, sets an irreducible limit on the ultimate sensitivity.

In the detection band accessible on the ground, the perturbations due to random forces fortunately all share the common property that their influence on the gravitational wave measurement are uncorrelated at the ends of the storage cavity. As a consequence their influence on a gravitational wave strain measurement is reduced in proportion to the length of the interferometer baseline. This is one of the principal arguments for the long baseline systems and enables the search for gravitational waves with frequencies less than a kHz, the region many believe is most promising for the detection of astrophysical sources.

Thermal noise (the fluctuations of a mechanical system about thermal equilibrium) enters the interferometers in two ways. The first is through center of mass motions imparted to the test masses by the suspension systems used to make the masses "free" at frequencies above the principal resonance of the suspension and, the second, through the thermal excitation of normal modes with motions about the center of mass of the test masses and mirrors below their resonance frequencies. In both cases, since mechanical resonance buildup of gravitational wave motions is not used in the interferometric systems (in contrast to acoustic bar systems), the thermal noise contribution comes from the off resonance behaviour of a thermally driven oscillator. There are subtleties depending on material properties which determine the actual spectrum of the thermal noise, some of which are discussed by P. Saulson at this conference; broadly the thermal noise contribution to the interferometer noise varies as the square root of the temperature divided by the product of the mass and the quality factor (Q) of the oscillator. Although not a severely limiting noise in the planned initial long baseline interferometers, the reduction of thermal noise, especially from the suspensions, will become a major research effort as the interferometers improve and are used at lower frequencies.

Seismic noise–random accelerations of the ground due to naturally occuring phenomena such as winds, ocean waves and the continuous micro relaxations of the earth as well as human activity–appears as a formidable noise source at low frequencies. Seismic noise precisely because it is not a direct force on the test masses but rather an acceleration with respect to the primary inertial frame, is amenable to isolation techniques which refer the test masses to the primary inertial frame. The simplest isolation techniques are cascaded passive stages of springs and masses which serve to reduce seismic noise above 100 Hz to below the levels of the other noises in the initial long baseline systems. The isolation of seismic noise between 10 to 100 Hz will involve passive stages with long periods (now being carried out by the Pisa research group) or more effectively (in my opinion) by active isolation systems which use the test masses as inertial references in active control of the test mass suspension point. This concept has been worked on by several of the research groups involved with gravitational wave interferometers. The improved isolation of seismic noise at low frequencies is another major research activity that will benefit improved interferometers after the initial long baseline interferometers are operating.

Gravitational gradients from seismically driven density changes in the earth and wind and acoustically driven atmospheric density fluctuations exert unshieldable Newtonian forces on the test masses. The spectrum of these forces was studied by P. Saulson and they set a lower boundary of approximately 10 Hz on the gravitational wave band observable from the ground. Fluctuating gravitational gradients as well as the possibility of achieving very long baselines to observe long period sources have been the main impetus for proposing the LAGOS, Laser Gravitational Observatory in Space, (P. Bender, J. Faller).

The noise sources discussed (short of gravitational gradients) can be reduced by improved technology but there is an irreducible limit (possibly for want of a new idea) to the minimum measurable gravitational wave strain imposed by the quantum nature of the radiation field and matter. The specific means by which the quantum fluctuations enter the interferometers was guessed and estimated qualitatively in early studies of such systems but a consistent and physically satisfying explanation was first given by C. Caves. A heuristic argument similar to the Heisenberg microscope (but with many quanta) can be made to estimate the quantum fluctuations. A fluctuating radiation pressure on the cavity mirrors arises from a cross term between the single mode laser field (assumed noise free) and the field of the uncorrelated vacuum fluctuations in the same mode generated in the two storage cavities. The radiation pressure fluctuations are proportional to the square root of the optical power while the noise in the optical phase measurement is proportional to the reciprocal of the square root of the optical power. The overall noise due to the quantum nature of the light is, therefore, minimized at a specific optical power determined by the mass of the mirrors, the optical wavelength and the gravitational wave frequency. The limiting gravitational wave sensitivity is approximately

$$h_{rms} \approx \sqrt{h/m}\left(\frac{\tau_{gwave}}{L\sqrt{\tau_{int}}}\right)$$

where h is Planck's constant, m the mirror mass, L the interferometer arm length, τ_{gwave} the period of the gravitational wave and τ_{int} the signal integration time. The quantum limit is far below the projected sensitivity of the initial long baseline interferometers and, if current estimates of the astrophysical gravitational flux are even approximately true, quantum noise will not play a role until gravitational wave astronomy is well underway and future gravitational wave astronomers want to push the limits of the new science.

Practical matters

It is all well and good to talk about fundamental limits and the bright future but the planning and hard work on more mundane but also critical technical developments to scale interferometers from the prototype to long baseline systems will occupy the bulk of the time and energy of the engineers and scientists now in this research for the next half decade. There is not much value in making an exhaustive list of technical issues which must be addressed, except to indicate by a few examples, the nature of some of the technical work that must be done and thereby anticipate why some of us on the experimental side of gravitational research will be hard to find and too busy to attend meetings and write papers.

The reduction of optical phase fluctuations due to variations in the forward scattering by the residual gas in the paths traversed by the interferometer beams requires a vacuum system which is a major cost of the large baseline systems. To maintain the costs and reliability will require scientific and engineering oversight. The modeling and control of phase fluctuations from scattered light by the use of low scatter optics, baffling and optical mode filtering is a substantial engineering and scientific task. Scaling from small aperture optics, used in the prototypes, to the large aperture optics needed in the long baseline systems requires design and test. The geometric stabilization of the optical beams by optical fibers and resonant cavity spatial filters must be scaled from the prototypes. The low noise servo systems that maintain beam position and control the positions of the test masses must be scaled..... Suffice it to say that there is a large and intensive but not impossible effort required to make the promises a reality.

Multiple interferometers and the need for an international effort

It is good fortune for the field that there is a world wide group of scientists willing to stake an appreciable part of their scientifically productive years to the development and operation of gravitational wave interferometers. Not only are there shared technical developments but it is critical for the science that will ultimately result. The complete characterization of the gravitational waves, and through them an understanding of the astrophysical processes at work in their sources, requires a global distribution of interferometers. At the minimum, in the initial stages of the search, multiple confirmations of events will lend credibility to the observations. As the science progresses a network becomes even more important. The measured time evolution of a gravitational wave signal in a single interferometer is a complex combination of the polarization states of the wave and the position of the source in the sky. The "antenna" pattern of a single interferometer is shown in Figure 1. (due to N. Christensen and "Mathematica"). As can be seen in the figure a single interferometer has virtually no directionality and one must resort to triangulation through timing the arrival of the waves at widely separated sites to gain information of the source position. A minimum of three, but for complete coverage four, interferometers at well separated sites is needed to gain adequate position to tie the gravitational wave sky to the better known electromagnetic one. The data analysis challenges for such a network are beginning to be worked on by people in this field in particular by B. Schutz, M. Tinto, Y. Gursel and N. Christensen.

In summary, the promise for a new field of gravitational wave astronomy is excellent and we stand to learn many new things about relativistic gravitation and the universe providing only that the engineers, scientists and funding agencies maintain their courage.

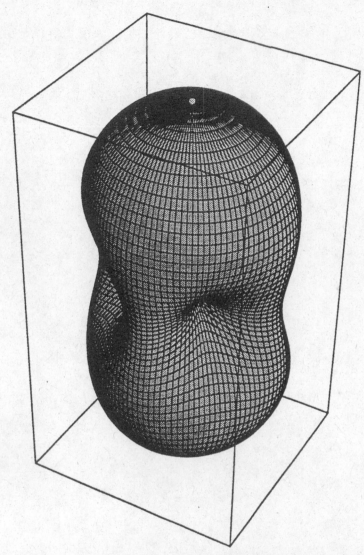

Fig. 1. The "antenna" pattern of an "L" shaped interferometer. The arms of the "L" are parallel to the edges of the prism in the horizontal plane and intersect the surface in the middle. The amplitude response (averaged over gravity wave polarizations) to a source at a specific polar and azimuthal angle is the radial distance from the center to the surface.

C1

Solar system and pulsar tests of general relativity

Ronald Hellings
Jet Propulsion Laboratory,
California Institute of Technology,
4800 Oak Grove Drive, Pasadena, CA 91109 USA

In the plenary talk on solar system tests of relativistic gravitation, Irwin Shapiro gave a summary of the present state of the experiments, including a new test of the geodetic precession effect using lunar laser ranging data. The contributed papers at this workshop, by contrast, seemed to look more to the future. What will be the next generation of gravitational experiments in space and what preparations are required now in order that these tests may be successfully accomplished?

The workshop began with a mini-workshop which discussed a new NASA program of technology development directed toward future experimental tests of relativistic gravitation. Ron Hellings of JPL began with an overview of the program and discussed two of the studies that are presently underway. Work to improve the accuracy of spacecraft ranging systems could pay immediate dividends of improved accuracy of tests of relativity on many planetary missions. Improvement from the accuracy that was available on the Viking landers (about 10 m) to a limit of 10 cm is the goal of work at JPL by Larry Young and collaborators. Also at JPL there is being pursued by Steve Macenka and Bob Korechoff a study of the capability of full-aperture metrology in optical systems. This technology seeks to use an auxiliary laser signal, sharing the optics in an astrometric telescope, to monitor small changes in the optical path, in order to allow the optics to be actively stabilized. This technology has application in highly accurate optical interferometers. Representatives of the three other studies presently being supported by this NASA project were present to describe their own work. Bob Vessot from the Harvard-Smithsonian Center for Astrophysics discussed the work that he has recently done to prepare to test hydrogen masers in the temperature environment provided by a space-qualified refrigerator. Allen Anderson of the University of California at Santa Barbara discussed the concept that he has pioneered that would use an existing Tracking and Data Relay Satellite System (TDRSS) along with two free-flying transponders to form a Michelson microwave interferometer that could have a sensitivity to strains as small as 10^{-19} for periodic signals at frequencies of several mHz. He also showed test data acquired from TDRSS that demonstrated sensitivity at a sub-micron level at 0.05 Hz. Finally, Peter Bender discussed work that he has done in conceptual design of a laser Michelson interferometer, with emphasis on the drag-free system. He presented a preliminary noise budget that would keep the sum of all noise sources below $10^{-17}g/\sqrt{Hz}$. The major contributor in this budget was the self-gravity produced by thermal expansion of the outer thermal insulation.

Following the report on the NASA advanced technology project, the contributed papers on solar system and pulsar tests of gravitation were presented. These were divided into four areas, with two representative papers being given in each area. The four areas were: 1) experimental results, 2) reports on ongoing design of future experiments, 3) discussions of new experiments or new effects to be tested, and 4) discussions of past or future pulsar observations.

Tim Krisher discussed results of an experiment performed by himself, John Anderson, and Jim Campbell, in which the frequency generated by the ultra-stable oscillator on board the Voyager spacecraft was monitored as the spacecraft passed through the gravitational potential well of Saturn. The gravitational redshift was measured at a level consistent with that predicted by General Relativity with < 1% error. Experimental results were also presented by Joe Weber for an Eötvös-type torsion balance. A positive result for an apparent violation of the equivalence principle was described and subsequently interpreted as being due to different absorption of solar neutrinos by the two different materials on the ends of the torsion balance. The interpretation relied on a new quantum-theory derivation of the neutrino cross-section which the author presented.

An update on the progress of Gravity Probe B (the Stanford gyroscope experiment) was presented by Keiser *et al.*. The construction of a major ground-based test facility was described and the results of some of the testing was reported. Using the test facility, a dynamical determination was made of the properties of the gyroscope rotor (mass imbalance, moments of inertia, and asphericity), allowing comparison with static measurements of the same properties. The two methods were shown to agree, and both indicated a gyroscope stability, in the presence of these imperfections, of 0.3 marcs/yr in a $3 \times 10^{-11}g$ environment. Ignazio Ciufolini discussed the results of a comprehensive error study of his proposed LAGEOS 3 mission. This mission seeks to measure the dragging of inertial frames produced by the rotating earth as it precesses two earth-orbiting spacecraft (LAGEOS 1 and LAGEOS 3) in the same direction, whereas other competing perturbations perturb the orbits in opposite directions, due to the two orbits having supplemental inclination angles. The result of the study indicates that the inertial frame dragging should be detectable at the 10% level. Finally, Bob Reasenberg discussed the state of studies of the POINTS measurement of second-order light deflection, with emphasis on the requirement for full-aperture metrology. This is a technique in which an auxiliary laser shares the same optics as the observed astronomical signal and allows changes in the optical path length to be monitored through a complex system of beams and sensors.

There were two papers presented that involved ideas for new tests of relativity. The first was a paper by Walter *et al.* in which it was suggested that the spin rates of the gyroscopes in the Stanford GPB experiment could be compared with earth-bound atomic clocks to give a redshift experiment using macroscopic mechanical clocks. Analysis of the torque disturbances lead to an expected accuracy of 10^{-4} of the effect predicted by General Relativity. A second paper, by Timothy Allison and Neil Ashby, recalculated the perturbation of an orbit produced by Lense–Thirring drag. They find small periodic discrepancies from the classical result (Lense and Thirring, *Phys. Z.* (1918), **19**, 156), derived from the fact that a Hamiltonian perturbing function must be used for the momentum-dependent Lense–Thirring perturbation, rather than the Lagrange planetary equations which have been used in the past.

The final topic of the workshop was analysis of timing data from binary pulsar systems. In the first paper, Zaglauer and Will used timing results from neutron star binaries to set limits on the coupling constant of Brans-Dicke type scalar-metric theories through limits on the existence of dipole radiation period damping in the binary. Data from the 11-minute binary, 4U1820-30, were most useful. A general lower limit on the coupling constant is 30, but, assuming the right stellar masses and equation of state of the neutron star, the limit could be as high as 600. Zhou and Haugan then presented arguments to show that gravitational time-delay effects on pulses passing out of the binary system PSR

1913+16 are detectable and can be used to determine the masses of the stars. These masses can then be used in orbit calculations to determine the gravitational fields of the stars, thereby constraining the nature of the companion to be a neutron star as well. The last paper was that of Blanchet and Schafer. In this paper a careful post-Newtonian expansion for the wave-generation problem was carried out to quadrupole-octopole order for the cases of elliptical, parabolic, and hyperbolic orbits and the results were compared with observations from PSR 1913+16.

In summary, the solar system remains one of the best laboratories for high-precision tests of relativistic gravitation via observations of the moon and planets and tracking of orbiting spacecraft. More recently, timing data from precise millisecond pulsars have added a new data type, and data from binary pulsars have defined a new laboratory where relativity produces measurable effects. All results to date continue to show that Einstein's 73-year-old theory is still in accord with observation, the accuracy of the tests now being generally a few tenths of a percent. The impressive accuracy of these tests has been the result of the last twenty years' unprecedented improvement in instrument precision and measurement techniques, though the improvement was most commonly obtained with goals other than gravitational physics in mind. This brings us to one final observation. We are concerned that this reliance on others to provide our measurement technology for us is probably not a wise policy. If the accuracy of the experiments is to improve, and if the experimental underpinnings of this basic branch of modern physical theory are to be thereby strengthened, we must pay more attention ourselves to development of the appropriate metrology. The new NASA program of technology development for relativity is an important step in the right direction.

C2

Earth-based
gravitational experiments

J. E. Faller
Joint Institute for Laboratory Astrophysics,
University of Colorado and National Institute of Standards and Technology
Boulder, CO 80309-0440 USA

Though historically, the solar system has been the principal area for testing theories of gravitation, we seem to be at the end of the golden age of solar-system tests (Reasenberg (1987)). The classical effects have now been measured within the limits of today's technology and further significant improvements cannot be expected in the near term. Future space-based experiments such as GPB, Gravity Probe B; LAGOS, Laser Gravitational-Wave Observatory in Space; and POINTS, the (proposed) Astrometric Optical Interferometer, await further technological as well as engineering developments and logistic (launch) support to deliver them to the laboratory of space.

Recent years have seen, however, ground-based gravitational experimentation undergo a resurgence, driven by new experimental capabilities and by new theoretical work. The question that was raised by Fischbach et. al. (1986) of a possible short-range gravity force, dubbed the "Fifth Force," has been particularly important. Though the experiments that gave rise to this suggestion have, in retrospect, turned out to be less compelling than was originally thought, the gravitational physics community has been forced to recognize the possibility of a short-range gravitational interaction.

This suggestion lent itself to fairly straightforward testing and the experimental community responded with great enthusiasm and ingenuity. Now some five years later, though it appears that this quite plausible theoretical suggestion has been ruled out by experiments at the level which initially was suggested. Nevertheless it gave rise to a rather exciting period in gravitational physics.

The Workshop C2 on Earth-Based Gravitational Experiments consisted of two oral-presentation sessions and one poster session. In all, 46 papers were presented during these three sessions. The papers broadly fell into two classes: Fifth-Force-related experiments and other ground-based experiments.

Table 1 lists the papers that were presented during the two oral presentation sessions of the Workshop.

In the first of these sessions we heard of how significant advances in precision either have been made or are being planned. The papers by Bollinger, Hils, and Ritter presented new experimental (null as it turned out) results.

The paper by Ni discussed some of the theoretical motivations behind the polarized test mass Eötvös experiment of Ritter et at. The other papers involved status reports on ambitious experiments presently under development to study various aspects of gravitation.

The second of the oral sessions was devoted to presentations about a possible "Fifth Force." This additional short-ranged component of gravity would manifest itself as an apparent violation of the weak Equivalence Principle for "nearby" test masses and/or the inverse-square law of gravity. The first two papers of this session were by Fischbach and Nieto who discussed various theoretical ideas that relate to the fifth-force hypothesis and also other theoretical aspects of non-Newtonian gravity. The subsequent experimental papers concentrated on Yukawa type interactions in the ranges between 1 m and 10^8 m using gravimeters and/or torsion balances as their main experimental tools. The paper by Kuroda discussed a cleverly conceived free-fall experiment between two objects whose centers-of-mass had been made to coincide (to avoid gravity gradient and tilt problems in the interpretation). Nelson reported on a very carefully conceived and carried out torsion balance fifth-force test. The papers by Thomas and Speake gave the results and further corrections to previously reported tower tests of the inverse-square law. These appear to be in agreement with the Newtonian prediction. Tuck discussed how errors in sampling the local topography may have affected the earlier Australian big G (mine) experiments. He also described plans for a new (modulated) water- mass determination of G.

An intriguing consequence of Newtonian gravity was presented by van Baak in the poster session. He showed how it is possible to have no gravitational field inside an asymmetric mass distribution and offered a heart shaped cavity inside a sphere as an example. At the same session

Tew and Bartlett displayed the progress of their cryogenic version of the Eöt vös experiment.

During all three sessions one had to be impressed with the creative use of advancing technologies to carry out the various ground-based gravitational experimental and by the evident enthusiasm of the experimenters. Though the fifth-force hypothesis appears to lack experimental confirmation at the sensitivity levels that have been achieved, it has nevertheless inspired a host of creative work that has significantly advanced our understanding of the experimental bases for gravitation.

Table 1.

Oral presentations at symposium on earth-based gravitational experiments.

Session A

J. J. Bollinger, *Search for Anomalous Long Range Interactions by ^{9}Be Nuclear Magnetic Resonance*

D. Hils, *Report on a Laser Version of the Kennedy-Thorndike Experiment*

R. C. Ritter, *Experimental Test of Equivalence Principle with Polarized Masses*

W.-T. Ni, *Polarized-Body Experiments to Test the Equivalence Principle for Spin-Polarized Electrons*

P. W. Worden, Jr., *A Satellite Test of the Equivalence Principle*

H. Terazawa, *Possible Measurement of the Gravitational Acceleration of Molecules, Atoms, Nuclei, Elementary Particles, and Antiparticles*

S. Vitale, *Progress on the Gyromagnetic Electron Gyroscope for a Ground Based Lense-Thirring Dragging Experiment*

Session B

E. Fischbach, *New Phenomenological Picture of the Fifth Force*

M. M. Nieto, *New Gravitational Forces and the Arguments Against Non-Newtonian Gravity*

K. Kuroda, *A Test of Composition-Dependent Force by a Free-Fall Interferometer*

N. Mio, *Dynamics Null Tests for a Possible Composition-Dependent Force*

P. Nelson, *A Search for an Anomalous Intermediate Range Composition Dependence in Gravity*

J. Thomas, *Gravity Measurements on the BREN Tower at NTS*

C. C. Speake, *A Test of the Gravitational Inverse Square Law Using the NOAA Tower in Erie, Colorado*

G. Tuck, *Gravity Gradients at Mount ISA and Hilton Mines*

G. Tuck, *A Test for Non-Newtonian Gravity Using a Hydroelectric Lake*

References

Fischbach, E., Sudarsky, D., Szafer, A., Talmadge, C. and Aronson, S. H. (1986), *Phys. Rev. Lett.* **56**, 3.

Reasenberg, R. D. (1987), in *Proceedings of the 11th International Conference on General Relativity and Gravitation*, ed. M. A. H. MacCallum (Cambridge University Press, Cambridge), p. 274.

C3

Resonant bar and microwave gravitational wave experiments

William O. Hamilton
Department of Physics and Astronomy,
Louisiana State University,
Baton Rouge, LA 70803 USA

In the 20 years after Weber's first publications there have been yearly announcements of new, better ways to detect the signals that everyone knows must be there. Drever, Billing, Douglass and Tyson, Garwin and Levine all built detectors that were of better sensitivity than that estimated by Weber in his original papers. None of those experiments were able to verify Weber's results. In almost every meeting of the past 20 years there have been the promises of the potential of the "second generation" detectors: cryogenic detectors using superconducting instrumentation that would be so sensitive that they would be able to see signals from as far away as the Virgo cluster. In many such meetings for the last 10 years these reports of the potential sensitivity have been tempered by descriptions of the experimental difficulties that were temporarily delaying the attainment of that sensitivity.

The exciting thing about this meeting was that–while there were still tales of wonders to come and of the experimental difficulties that some detectors were confronting–we also had reports of coordinated runs between groups looking for coincidental excitation of their antennas. We had informal meetings outside of the workshops to coordinate data taking and exchange of data. In short, this workshop was the first meeting where three or more groups were confident enough about their detectors that they could talk of coordinating their experiments and of coordinating their existing experiments with the upcoming third generation resonant bar detectors. In addition we had a report of data analysis from the deep space network using the Pioneer 10 spacecraft that established a new upper limit on very long wavelength gravitational waves.

The operating resonant bar detectors are now reporting interesting sensitivities to gravitational waves. Rome finds periods when their noise level is 7×10^{-19}. LSU operated for 6 months with a best noise level of 2.2×10^{-18}. Stanford had operated for over one year in the early 1980's with a gravitational wave noise level approximately equal to 10^{-18}. All of these detectors should be sensitive to any of the signals predicted by theory to come from events in our galaxy.

The experimental workshop was divided into two afternoon sessions. The first session had one paper that presented experimental results from an analysis of tapes from NASA's deep space network. J. D. Anderson reported on the magnitude of the residuals from computer analysis of the ranging data to the pioneer 10 spacecraft. The data puts a limit on the magnitude of long wave gravitational radiation.

Bocko also had results to report. The Rochester group has constructed a test stand for transducer development which uses a vertical dumbbell antenna with a very respectable low temperature Q of 1.4×10^7. He is using this test stand to test a parametric transducer of a novel design. There were only room temperature results to report at this meeting.

Richard has continued his investigation of the multimode transducer that he invented in 1982. He is now using a Fabry-Perot cavity as the transducer output. Optical readout is also being investigated by the Tokyo group in conjunction with disk antennas. This was reported by K. Tsubono.

Some potentially very important results for transducer development were presented by Carroll. He measured high electrical quality factors in a niobium diaphragm that was also loaded by pressure from an applied magnetic field. There are some differences between the results obtained in this Maryland investigation and the results reported by Solomonson and the LSU group in the poster session but taken together they indicate that the low electrical Q problem in niobium is not insurmountable.

Noteworthy results reported in the second session included Bill Duffy's report of steadily increasing Q values for aluminum alloys as the temperature was reduced from 4 K to below 50 mK. It has been a matter of faith among experimenters that internal friction losses would decrease as the temperature was lowered and Duffy's work is evidence to justify that faith. It is important that Duffy and Griffin's measurements were made at 1 kHz, the frequency close to that used by all resonant mass antennas except the Japanese torsional antenna, and thus should be applicable to all of the bar antennas now constructed or under consideration. Aluminum 5056 alloy gave the best low temperature performance. When this work is combined with the previous work of Marsden and Douglass it makes Q values of 10^8 seem reasonably within reach.[1] It is important to keep in mind, however, that the Q of the antenna is not the only consideration. A high antenna Q reduces the Langevin forces due to losses in the antenna but there are many other sources of system losses. A transducer must be attached to the antenna without introducing additional loss and that transducer must in itself have low dissipation, both in its electrical and in its mechanical modes.

None of the remaining papers about the operation of gravitational wave antennas can be considered outside of the context of the plenary paper delivered by Pizzella. There can be no question of the fact that the Italian effort in gravitational wave detection is the strongest in the world at this time. The Rome antenna at CERN ("Explorer") is operating with a noise temperature of 7 mK. An ultra low temperature antenna ("Nautilus") has been constructed and a dilution refrigerator for that antenna has been purchased and tested. A new resonant bar group is building an antenna near Venice and a new group has started operation of an antenna in Frascati. All of the experiments are designed to use 5056 aluminum as the antenna material.

Johnson focussed on the steps that must be taken to bring an experiment to observatory status. He reported on the joint operation between the LSU antenna and the Rome antenna "Explorer." The two groups operated in coincidence for two months in 1988. In addition, they also conducted separate experiments designed to optimize and improve their individual detectors. Previous coincidence experiments between Stanford, Rome, and LSU have already been published.[2] The LSU experiment ran for long periods with a rms noise level corresponding to a gravitational wave pulse amplitude of 2.2×10^{-18}. During this time there were very few events which could not be considered to be due to normal noise processes of the antenna. The 1988 data was still being analyzed at the time of the talk and data exchange protocols were still being worked out. Hamilton reviewed the breakthrough in transducer design that was detailed in the poster session. This transducer will give noise performance that is limited by squid noise to a noise temperature of 3 mK. He spoke briefly on the possibility of implementing a spherical antenna, originally suggested by Paik and Wagoner[3] and revived by Johnson, which would gain omni-directional capability as well as additional effective mass for the antenna modes. Such an antenna, operated at 50 mK, could probably be constructed relatively inexpensively and promises rms noise performance of 10^{-22}. It is certainly an interesting complement to the interferometer detectors which were discussed in another session of the experimental workshop.

Before going further it is worthwhile to restate some obvious facts about statistics. While all the experimenters characterize their individual detectors in terms of the noise temperature or the rms strain sensitivity corresponding to that noise temperature, the noise temperature is only part of the story. The noise temperature is the parameter that describes the exponential distribution of unmodeled fluctuating forces that pass the computer filter on the experiment. Vibration isolation is extremely important. If there is vibrational excitation that passes the vibration filters and is not vetoed, that excitation will appear as an unmodeled fluctuating force at the output of the computer filter. For that reason the exponential character of the sampled and filtered data must be well established. There must be very few events which are outside of the established statistical distribution.

Assuming that an exponential probability distribution for the filtered data has been established there is an easily calculatable probability that a single detector will experience a force greater than that given by the noise temperature. The probability that two or more detectors will experience simultaneous excitations is much reduced.[4] As an example, for a single detector which is measuring the fluctuating force amplitude ten times each second for one year, an energy threshold must be established at 22 times the noise temperature in order that there be less than a 10% probability that a detected signal that exceeds that threshold at some random time during the year be due to the Langevin forces internal to the antenna. If two identical antennas operate in coincidence and we use the Poisson approximation for the coincidence statistics we can show that a threshold of 11 times the noise temperature will suffice. Three detectors enable one to set a threshold of 7.3 times the noise temperature while 4 detectors will allow a threshold of 5.5 times the noise temperature. We thus see that it is important to obtain good noise temperature but that it must be obtained in a well characterized antenna and hopefully, in a multitude of well characterized antennas all operating in coincidence.

The Italian group all spoke of reducing the noise temperature and characterizing the performance of their antennas. Coccia discussed the extensive design studies on an ultra-low temperature antenna and the results of testing a dilution refrigerator for an ultra cryogenic antenna ("Nautilus"). The dewar is constructed and awaiting first testing. The dilution refrigerator has shown itself capable of attaining 40 mK. Bonifazi reported that the 1988 performance had a noise temperature of 7 mK. There were some difficulties however that were under investigation. The noise level seemed to increase as the helium level became low, leaving them to question the existence of an unfiltered noise path. Also, they report occasional long time excitation of only one normal mode in their tuned mass system when both modes should be equally excited. They are investigating this phenomenon.

Pallottino reported a calibration experiment whereby an oscillating quadrupolar field was created by rotating a massive dumbbell outside the dewar and measuring its effect on the antenna. Such a calibration procedure is very dangerous because of the high speed required for the rotating dumbbell but avoids many of the problems experienced in early direct calibration experiments such as Sinsky's where very large electric fields were created at the excitation frequency of the antenna.[5]

Mann and Chiang reported on a different approach Stanford has taken to creating a ultra low temperature experiment. While Rome has designed their experiment to cool the antenna from a support at the center of the antenna, Stanford is intending to cool their antenna from its end and to attach the cooling to the transducer through solid metal vibration isolators modeled after the original Taber isolators that everyone now uses. Extensive finite element analysis was reported to demonstrate that the spring concept was workable. They reported the delivery of a large dilution refrigerator. They also reported that extensive damage was suffered by the original Stanford experiment during the 1989 San Francisco earthquake. Their manpower situation probably will not allow them pursue both the repair of the current system and the construction of the ultra-low temperature experiment. That ultra-low temperature cryostat is designed and construction is starting.

Perhaps the most memorable remark of the meeting was made by Blair who, in describing the Western Australia experiment, said "We know that we have all of the right ingredients but we just haven't been able to bake the cake." Every experimentalist understood exactly how he felt.

There were a number of important results given in the experimental poster session. The Stanford group gave more details of the design of their proposed ultra low temperature experiment. The LSU group reported on two transducers, the breakthrough in the mushroom transducer of Solomonson and a progress report on the development of a parametric transducer by Aguiar and Johnson.

Two theoretical papers were also included in the poster session. Weber again presented his theory of gravitational wave cross sections and Peng, arguing by analogy, presented his ideas concerning the existence of an electric field in the interior of a superconductor when it is being subjected to the time varying stress of incident gravitational waves.

Thanks are due to all of those individuals who shared their notes when the unfortunate duty of writing this review came to me. Odyno Aguiar, Norbert Solomonson and Warren Johnson were very helpful. Rich Isaacson sent his copy of the much changed schedule for the workshops. Any mistakes or misstatements remaining are mine alone.

References

1. R. Marsden (1982), Ph. D. Thesis, University of Rochester (unpublished).
2. E. Amaldi et al. (1989), *Astronomy and Astrophysics* **216**, 325-332.
3. H. J. Paik and R. Wagoner (1976), in *Gravitazione Sperimentale, Pavia 17-20 Sett.* 257-265, (Academia Nazionale dei Lincei, Rome, (1977)).
4. W. O. Hamilton and W. W. Johnson, LSU Technical Memo (unpublished).
5. J. A. Sinsky, (1967), Ph. D. Thesis, University of Maryland (unpublished).

Table 1.

The following table gives the authors and titles of the papers delivered orally in the workshop. The individuals listed in bold type presented the paper.

J. D. Anderson, J. W. Armstrong and E. L. Lau, *December 1988 Microwave Gravitational Wave Experiment*

J.-P. Richard, Y. Pang and J. J. Hamilton, *Development of Laser Sensors and Wide-Band Detectors*

K. Tsubono, N. Mio and A. Mizutani, *Disk Antennas Instrumented with Laser Interferometer*

K. R. Carroll and H. J. Paik, *Electrical Quality Factor in a Gravitational Wave Transducer*

M. Bocko, M. Fisher, L. Marchese and M. Karim, *Development of a Parametric Transducer for a Resonant Bar Gravitational Wave Detector*

B. X. Xu, N. Solomonson, B. Price, **W. W. Johnson, W. O. Hamilton** and O. Aguiar, *Operation of the LSU Gravitational Wave Observatory - 1988*

T. Aldcroft, **J. Chiang**, W. M. Fairbank, J. Henderson, Hu En-ke, **L. D. Mann**, P. F. Michelson, J. C. Price, T. Stevenson, B. Vaughan, and Z. Zhou, *Stanford MilliKelvin Gravitational Wave Antenna*

E. Amaldi, P. Astone, M. Bassin, **P. Bonifazi**, P. Carelli, M. G. Castellano, G. Cavallari, **E. Coccia**, C. Cosmelli, S. Frasca, I. Modena, G. V. Pallottino, G. Pizzella, P. Rapagnani, F. Ricci, and M. Visco, *Progress Report on the Rome Gravitational Wave Experiment and Preliminary Results of the Dilution Refrigerator for the new MilliKelvin Antenna*

P. J. Veitch, **D. G. Blair**, N. P. Linthorne, M. J. Buckingham and C. Edwards, *Progress Report on the UWA Gravitational Wave Antenna*

W. Duffy, Jr. and Garrett Griffin, *Ultralow Temperature Quality Factor of Aluminum*

E. Amaldi, R. Bizzari, R. Cardarelli, D. DePedis, A. Degasperis, E. G. Muratori, **G. V. Pallottino**, G. Pizzella, M. Price, F. Ricci, *Upper Limit of Gravitational Waves and A Rotating Source of Gravitational "Near Field"*

Y. Tomozawa, *Gravitational Waves, Supernovae and Quantum Gravity*

C4

Laser gravitational
wave experiments

J. Hough
Department of Physics and Astronomy,
University of Glasgow,
Glasgow G12 8QQ, UK

Introduction

After 20 years of careful and innovative experimenting most of the researchers in the field of gravitational wave detection believe that success is on the horizon. Theoretical predictions of source strengths and source types have been steadily evolving and it is now clear that we should be aiming to build gravitational wave detectors which have strain sensitivities of $\sim 10^{-22}$ over kilohertz bandwidths.[1] Such sensitivity should allow the detection of signals from, for example, supernova events at distances out to the Virgo cluster, coalescing compact binary systems and continuous and stochastic background sources.

One of the most promising ways of achieving the required sensitivity and bandwidth is to use laser interferometry between freely suspended test masses placed several kilometers apart,[1] and the majority of contributions to the workshop on laser interferometer gravitational wave detectors were related to the large laser interferometer projects currently well advanced in planning. These instruments rely on searching for changes in the relative length of two paths, usually at right angles to each other, and formed between the test masses suspended as pendulums.

As will be mentioned again later, a number of interferometers are required around the world to obtain useful astrophysical information from the strength, polarization and timing of signals detected, and currently the belief is that at least three separate detector systems at different sites are necessary.

Laser interferometer projects

There are three main projects at the planning stage:

1) the American Laser Interferometer Gravitational Wave Observatory (LIGO) which is intended to consist of two detector systems with arm lengths up to 4 km;

2) the Italian/French project, named VIRGO after the galactic cluster which may contain many interesting sources, for an instrument of 3 km arm length; and

3) the proposed German/British interferometer project for a detector again of 3 km arm length.

These projects are complemented by a proposal for an Australian interferometer project of similar size.

The LIGO project is the most ambitious of those presented, being designed to allow observations and experimental development to be conducted simultaneously. R. Vogt (Caltech) outlined plans for the installation of up to nine interferometers between the two detector sites. He also described special vacuum housings for the test masses, designed such that some masses can be removed or inserted with only minimal disturbance to other interferometers in the system. The vacuum pipes joining the tanks containing the test masses would be 48 inches in diameter and the pressure in the system would be less than 10^{-8} torr of residual hydrogen. It is planned that green light from argon lasers or from frequency doubled neodymium YAG lasers would be used to illuminate the interferometer. The apparent changes in length of the arms would be magnified by effectively reflecting the light in the arms back and forth between mirrors mounted on the test masses. These mirrors would form Fabry-Perot interferometers in the arms, and certain of the light recycling schemes proposed in recent years[2,3,4] for interferometers to make the most use of the input light would be implemented.

The other projects are somewhat simpler in concept although similar in nature. The VIRGO system, presented by A. Giazotto (Pisa) and A. Brillet (Orsay), would contain one interferometer (perhaps two at a later stage) optimized for low frequency operation near a few tens of Hertz. Again Fabry-Perot cavities would be formed in the arms but it is intended that light of wavelength 1.06 microns from a neodymium YAG laser be used. The longer wavelength would decrease the amount of unwanted scattered light in the interferometer with a possible reduction in related noise, but would require the use of larger mirrors and optical components in the system. The German/British proposal, described by H. Ward (Glasgow), is for two interferometers in the one vacuum envelope, one aimed at high frequency and one for lower frequency, the illumination being with green light. The German/British project is somewhat different from the others as the proposal is to use optical delay lines[5] in the 3 km arms at least in the first instance. The Fabry-Perot approach is reverted to in the Australian proposal described by D. Blair (University of Western Australia) and H. Bachor (Australian National University).

Technology development for the long baseline detectors

The various technical requirements for such detectors and the noise sources which limit the operation of the instruments have been fully discussed by R. Weiss in his invited talk and by Kip Thorne.[1] The main sources of limitation to sensitivity are expected to include photon noise in the detected light, thermal noise from the resonant modes of the suspension systems and the test masses themselves, seismic noise, and fluctuations in the refractive index of the residual gas in the vacuum system.

Laser light sources and associated issues

In order to reduce the limitation to sensitivity imposed by photon noise in the detected light, high power continuous wave lasers providing up to 100 Watts of single frequency light are needed for illumination. In the prototype detectors built at present argon ion lasers have been used as the source of illumination. However these lasers are limited to providing several watts of light, exhibit fairly high levels of amplitude and frequency noise and are very inefficient. The situation is changing. A. Brillet (Orsay) presented new results[6] on the development of a reasonably efficient flash lamp pumped ring YAG laser producing 18 W of single mode light at 1.06 microns. This laser was injection locked to a much smaller frequency stabilized ring YAG laser and the combination was shown to retain the excellent noise characteristics of the small laser. Flashlamp pumping is still not very efficient and it is a generally held belief that YAG lasers pumped by semiconductor diode lasers are the best candidates for the future. Work on such systems is being undertaken in a number of laboratories in Europe and the USA and the developments required by the LIGO project are being carried out by R. Byer and his group at Stanford. Byer was particularly enthusiastic about the prospects for diode pumped frequency doubled YAG lasers reaching the required power levels in step with the building and development of the large gravitational wave detector systems.

A potential problem in optical systems, which appears with the use of high powers, is the effect of heat dumped in the materials of optical components. For example, surfaces of dielectric coated mirrors may become distorted, or variations in the refractive index with temperature distribution in a substrate may distort wavefront geometry. Both effects lead to degradations of the quality of interference possible and hence to impaired noise performance. Such problems have been observed on the prototype at Caltech for the LIGO detector. F. Raab (Caltech) presented some experimental findings that transmission through optical cavities, used to clean the spatial mode structure of the input light, seemed to saturate at a few Watts of input power. The exact level of saturation seemed to depend on the particular mirrors used in the cavity and the effect was consistent with thermal changes in the refractive index of the input substrate affecting the coupling of the incoming light to the cavity. However the mirrors used were not of the lowest loss available, resulting in more heat being dumped than is necessary, and the effect will be reduced by the use of materials of lower optical absorption and higher thermal conductivity both for substrates and coatings. Materials of more suitable properties are being investigated. Further R. Drever (Caltech) pointed out in his talk that, apart from choosing more suitable materials there are a number of ways around these problems. For example it is possible to inject light into an optical cavity by directing it on to a very thin wedge of transparent material placed inside the cavity and oriented at Brewster's angle to the resonating light. This avoids the transmission of light into the arms of the interferometer through the mirror substrates. It is clear that development work is needed in the area of thermal effects; and the computer simulation techniques being developed by J.-Y. Vinet (Paris) for studying the effects of optical imperfections in laser interferometers will be very useful for this.

Another subject relevant to the use of high light powers in long baseline interferometers with Fabry-Perot cavities in their arms – that of possible instability and chaotic behaviour induced by light pressure effects – has been a subject of discussion for a number of years.[7] This was addressed by B. Meers (Glasgow) and he clearly emphasized that for the designs of detectors envisaged there should be no problems associated with such effects.

Suspension systems - seismic isolation

Design of suspension systems is a critical area for gravitational wave detectors. Adequate filtering for seismic and local mechanical noise has to be provided over the operating range of frequency of the detector, and the mechanical quality factor, Q, of the pendulums has to be high enough that significant levels of thermal motion do not appear in the frequency band of interest. Most research groups involved in the planning of large gravitational wave detectors are carrying out some suspension development, the most ambitious experimental programme being that of A. Giazotto and colleagues at Pisa. In this case a seven stage system[8] to carry a test mass of about 400 kg is being developed for the VIRGO project. The suspension operates in three degrees of freedom using gas springs for the vertical direction and normal pendulums for the horizontal directions. Calculations indicate that isolation factors of 10^{11} should be attainable at 10 Hz, allowing useful operation of the VIRGO detector at frequencies as low as this. Initial experimental tests, limited at present by the sensitivity of the motion sensors, have indicated an isolation factor of about 10^9 at 10 Hz. Unwanted modes of the pendulum systems are controlled by active (electronic) damping techniques. It seems that similar isolation performance might be achieved by a more compact system based on the application of active feedback systems to sense and control the effects of ground borne noise. A proposal to build such a system for operation in six degrees of freedom for use in the LIGO project was described by R. Stebbins (JILA).

A very interesting contribution, relevant to thermal noise in suspension systems was presented by P. Saulson (MIT). He pointed out that the normally used theory for thermal noise in pendulums assumes that the equivalent noise force is white in frequency. This is only true if the damping force is proportional to the velocity of the pendulum and Saulson pointed out that this is not the case for the normally used suspension materials. In fact it appears that the equivalent noise force should fall towards higher frequencies, thus making the thermal motion of the test masses in the frequency band of interest for gravitational wave detectors smaller than might have been expected.

Finding the signals in the noise

In order to uniquely identify the position of a source of gravitational waves from the time of arrival of the signals at a number of detectors in a network it was often stated that a minimum of four detectors were required in the network. However the work of Gürsel and Tinto[9] has lead to a change in thinking. They have pointed out that from three independent functions of time which are detector responses and from two independent time delays between events as measured with respect to one receiver, it is possible to uniquely determine the parameters of the wave - thus three detectors are sufficient. Further they have shown from numerical simulations that with optimal filtering techniques and a signal to noise ratio of 10 it should be possible to determine the source direction with an average uncertainty of 2×10^{-5} steradians. While this essentially means that the minimum number of interferometric detectors required worldwide is three, extra detectors would be a great advantage in overcoming duty cycle problems and in handling the realistic situation where signal to noise ratios are somewhat lower.

There is a situation where even one detector can be useful on its own and this is in the search for continuous radiation from pulsars. In the case of a pulsar of known frequency and known position, data analysis for integration over times of 10^7 seconds can be seen to be relatively straightforward. The data analysis power required for pulsars of unknown position and frequency is considerably larger - each position in the sky requires a different Doppler correction, for example - but in a short review of these matters I. Pinto (Salerno) showed that, even with present computational limits, up to 250 patches of sky could be searched for pulsars in a frequency band up to 1 kHz with a resolution of 10^{-7} Hz.

Space experiments

A laser interferometer detector is also eminently suited as the base for a gravitational wave observatory in space (LAGOS as discussed by P. Bender, JILA) aimed at detecting the stochastic superposition of signals from many galactic binary systems. A right triangular cluster of three spacecraft is envisaged with spacecraft separations of 10^6 or 10^7 meters. The frequency range over which such a detector is sensitive - $\sim 10^{-5}$ Hz to $\sim 10^{-2}$ Hz or 10^{-3} Hz depending on spacecraft separation - is very different from that of ground based instruments. Thus such an instrument will be a valuable addition to the proposed ground based networks. There is an advantage in the shorter baseline as it allows the slightly higher frequency of 10^{-2} Hz to be reached, thus allowing signals from the formation of black holes down to 10^4 solar masses to be searched for. However this is offset to some extent by the need to reduce test mass disturbance limits to below 2×10^{-18} g/$\sqrt{\text{Hz}}$ over the operating bandwidth to achieve the designed performance.

Prototype detectors

There are currently four prototype interferometer detectors under development: at Caltech (40 m arm length with Fabry-Perot cavities in the arms); at the Max-Planck-Institut für Quantenoptik, Garching (30 m arm length with delay lines in the arms); at the University of Glasgow (10 m arms, Fabry-Perot) and at ISAS in Japan (10 m arms, delay line). Three of these detectors have been used to take data for a period during the last year. The Garching and Glasgow prototypes were run for 100 continuous hours in coincidence just after the observation of the pulsar in SN 1987A and each system achieved a duty cycle of about 90%. Results from this run are at present being analysed and T. Niebauer (MPQ) presented some very preliminary data, setting an upper limit of $h \sim 10^{-20}$ on radiation from the pulsar. N. Kawashima (ISAS) presented data for 46 hours of high quality operation of his detector over a 126 hour period between 20 February and 12 March 1989, some of which coincided with the period of operation of the Glasgow and Garching detectors. He also set an upper limit on radiation from the pulsar, at an amplitude level below 10^{-19}.

Conclusion

There were many other contributions to this workshop covering many areas of work ranging from servo control of optical cavities to signals associated with the ringing of black holes and a list of contributions is given in the appendix.

In summary, the impression given was one of considerable enthusiasm and progress. It is clear that the preparations for the large laser interferometer experiments are well advanced and it is to be expected that decisions as to how these experiments will go ahead will be made in the next year.

References

1. K. S. Thorne (1987), Gravitational Radiation, in *300 Years of Gravitation*, eds. S. W. Hawking and W. Israel (Cambridge University Press, Cambridge).
2. R. W. P. Drever (1983), Interferometric Detectors of Gravitational Radiation, in *Gravitational Radiation, Les Houches 1982*, eds. N. Deruelle and T. Piran (North Holland, Amsterdam) pp. 321-338.
3. J.-Y. Vinet, B. J. Meers, C. Nary Man and A. Brillet (1988), Optimisation of long-baseline optical interferometers for gravitational wave detection, *Phys. Rev.* **D38**, 433-447.
4. B. J. Meers (1988), Recycling in laser interferometric gravitational wave detectors, *Phys. Rev.* **D38**, 2317-2326.
5. For a description of a delay line system see D. Shoemaker, R. Schilling, L. Schnupp, W. Winkler, K. Maischberger and A. Rüdiger (1988), Noise behaviour of the Garching 30 meter prototype gravitational wave detector, *Phys. Rev.* **D38**, 423-432.
6. O. Cregut, C. N. Man, D. Shoemaker, A. Brillet, A. Menhert, P. Peuser, N. P. Schmitt, P. Zeller and K. Wallmeroth (1989), 18 W single-frequency operation of an injection-locked, CW, Nd:YAG laser, *Phys. Lett. A (Netherlands)* **140**, 294-298.
7. J. M. Aguirregabiria and L. Bell (1987), Delay-induced Instability in a Pendular Fabry-Perot Cavity, *Phys. Rev.* **A36**, 3768-3770.
8. R. Del Fabbro, A. Di Virgilio, A. Giazotto, H. Kautsky, V. Montelatici and D. Passuello (1988), Low Frequency Behaviour of the Pisa Seismic Noise Super-Attenuator for Gravitational Wave Detection, *Phys. Lett. A (Netherlands)* **133**, 471-475.
9. Y. Gürsel and M. Tinto (1989), A Near Optimal Solution to the Inverse Problem for Gravitational Wave Bursts, Caltech LIGO Project preprint **89-1**.

Table 1.

Papers presented at the workshop on laser gravitational wave experiments. Presenting authors are indicated in boldface.

A. Brillet, O. Cregut, C. N. Man, A. Marraud, M. Pham-Tu, D. Shoemaker, J. Y. Vinet, **A. Giazotto**, C. Bradaschia, A. di Virgilio, D. Passuello and V. Montelatici, *The VIRGO project - a progress report*

J. Y. Vinet and P. Hello, *Numerical Simulation of the VIRGO Project's Optical System*

H. Ward, J. Hough, G. P. Newton, N. A. Robertson, B. J. Meers, N. L. MacKenzie, D. I. Robertson, C. A. Cantley, K. A. Strain and A. Carmichael, *Developments at Glasgow including brief discussion of a 100 hour Data Run and Proposed German/British Collaborative Project*

T. M. Niebauer, A. Rüdiger, R. Schilling, L. Schnupp, W. Winkler and G. Leuchs, *Some results from a 100 hour Data Run with the Garching 30 m Prototype*

N. Kawashima, S. Kawamura, J. Hirao, J. Mizuno and L. Yong - gui, *A Search for a Sub-Millisecond Pulsar in SN 1987A using the ISAS Laser Interferometer Gravitational Wave Antenna*

D. Blair and **H. Bachor**, *AIGO - Australian International Gravitational Wave Observatory*

Y. Gürsel and M. Tinto, *An Optimal Solution to the Inverse Problem for Gravitational Wave Bursts*

B. J. Meers, *Radiation Pressure Instabilities in Interferometers*

M. Longo and **I. M. Pinto**, *Perspectives in Signal Processing for Detecting Gravitational Waves from Pulsars*

P. L. Bender, J. E. Faller, D. Hils and R. T. Stebbins, *Disturbance Reduction Techniques for a Laser Gravitational Wave Observatory in Space*

M. Tinto and J. W. Armstrong, *Search for Gravitational Waves from Massive Coalescing Binaries with a Network of Spacecraft*

R. Price and Y. Sun, *Gravitational Waves and Black Hole Ringing*

F. Echeverria, *Gravitational Wave Measurements of the Mass and Angular Momentum of a Black Hole*

R. Vogt, *The LIGO Project - a progress report*

R. Drever, *Developments in Interferometric Gravity Wave Detectors for Large Scale Operation*

R. Byer, *Solid State Lasers for Gravitational Wave Interferometry*

R. Spero, *Status Report on the Caltech 40 m Fabry-Perot Prototype Gravitational Wave Detector*

F. J. Raab, *Observation and Analysis of a Mirror Heating Threshold in Fabry-Perot Cavities*

N. Christensen, **P. Fritschel** and J. Giaime, *Tests of Recombination Schemes in a Fixed Arm Fabry-Perot Laser Interferometer*

C. A. Cantley, **N. A. Robertson** and J. Hough, *A Suspension System for use in a Laser Interferometer Gravitational Wave Detector*

J. Kovalik, P. Saulson and **M. Stephens**, *The Performance of a Double Pendulum Vibration Isolation System*

R. T. Stebbins, P. L. Bender, J. E. Faller, D. B. Newell, and C. C. Speake, *A 1 to 10 Hz Prototype Isolation System For Gravitational Wave Interferometers and Thermal Noise Measurements*

P. Saulson, *Thermal Noise in Suspensions For Interferometric Gravitational Wave Antennas*

A. Krolak, *Application of the Classical Theory of Hypothesis Testing to Detection of Gravitational Waves*

D. Anderson *Comments on Auto Aligning systems for Laser Interferometers*

Poster: L. Schnupp. *Signal to Noise Ratio of a Modulated Laser Interferometer Gravitational Wave Detector*

Part D.

Quantum gravity, superstrings, quantum cosmology

14

Self-duality, quantum gravity, Wilson loops and all that

Abhay Ashtekar
Physics Department, Syracuse University,
Syracuse, NY 13244-1130 USA

1. Introduction

At the Padova GRG conference, a new avenue to non-perturbative, canonical quantum gravity was suggested (Newman (1984)). By the time the Stockholm conference was held, these preliminary ideas had blossomed into a broad program aimed at analysing the structure of classical and quantum general relativity from a somewhat unusual standpoint (Ashtekar (1986), (1987)). By now, over two dozen individuals have contributed to this program. The purpose of this chapter is to present a brief status report of this body of ideas. Although I will try to be objective, it is inevitable that not everyone who is working in this field will agree with all the views expressed here. Also, since my space is limited, I will have to leave out several interesting results; I apologize in advance for these omissions.

The key idea underlying this program is to shift the emphasis from geometrodynamics to connection dynamics. In the classical theory, the new viewpoint merely complements the traditional one in which the metric, rather than a connection, is taken as the fundamental variable. We do obtain a fresh perspective that simplifies certain issues and suggests new ways of tackling unresolved problems. However, as far as the basic features of the theory are concerned, nothing is really altered conceptually. It is in the quantum regime that the shift of emphasis plays a major role. More precisely, there are indications that connection dynamics is indeed a better tool to analyze the micro-structure of space-time in a non-perturbative way. The emphasis on connections opens up new windows; a number of unanticipated concepts begin to play a major role in the formulation of the basic questions and a new technical machinery becomes available to analyze these issues. Finally, the resulting mathematical framework is closely related to the one which has been successful in the study of other basic interactions in physics. Thus, if these ideas succeed, quantum general relativity would no longer remain isolated; a unifying mathematical framework would bind it to other theories of Nature. It is interesting to note that, in a certain sense, this approach strikes a chord with views that have been expressed both by relativists and quantum field theorists. Relativists have long felt that a *non-perturbative* approach is needed; any attempt that begins by "steam-rollering general relativity into flatness and linearity" cannot ultimately succeed (Penrose (1976)). Field theorists, on the other hand, have felt that the emphasis on space-time metric and its geometrical properties have "driven a wedge between general relativity and the theory of elementary particles" in which quantum mechanics (of connections) plays a deep role (Weinberg (1972)). A non-perturbative analysis of connection dynamics appears to bridge this gap.

Why does one need a non-perturbative framework? After all, as field theorists are apt to point out, perturbation expansions have been highly successful in quantum theories of other interactions. In the gravitational case, however, these methods have failed even by their own criteria. To emphasize this point, let me begin with a sketch of the history of these ideas in quantum gravity.

It has been known for some time now that Einstein's theory itself is perturbatively non-renormalizable at two loops for pure gravity and at one loop for gravity interacting with matter. One can change the theory by adding higher derivative terms to the action. Since all the classical tests of general relativity refer to the large-distance behavior of the theory, from a strict experimental viewpoint, we are free to add terms that leave this behavior untouched. The addition of suitable higher derivative terms *does* improve the ultraviolet behavior of the quantum theory. Not only can one make the theory renormalizable but one can even achieve asymptotic freedom; at high energies, the theory behaves as if it were free. However, now, the theory is not unitary. In the perturbative treatment, the Hamiltonian is in fact unbounded from below, signaling a dramatic instability. One might consider supersymmetric extensions of general relativity. The hope here is that the bosonic infinities of gravity would be cancelled by the fermionic infinities of matter, giving us an acceptable theory. Again, the the resulting theory, supergravity, *is* better behaved. It has a positive definite Hamiltonian and even in the presence of (supersymmetric) matter, it is two-loop renormalizable. Unfortunately, however, the renormalizability fails at the third loop. A more radical revision is suggested by string theory. Here one abandons local field theories altogether and considers extended objects–strings–as the fundamental entities. It is a very attractive strategy because, unlike in any other attempt, all matter couplings are now fixed automatically; very little is fed in by hand. There is also a tremendous technical improvement. The theory is unitary and, although a conclusive proof is yet to emerge, by now a general consensus has developed among experts that the theory is perturbatively finite. To those who are unfamiliar with the technical jargon, however, this terminology is misleading: finiteness here refers only to each term in the perturbation expansion. To provide finite answers to physical questions, the entire sum must also converge. Unfortunately, it is known that the *sum diverges* and does so rather badly.* Now, one might wonder why this is a criticism since even in quantum electrodynamics the renormalized perturbation series, when summed, gives a divergent result. The answer is that theories such as quantum electrodynamics are inherently incomplete. They ignore the micro-structure of space-time and assume that space-time can be represented by Minkowski space at arbitrarily small distance scales. They shift the burden of infinities to the quantum theory of space-time structure. A "theory of everything," on the other hand, has no such escape route. It must face the Planck regime squarely. It cannot plead ignorance and shift the burden of infinities on yet another theory.

The general consensus now is that the root of the problem lies in the use of perturbative methods. More precisely, it is felt that the imaginative field theoretic attempts at quantizing gravity have failed because they assume that even at small distances, where by simple dimensional arguments, gravity should dominate, space-time geometry is smooth. To obtain a viable theory, one must drop this assumption and let the theory itself determine the micro-structure of space-time. For this, irrespective of whether one prefers general relativity or higher derivative theories, strings or membranes, one must face the problem of quantization *nonperturbatively*. My own view is that once one accepts this premise, the

---

*For the case of the bosonic string, this result was first proved by Gross and Periwal (1988). They argued that the superstring expansion should behave in the same way. By now there are independent calculations supporting this view (Jan Ambjorn ((1989)).

initial rationale for abandoning general relativity as the point of departure for quantization loses much of its force. Furthermore, the question of whether quantum general relativity exists as a consistent theory is of considerable interest in its own right in mathematical physics. Therefore, our program focuses on this issue. Note that I do *not* wish to imply that quantum general relativity would necessarily be the correct physical theory. Indeed, I believe that one should aim at constructing a robust framework so that the broad qualitative conclusions would continue to hold should general relativity be ultimately replaced by a more complete theory. Rather, the viewpoint is that general relativity is as good a starting point as any to address the key conceptual problems in quantum gravity that arise because of the absence of a background space-time geometry.

The material in this report is divided as follows. Section 2 summarizes the new Hamiltonian formulation of classical general relativity on which the program is based. I also point out the close relation between this formulation and the standard canonical framework for Yang-Mills fields. Section 3 outlines the quantization program. In particular, I will discuss the "loop representation" introduced by Rovelli and Smolin (1989a,1989b) in which quantum states arise as suitable functions of closed loops on a 3-manifold and present the infinite dimensional space of their solutions to all quantum constraints. Although these developments represent substantial progress, a number of important problems remain unresolved. To gain insight into the remaining problems, the program has been carried to conclusion in some simplified models. I will conclude with a brief discussion of these results.

2.   Hamiltonian framework

For simplicity, I shall restrict myself to source-free general relativity. The framework is, however, quite robust: all its basic features remain unaltered by the inclusion of a cosmological constant and coupling of gravity to Klein-Gordon fields, (classical or Grassmann-valued) Dirac fields and Yang-Mills fields with any internal gauge group (Jacobson (1988); Ashtekar et al. (1989a)). For brevity of presentation, I will begin with complex general relativity–i.e. by considering complex Ricci-flat metrics $g_{\mu\nu}$ on a real 4-manifold M–and take the "real section" of the resulting phase-space at the end.

The idea is to use a first-order formalism. The basic space-time fields will consist of a pair, $(e^\mu,\ {}^4A_\mu{}^\nu)$, of a (complex) co-tetrad e^μ and a connection 1-form ${}^4A_\mu{}^\nu$ which is *self-dual* in the "internal" (or Lorentz) indices '$\mu\nu$'; we have: ${}^4A^{\mu\nu} = \frac{i}{2}\epsilon^{\mu\nu}{}_{\sigma\tau}{}^4A^{\sigma\tau}$. (For simplicity, I have suppressed the space-time indices. If one is ultimately interested in Euclidean general relativity, one should drop the i in the right hand side while defining self-duality.) The action is given by (Samuel (1987); Jacobson and Smolin (1987, 1988)):

$$S(e,{}^4A) := \int e_\mu \wedge e_\nu \wedge {}^4F^{\mu\nu}, \tag{1}$$

where ${}^4F^{\mu\nu}$ is the curvature of ${}^4A_\mu{}^\nu$. This action is rather like the one introduced by Palatini except that, since the connection $A^{\mu\nu}$ is now required to be self-dual in the internal indices, so is the curvature tensor in Eq. (1). By setting the variation of the action with respect to ${}^4A_\mu{}^\nu$ to zero we obtain the result that ${}^4A_\mu{}^\nu$ is the self-dual part of the (torsion-free) connection ${}^4\Gamma_\mu{}^\nu$ compatible with the co-tetrad e^μ . Thus, ${}^4A_\mu{}^\nu$ is completely determined by e^μ . Setting the variation with respect to e^μ to zero and substituting for the connection from the first equation of motion, we obtain the result that the space-time metric $g = e_\mu e_\nu \eta^{\mu\nu}$, where $\eta_{\mu\nu}$ is the fixed, Minkowski metric on the internal space, satisfies the vacuum Einstein's equation. Thus, even though we are using a self-dual connection, Eq. (1) is completely equivalent to the standard Palatini action as far as the classical equations of motion are concerned. The reason behind this can be traced back to Bianchi identities.

One can carry out the Legendre transform of this action by carrying out a 3+1-decomposition in a straightforward way. The resulting canonical variables are then complex fields on a ("spatial") 3-manifold Σ. The configuration variable turns out to be an $SO(3)$-connection 1-form $A_a{}^i$ and its canonical momentum turns out to be a triad $\widetilde{E}^a{}_i$ with density weight one, where 'a' is the (co)vector index and 'i' is the triad or the $SO(3)$ internal index. The (non-vanishing) fundamental Poisson brackets are:

$$\left\{ \widetilde{E}^a{}_i(x),\ A_b{}^j(y) \right\} = -i\delta^a{}_b\delta_i{}^j\,\delta(x,y). \tag{2}$$

The geometrical interpretation of these canonical variables is as follows. In any solution to the field equations, $A_a{}^i$ turns out to be a potential for the self-dual part of the Weyl curvature and $\widetilde{E}^a{}_i$ the "square-root" of the 3-metric (times its determinant) on Σ. The relation of these variables to the familiar geometrodynamical variables, the 3-metric q_{ab} and the extrinsic curvature k_{ab} on Σ, is as follows:

$$GA_a{}^i = \Gamma_a{}^i - ik_a{}^i \quad and \quad \widetilde{E}^a{}_i \widetilde{E}^{bi} = (det\, q)q^{ab} \tag{3}$$

where G is Newton's constant, $\Gamma_a{}^i$ is the spin connection determined by the triad $\widetilde{E}^a{}_i$, $k_a{}^i$ is obtained by transforming the space index 'b' of k_{ab} into an internal index by the triad and $(det\, q)$ is the determinant of q_{ab}. Note, however, that, as far as the mathematical structure is concerned, we can also think of $A_a{}^i$ as a (complex) $SO(3)$-Yang-Mills connection and $\widetilde{E}^a{}_i$ as its conjugate electric field. Thus, the phase space has a dual interpretation. It is this fact that enables one to import into general relativity and quantum gravity ideas from Yang-Mills theory and quantum chromodynamics and may, ultimately, lead to a unified mathematical framework underlying the quantum description of all fundamental interactions. In what follows, we shall alternate between the interpretation of $\widetilde{E}^a{}_i$ as a triad and as the electric field canonically conjugate to the connection $A_a{}^i$.

Since the configuration variable $A_a{}^i$ has nine components per space point and since the gravitational field has only two degrees of freedom, we expect seven first class constraints. This expectation is indeed correct. We have:

$$\begin{aligned}
\mathcal{G}_i &:= \mathcal{D}_a \widetilde{E}^a{}_i = 0, \\
\mathcal{V}_a &:= F_{ab}{}^i \widetilde{E}^b{}_i = 0, \\
\mathcal{S} &:= \epsilon^{ijk} F_{abk} \widetilde{E}^a{}_i \widetilde{E}^b{}_j = 0,
\end{aligned} \tag{4}$$

where $F_{ab}{}^i := 2\partial_{[a}A_{b]}{}^i + G\epsilon^{ijk}A_{aj}A_{bk}$ is the field strength constructed from $A_a{}^i$. Note that all these equations are simple polynomials in the basic variables; the worst term occurs in the last constraint and is only quadratic in each of $\widetilde{E}^a{}_i$ and $A_a{}^i$. The three equations are called, respectively, the Gauss constraint, the vector constraint and the scalar constraint. The last two are familiar; they arise also in the older canonical formulation of general relativity based on 3-metrics q_{ab} and their conjugate momenta $\widetilde{p}^{ab}$. The first, Gauss law, arises because we are now dealing with triads rather than metrics. It simply tells us that the internal $SO(3)$ triad rotations are "pure gauge."

From geometrical considerations we know that the "kinematical gauge group" of the theory is the semi-direct product of the group of local triad rotations with that of spatial diffeomorphisms on Σ. This group has a natural action on the canonical variables $A_a{}^i$ and $\widetilde{E}^a{}_i$ and thus admits a natural lift to the phase-space. It turns out that this is precisely the group formed by the canonical transformations generated by the Gauss and the vector constraints. Thus, six of the seven constraints admit a simple geometrical interpretation. What about the scalar constraint? Note that, being quadratic in momenta, it is of the form $G^{AB}P_A P_B$ where, the connection supermetric $\epsilon^{ijk}F_{abk}$, plays the role of G^{AB}. Consequently, the canonical transformations generated by the scalar constraints correspond precisely to the *null geodesics of the connection supermetric*. It is well-known that the space-time interpretation of these canonical transformations is that they correspond to "multi-fingered" time-evolution. Thus, we now have an attractive representation of the Einstein evolution as a null geodesic motion in the (connection) configuration space.

In the asymptotically flat situation, asymptotic (space and time) translations are generated on the phase-space by Hamiltonians. As in geometrodynamics, these are obtained by adding suitable surface terms to constraints. Given a lapse $\underset{\sim}{N}$ and a shift N^a, the Hamiltonian is given by:

$$H(A, \widetilde{E}) = i \int_\Sigma d^3x \, (N^a F_{ab}{}^i \widetilde{E}^b{}_i - \tfrac{i}{2} \underset{\sim}{N} \epsilon^{ijk} F_{abk} \widetilde{E}^a{}_i \widetilde{E}^b{}_j)$$
$$- \oint_{\partial\Sigma} d^2 S_a \, (\underset{\sim}{N} \epsilon^{ijk} A_{bk} \widetilde{E}^a{}_i \widetilde{E}^b{}_j + 2i N^{[a} \widetilde{E}^{b]}{}_i A_b{}^i).$$

$$(5)$$

So far, we have discussed *complex* general relativity. To recover the Lorentzian theory, we must now impose reality conditions, i.e., restrict ourselves to the real, Lorentzian section of the phase-space. Let me explain this point by means of an example. Consider a simple harmonic oscillator. One may, if one so wishes, begin by considering a complex phase-space spanned by two complex co-ordinates q and p and introduce a new complex co-ordinate $z = q - ip$. (q and p are analogous to the triad $\widetilde{E}^a{}_i$ and the extrinsic curvature $k_a{}^i$, while z is analogous to $A_a{}^i$.) One can use q and z as the canonically conjugate pair, express the Hamiltonian in terms of them and discuss dynamics. Finally, the real phase-space of the simple harmonic oscillator may be recovered by restricting attention to those points at which q is real and $ip = z - q$ is pure imaginary (or, alternatively, $\dot{q}$ is also real.) In the present phase-space formulation of general relativity, the situation is analogous. In terms of the familiar geometrodynamic variables, the reality conditions are simply that the 3-metric be real and the extrinsic curvature–the time derivative of the 3-metric–be real. If these conditions are satisfied initially, they continue to hold under time-evolution. In terms of the present canonical variables, these become (Ashtekar, et al. (1989a)): i) the (densitized) 3-metric $\widetilde{E}^a{}_i \widetilde{E}^{bi}$ be real; and ii) its Poisson bracket with the Hamiltonian H be real, i.e.,

$$(\widetilde{E}^a{}_i \widetilde{E}^{bi})^\star = \widetilde{E}^a{}_i \widetilde{E}^{bi}$$
$$[\epsilon_{ijk}\widetilde{E}^a{}_i \mathcal{D}_c(\widetilde{E}^c{}_j \widetilde{E}^b{}_k)]^\star = -\epsilon_{ijk}\widetilde{E}^a{}_i \mathcal{D}_c(\widetilde{E}^c{}_j \widetilde{E}^b{}_k)). \qquad (6)$$

(In Euclidean relativity, these conditions can be further simplified; they only require that we restrict ourselves to real triads and real connections.) As far as the classical theory is concerned, we could have restricted to the "real slice" of the phase-space right from the beginning. In quantum theory, on the other hand, it is simpler to first consider the complex theory, solve the constraint equations and then impose the reality conditions as suitable Hermitian-adjointness relations. Thus, the quantum reality conditions would be restrictions on the choice of the inner-product on physical states.

A number of remarks are in order.

   i) Note that all equations of the theory–the constraints, the Hamiltonian and hence the evolution equations and the reality conditions–are simple polynomials in the basic variables $\widetilde{E}^a{}_i$ and $A_a{}^i$. The framework may therefore be of considerable interest to numerical relativists (Ashtekar and Romano (1989)). In this connection, it is especially worth noting that, since the Boulder conference, a simple approach has become available to obtain the "free data," i.e. solutions to the constraint equations (4) (Capovilla et al. (1989a)). The strategy is the following. Choose any connection $A_a{}^i$ such that its magnetic field $\widetilde{B}^a{}_i := \widetilde{\eta}^{abc}F_{bci}$, regarded as a matrix, is non-degenerate. (We shall denote by $\widetilde{\eta}^{abc}$ the metric independent Levi-Civita density.) A "generic" connection $A_a{}^i$ will satisfy this condition; it is not too restrictive an assumption. Now, we can expand out $\widetilde{E}^a{}_i$ as $\widetilde{E}^a{}_i = M_i{}^j \widetilde{B}^a{}_j$ for some matrix $M_i{}^j$. The pair $(A_a{}^i, \widetilde{E}^a{}_i)$ then satisfies the vector and the scalar constraints *if and only if* $M_i{}^j$ is of the form $M_i{}^j = [\phi^2 - \frac{1}{2}\operatorname{tr}\phi^2]_i{}^j$, where $\phi_i{}^j$ is an arbitrary trace-free, symmetric field on Σ. Thus, as far as these four constraints are concerned, the free data consists of $A_a{}^i$ and $\phi_i{}^j$. It only remains to solve the Gauss law which simply reduces to: $\widetilde{B}^a{}_i \mathcal{D}_a M_i{}^j = 0$. Although it is straightforward to invent procedures to solve this remaining equation, a simple, elegant method has not yet emerged.

ii) The phase-space of general relativity is now identical to that of complex-valued Yang-Mills fields (with internal group $SO(3)$). Furthermore, one of the constraint equations is precisely the Gauss law that one encounters on the Yang-Mills phase-space. Thus, we have a natural imbedding of the constraint surface of Einstein's theory into that of Yang-Mills theory: Every initial data $(A_a{}^i, \tilde{E}^a{}_i)$ for Einstein's theory is also an initial data for Yang-Mills theory which happens to satisfy, in addition to the Gauss law, a scalar and a vector constraint. From the standpoint of Yang-Mills theory, the additional constraints are essentially the simplest diffeomorphism and gauge invariant expressions one can write down in absence of a background structure such as a metric. Note that the degrees of freedom match: the Yang-Mills field has 2 (helicity) $\times 3$ (internal) $=$ 6 degrees and the imposition of four additional first-class constraints leaves us with $6 - 4 = 2$ degrees of freedom of Einstein's theory. I want to emphasize, however, that in spite of this close relation of the two initial value problems, the Hamiltonians of the two theories are *very* different. Nonetheless, the similarity that does exist can be exploited to obtain interesting results relating the two theories (Ashtekar and Renteln in Ashtekar (1988); Mason and Newman (1989); Samuel (1988); S. Koshti and N. Dadhich (1989); Capovilla et al, (1989b)).

iii) Since all equations are polynomial in $A_a{}^i$ and $\tilde{E}^a{}_i$ they continue to be meaningful even when the triad (i.e. the "electric field") $\tilde{E}^a{}_i$ becomes degenerate or even vanishes. Results obtained in 2+1 gravity (Witten (1988)) indicate that this fact would play a significant role in the "unbroken," diffeomorphism invariant phase of quantum gravity where the vacuum state would correspond to just the zero, rather than Minkowskian, metric. In fact, recently, Capovilla, Dell and Jacobson (1989a) have introduced a Lagrangian framework which reproduces the Hamiltonian description discussed above but which *never even introduces a space-time metric*! This formulation of "general relativity without the metric" lends strong support to the viewpoint that the traditional emphasis on metric-dynamics, however convenient in classical physics, is not indispensible.

3.   Quantum theory

I will now outline the quantization program and summarize its status in full quantum gravity in $3+1$ dimensions. We will see that although significant progress has been made on a number of issues, several important and difficult problems remain to be investigated; even as a mathematical framework, the theory is incomplete. To gain insight into these problems, we have carried out the entire program to its conclusion in certain simplified contexts. I will simply state these results and provide references where detailed derivations can be found.

The program

We shall use a mixture of an algebraic approach and geometric quantization methods. The key steps may be stated as follows.

1. Select a subspace S of functions on the classical phase-space, closed under the operation of taking Poisson-brackets. Each element of S is to be promoted to a quantum operator unambiguously. S has to be "small enough" so that this Dirac-quantization procedure can be carried out unambiguously and yet "large enough" so that they can serve as (complex) coordinates on the phase-space. For the harmonic oscillator discussed in Section 2, for example, S can be taken to be the 2-dimensional space spanned by (complex-valued) q and p. In general relativity, S could be the space of linear functionals on the complex phase-space.

2. Associate with each element f of S an abstract operator $\hat{f}$ and construct the (free) algebra generated by these elements subject to the condition: $[\hat{f}, \hat{g}] = i\hbar\widehat{\{f, g\}}$. For later use, we also introduce a $\star$-relation (i.e. involution) on this algebra by requiring that $\hat{g} = (\hat{f})^\star$ if and only if the restriction of g to the real slice of the phase-space is the complex-conjugate of the restriction of f to this slice. Denote the resulting $\star$-algebra by **A**.

3. Find a representation of this algebra **A** by operators on a complex vector space V. Note that V is *not* equipped with any inner-product and that the $\star$-relations are ignored at this stage. This representation may be obtained by any convenient means. The leading candidates are: geometric quantization techniques (see, e.g., Woodhouse (1980)) and group theoretical methods (Isham (1984)).

4. Obtain the quantum analogs of the classical constraints. This requires the choice of a factor-ordering *and* regularization. Find

the linear subspace $\bar{V}$ of V which is annihilated by all quantum constraints. This is the space of physical quantum states.

5. Introduce an inner-product on $\bar{V}$ such that the $\star$-relations–ignored so far–become Hermitian adjoint relations on the resulting Hilbert space. Note that the full $\star$-algebra **A** itself does *not* have a well-defined action on the physical subspace $\bar{V}$; a general element of **A** would not weakly commute with the constraints and would therefore throw physical states out of $\bar{V}$. However, the $\star$-relations on **A** do define, in particular, $\star$-relations on physical operators–i.e. those that weakly commute with constraints–and it is these relations that should be transformed into Hermitian-adjointness by the inner-product. This would be a rather involved procedure. In simple quantum mechanical systems where constraints are either absent or play a minimal role, the $\star$-relations–quantum versions of the classical reality conditions–essentially determine the inner-product. In particular, this is the case in $2+1$-gravity and the weak field limit of the $3+1$ theory. It is not known if the situation in the full $3+1$ theory is analogous.

6. Interpret a sufficiently large class of self-adjoint operators; devise methods to compute their spectra and eigenvectors; analyze if there is a precise sense in which the 1-parameter family of transformations generated by the Hamiltonian can be interpreted as "time evolution;"....

The connection representation

The obvious candidate for the subspace S of functions on the phase-space is the space of functionals which are linear in $A_a{}^i$ and $\tilde{E}^a{}_i$. With this choice, it is straightforward to complete the first three steps given above using, as the representation space, functionals of either $A_a{}^i$ or $\tilde{E}^a{}_i$. The problem arises in step 4. The quantum Gauss constraint poses no problem. However, the vector and the scalar constraints do. As in the traditional canonical framework, while it is straightforward to write down a large class of solutions to the vector constraint in the E-representation, not a single solution is known to the scalar constraint. In the A-representation, the situation is just the opposite. An infinite dimensional space of solutions to the scalar constraints is available (Jacobson and Smolin (1988)). These are the "Wilson-loops," i.e., traces of holonomies, $h[\gamma, A]$, (regarded as functions of $A_a{}^i$, parametrized by

γ.)* However, none of these is diffeomorphism invariant since, under the action of diffeomorphisms, each γ is mapped to another loop. One can *formally* integrate these functionals over the diffeomorphism group of the 3-manifold and the resulting "functional" of $A_a{}^i$ *will* satisfy all constraints. But this construction is only formal since one does not know how to carry out the required integration.

Significant progress has, however, been made in a "truncated theory" in which one expands the constraints around a classical background (both $\widetilde{E}^a{}_i$ and $A_a{}^i$ flat) and keeps terms up to *second* order (Ashtekar (1989)). In this theory, not only can one solve all constraints but it is also possible to identify the component of the connection, A_T, that is to play the role of time; the true, dynamical degrees of freedom of the theory, A_D, evolve with respect to this "internal clock." Let me focus on the scalar constraint. Recall that, in the classical theory, this constraint plays a dual role. On the one hand, it restricts the physically allowable states and, on the other, it generates time evolution. In the truncated theory, the situation is completely analogous in the quantum picture as well. Physical states are annihilated by the scalar constraint. In this sense, the constraint is just a restriction on the physical states. Nothing "happens." This is analogous to the relativists' view that a solution of Einstein's equation in the classical theory is simply an entire space-time; intrinsically, there is no dynamics. Things begin to "happen" when we identify a time variable and slice space-time into space and time. Again, the situation in quantum theory is similar. If we identify one of the connection components, A_T, as "internal time"–and the choice is suggested by the form of the scalar constraint itself–then, the quantum constraint equation can be regarded as an "evolution equation" in the infinite-dimensional space of connections; the constraint simply tells us how to "evolve" the wave function $\Psi(A_D, A_T)$, given its value, $\Psi(A_D, A_T = 0)$, at the "initial time" $A_T = 0$. It turns out that the generator of this "evolution" is *precisely* the standard Hamiltonian of linearized gravity plus matter. Therefore, the "evolution," which takes place in the space of connections, can be re-interpreted as taking place in Minkowski space. Thus, in this truncated theory, the quantum scalar constraint actually reproduces the familiar Schrödinger equation for evolution! It is not yet clear whether this "derivation" of the Schrödinger

---

*Thus, $h[\gamma, A] := \operatorname{tr} \mathcal{P} \operatorname{Exp} \oint_\gamma A_a \, dS^a$ is the trace of the path ordered exponential of the integral of the connection $A_a{}^i$ around the closed loop γ.

equation from quantum gravity will extend to the full, non-truncated theory. Nonetheless, it is gratifying that at least in the weak field limit, the general framework outlined here reproduces the familiar, flat space quantum mechanics. While the flatness of the chosen background point $(A_a{}^i, \widetilde{E}^a{}_i)$ in the phase space simplifies calculations, I believe it is not essential. The main ideas underlying this truncated theory should provide new insight into the nature of quantum fields on curved backgrounds, including the origin of Hawking radiation. In this picture, quantum field theory in curved space-times would "descend from above" as an interesting approximation to full quantum gravity; traditionally, it has "emerged from below" as a first order correction to the classical description.

The loop representation

In the full 3+1 theory, the most promising approach to date is to use the "loop representation" in which quantum states are represented by functions of loops on the 3-manifold Σ (Rovelli and Smolin (1989a,1989b)). This construction rests heavily on the shift of emphasis in the classical canonical theory from geometrodynamics to connection dynamics. The motivation comes from the fact that the Wilson loops solve the scalar constraint. The intuitive idea is the following. If one formally writes the integral:

$$\mathcal{A}(\gamma) := \int d\mu[A] \, h[\gamma, A] \Psi[A] , \qquad (7)$$

(where "$d\mu[A]$" is a diffeomorphism invariant "measure" on the space of connections), one finds that, because the kernel $h[\gamma, A]$ satisfies the scalar constraint in the A-representation, the left hand side, $\mathcal{A}(\gamma)$ satisfies the scalar constraint for *any* choice of $\Psi(A)$ in the integral on the right. This is analogous to the fact that, in the Klein-Gordon theory in Minkowski space, the function $\mathcal{A}(x)$ in space-time, given by

$$\mathcal{A}(x)(x) := \int d^4k \, [\delta^4(k \cdot k - \mu^2) e^{ik \cdot x}] \Psi(k) , \qquad (7')$$

satisfies the Klein-Gordon equation automatically for *any* $\Psi(k)$ because the kernel in the square-brackets satisfies the equation in the k-space. In general relativity, then, it only remains to impose the vector constraint on the loop-functions $\mathcal{A}(\gamma)$ of (7). Since under the action of the diffeomorphism group of Σ, a loop is mapped to another loop in the same knot class, it follows that loop functions will satisfy *all* constraints if they are obtained via (7) *and* they depend only on the knot-classes of loops (rather than on individual loops.)

With these formal constructions as motivation, Rovelli and Smolin carry out the first four steps in the quantization program as follows. First they introduce a brand new space of functions on the phase-space which is to serve as the space S of step 1. Associated with a loop γ, there is an element of S called $T^0[\gamma](A)$, and associated with a loop γ, and a point x on γ there is another element, $T^1[\gamma, x](A, \widetilde{E})$ (which takes values in the space of vectors of density weight one at the point x). These are defined as follows:

$$T^0[\gamma](A) = h[\gamma, A] \equiv \operatorname{tr} \mathcal{P}(\operatorname{Exp} \oint_\gamma A_a ds^a),$$

and

$$T^1[\gamma, x](A, \widetilde{E}) = \operatorname{tr} \widetilde{E}^a(x) \mathcal{P}(\operatorname{Exp} \oint_\gamma A_a ds^a), \tag{8}$$

where $A_a = A_a{}^i \sigma_i$ and $\widetilde{E}^a = \widetilde{E}^a{}_i \sigma^i$ with σ^i the 2×2 Pauli matrices. (One can define in a similar way functions $T^{(n)}[\gamma, x_1, ... x_n]$ which are of order n in the momenta $\widetilde{E}^a{}_i$. These are useful but not essential to the construction.) One can check that the vector space S spanned by T^0 and T^1 is closed under the Poisson bracket. This is called the (classical) T-algebra.

Step 2 of the quantization program is now easily carried out by promoting T^0 and T^1 to quantum operators $\hat{T}^0$ and $\hat{T}^1$. To carry out step 3, one uses for V the space of functions $\mathcal{A}(\alpha)$ on the loop space of Σ satisfying certain properties. $\hat{T}^0$ and $\hat{T}^1$ are then represented as linear operators on V. Although their expressions are rather simple, I would have to introduce new notation to write them out explicitly. Therefore, I will just explain their general structure. The action of $\hat{T}^0[\gamma]$ on a state $\mathcal{A}(\alpha)$ gives the state $\mathcal{A}(\gamma \cup \alpha)$ where the loop $\gamma \cup \alpha$ is obtained from the individual loops α and γ by a specific prescription. $\hat{T}^1[\gamma, x]$ operating on $\mathcal{A}(\alpha)$ is zero unless α intersects γ at the point x. If it does, then the result is a linear combination of the value of $\mathcal{A}$ on the loops obtained by composing (at x) the two loops γ and α in two different ways. Using their definitions, one can explicitly check that the commutator of the two operators mirrors the Poisson algebra of their classical counterparts, T^0 and T^1.

The next task is to find regularized operators corresponding to the constraints. Since the entire T-algebra is gauge-invariant, the Gauss constraint has been taken care of already. The vector constraint now requires that the states $\mathcal{A}(\alpha)$ should depend only on the knot-class of α. Thus, in the loop representation, we know the *general* solution to the vector constraint. In the traditional metric representation, by contrast, while we know that the imposition of the quantum vector constraint is equivalent to demanding diffeomorphism invariance of the wave functional, the problem of explicitly finding the *general* solution is difficult and unresolved. The fact that one is led to represent quantum states by functions of a *discrete* space–the space of knots–may cause some discomfort since in field theories states normally arise as functionals of fields. After all, in some sense, the quantum states should carry the information that the gravitational field has two degrees of freedom per space point. Is that not intrinsically impossible if states are functions on the discrete set of knots? I believe that this discomfort is not justified. Consider the quantum theory of the hydrogen atom. In the position or momentum representation, states are functions of a continuous variable, x or k, reflecting the fact that the classical system has 3 degrees of freedom. The fact that these variables take values in a continuum is comforting because the classical configuration space is continuous. However, in quantum theory, we can also go to the basis in which the Hamiltonian, the total angular momentum and the z-component of the angular momentum are all diagonal. Then, a generic state is given by a complex-valued function, $\psi(n, l, m)$ of three integers! A similar representation dependence arises also in the quantum theory of a free field confined to a box. Therefore, in quantum gravity, the discreteness of the space on which physical states are defined may just be a reflection of the choice of basis that happens to be well-adapted to the loop picture.

It remains to impose the scalar constraint. In the loop representation, one can first construct, at the classical level, a function $T^{(2)}(\gamma; x_1, x_2)$ whose appropriately taken limit, as the loop γ shrinks to a point x, gives us the scalar constraint evaluated at x. One can promote $T^{(2)}$ to a quantum operator $\hat{T}^{(2)}$ on V in a precise way and require that the physical quantum states be those elements $\mathcal{A}(\alpha)$ of V which satisfy: $\mathrm{Lim}\,(\hat{T}^{(2)} \cdot \mathcal{A}(\alpha)) = 0$. The availability of $\hat{T}^{(2)}$ provides a way of both regularizing and imposing the scalar constraint on quantum states.* It

---

*This regularization *is* unconventional in that it does not use an inner-product on the space V on which $\hat{T}^{(2)}$ operates. As a result, there is some controversy as to whether this procedure is fully satisfactory. However, in every instance where this issue has come up so far, a background geometry is used in an essential way to construct the inner-product on the space of unconstrained states. This would be unacceptable in non-perturbative quantum gravity. Is there a more satisfactory procedure?

is known that if the state $\mathcal{A}(\alpha)$ has support on only those (differentiable) loops which contain no self-intersections, it is annihilated by the quantum scalar constraint given above. Thus, a function $\mathcal{A}(\gamma)$ which has support only on non-intersecting loops γ *and* which depends only on the knot class of these loops is in the physical subspace $\bar{V}$. While this space is infinite-dimensional, there are several reasons to believe that this is *not* the complete set of physical states. (In fact, there is some question as to whether this sub-space of $\bar{V}$ is physically interesting at all.) In the case when the spatial topology is non-trivial, additional solutions *are* known (Blencowe (1989)). However, one would expect entire sectors of new solutions even when the topology is trivial. For example, there must exist loop space analogs of the solutions discovered by Jacobson and Smolin in the connection representation which are parametrized by self-intersecting loops. Are there more general solutions? Unfortunately, so far attempts to find these have led only to negative results which state that there are no solutions of a certain type (see, e.g., Husain (1988)).

The loop representation is quite unconventional. In particular, the space of states is not constructed by introducing a polarization on the phase-space; states are functions on the loop-space of the 3-manifold rather than functionals of classical fields. Consequently, interesting quantum operators have to be constructed by indirect means. For example, the quantum scalar constraint operator was not obtained by simply replacing each classical field in the expression of the scalar constraint by an operator-valued distribution and then regularizing the resulting object. In this sense, the procedure adapted in the loop representation takes the suggestions that Dirac made in his last paper* very seriously: in the passage to quantum theory, one does not copy the classical expressions but allows more general dynamical variables. It is for this reason that the final physical variables–operators on the space of physical quantum states–do not have an obvious physical interpretation in terms that we are familiar with from the classical theory. As advocated by Dirac, the task is to first construct a consistent mathematical framework and

---

* "The theory of Heisenberg is more powerful than classical mechanics because its dynamical variables can be of a more general nature. ... work should be concerned with finding the correct Hamiltonian, making use of the vast possibilities of non-commuting quantities which need not be suggested by classical mechanics. That would mean some kind of degrees of freedom occurring in a fundamental way in the equations of quantum theory. The trend followed by most physicists of keeping ideas suggested by classical mechanics and then supplementing them by certain groups is a very restricted one." (Dirac, published posthumously (1987)).

then let it guide us to the appropriate interpretation of the dynamical variables.

This is an unconventional viewpoint and one might therefore wonder if such a description is viable at all. Are there some intrinsic difficulties associated with this unfamiliar procedure which are masked merely by the technicalities of 3+1-dimensional general relativity? To ensure that there are no inherent problems, the loop representation was constructed in two tractable, complementary cases, each of which captures an aspect of general relativity. We shall conclude this section with a brief discussion of these models.

The first model is general relativity in 2+1-dimensions (Ashtekar et al. (1989b)). In 3 dimensions, the vanishing of the Ricci tensor implies the vanishing of the Riemann tensor, whence every solution to the vacuum equations is flat. Thus, there are no gravitational waves or "gravitons." Yet the theory is not empty: it has a finite number of topological degrees of freedom. Furthermore, structurally it is identical to the 3+1 theory. It therefore serves as a useful toy model in which one can test various ideas. Since the notion of self-duality of the curvature tensor does not go over to 2+1-dimensions, one might wonder if the ideas introduced in Section 2 have any place at all in this model. It turns out that a completely analogous Hamiltonian framework can in fact be constructed (Witten (1988); Bengtsson (1989)). The basic configuration variable is again a connection $A_a{}^i$ on a Cauchy surface (which, however, is real and takes values in the $SO(2,1)$ Lie-algebra). It is therefore possible to introduce both the connection and the loop representations explicitly and exhibit the Rovelli-Smolin transform relating them. (In fact, in asymptotically flat space-times, the 2+1-analogs of T^0 and T^1 have a direct physical interpretation: they are related, in a simple way, to the total mass and angular momentum of the space-time.) In the connection representation, physical states turn out to be functions on the (finite-dimensional) moduli-space of flat connection. In the loop representation, they are suitable functions of *homotopy classes* of closed loops on the spatial 2-manifold. Since the homotopy classes in 2-dimensions are the analogs of knot classes in 3-dimension, this picture is perhaps not too surprising. In both representations, the reality conditions suffice to pick out the inner-product on physical states. In the loop representation, three distinct, concrete procedures are available to make this selection. One of these, based on the Gelfand-Naimark-Segal construction, appears to be best suited for extension to the 3+1-theory. Work is now in progress to see if this is indeed the case.

While the 2+1 theory captures many features of 3+1 general relativity, it does have only a finite number of degrees of freedom. To gain insights into the infinite dimensional problems, the loop representation was constructed also for source-free Maxwell theory in 3+1-dimensional Minkowski space (Ashtekar and Rovelli (1989)). Again, it was possible to show rigorously that the Rovelli-Smolin transform exists and that the reality conditions essentially suffice to select the inner-product. Once again, one can introduce the inner-product directly in the loop representation, without any reference to the transform, using the Gelfand-Naimark-Segal construction.

4. Discussions

In this section, I shall return to the key feature of the present canonical framework: The shift of emphasis from geometrodynamics to connection dynamics.

We saw in Section 2 that the scalar constraint S is of the form

$$S = G_{ab}{}^{ij} \widetilde{E}^a{}_i \widetilde{E}^b{}_j = 0, \tag{9}$$

where the "connection supermetric," $G_{ab}{}^{ij}$ is given by $\epsilon^{ijk} F_{abk}$. Since S is purely quadratic in momentum, it follows that the Hamiltonian flow it generates on the phase-space corresponds precisely to null geodesics of $G_{ab}{}^{ij}$ on the space of connections. By contrast, in terms of geometrodynamical variables, the 3-metric q_{ab} and its conjugate momentum $\widetilde{p}^{ab}$, the scalar constraint has the form:

$$S' := G'_{abcd} \widetilde{p}^{ab} \widetilde{p}^{cd} + \frac{\sqrt{(\det q)}}{G} \, {}^3R = 0, \tag{9'}$$

where $G'_{abcd}(q) = \frac{G}{2\sqrt{\det q}}(q_{ac}q_{bd} + q_{bc}q_{ad} - q_{ab}q_{cd})$ is the Wheeler-DeWitt supermetric and G, Newton's constant. Not only is the functional dependence of S' on q_{ab} and $\widetilde{p}^{ab}$ more complicated than that of S on $\widetilde{E}^a{}_A{}^B$ and $A_a{}^i$, but S' also contains a "potential term" ${}^3R$. Consequently, the trajectories in the space of metrics defined by the Einstein evolution do not have a natural geometrical interpretation. (One may divide S' by ${}^3R$ and regard $({}^3R)^{-1} G'_{abcd}$ as the new supermetric to make S' resemble S. However, the new supermetric is ill-defined at points where ${}^3R$ vanishes and this leads to problems even in simple models such as Bianchi cosmologies.) In particular, Misner (1989) has pointed out that, since dynamical orbits in the space of 3-metrics can start out in, say, a time-like direction of G'_{abcd} and then become space-like in the course of evolution, the supermetric G'_{abcd} appears to play a only secondary role in the geometrodynamic description of the Einstein evolution. In the space of connections, on the other hand, it is the supermetric $G_{ab}{}^{ij}$ that completely determines the evolution. It is therefore possible that many of the interesting properties of space-times are captured in the intrinsic, geometric structure of the connection supermetric.

An example is provided by Bianchi cosmologies. In the type I model, the connection supermetric turns out to be flat except at certain points where it is degenerate. Points at which it vanishes correspond to the flat Kasner solutions. If we excise degenerate points, the remainder of the connection space is isometric to the three dimensional Minkowski space. It is trivial to integrate null geodesics in this portion. These correspond to general Kasner space-times. The end-points of these null geodesics (at which the gauge-fixed connection diverges) correspond to space-time singularities. Thus, singularities in space-time are represented by points at (future) null-infinity of the connection supermetric! The connection supermetric has a rather simple form also in the type IX model (Kodama (1988)). In the diagonal gauge, the space of connections is again 3-dimensional and the diagonal connection components, A_I, provide a natural chart. The A_I turn out to be the null coordinates of the supermetric. Consequently, the $A_I =$ constant curves trivially provide a family of null geodesics, i.e., solutions to Einstein's equations. A number of other features could be similarly explored. For example, it is of considerable interest to find out if the type IX supermetric is asymptotically flat at null infinity. If it is, one could analyze the asymptotic behavior of solutions near space-time singularities using local, differential geometric techniques at null infinity of the connection space. In this case, Lyapounov exponents associated with these geodesics will almost certainly go to zero as one approaches null infinity. Therefore, it is likely that the chaotic behavior of the exact type IX solutions–if it exists–will be coded in the lack of asymptotic flatness of the supermetric at null infinity. Thus, the emphasis would be shifted: *One would investigate invariant properties of the connection supermetric rather than the space-time behavior of specific solutions to type IX equations.* Similarly, in the full theory, the asymptotic behavior of the null geodesics of $G_{ab}{}^{ij}$–i.e. of space-time geometries near singularities–is fully coded in the conformal structure of $G_{ab}{}^{ij}$. Therefore, a detailed analysis of this structure will shed new light on a number of issues in classical relativity.

The use of the space of connections as the arena for dynamics is especially attractive in quantum gravity. Already in classical general relativ-

ity, since we do not have access to a background geometry, the issue of
the kinematical arena becomes more subtle than it is in other field theories. In other theories, the presence of a fixed, kinematical space-time
metric makes it relatively straightforward to introduce notions such as
time evolution, causal propagation, and scattering matrices. In general
relativity, on the other hand, one can introduce such notions only *after* one has a complete solution to Einstein's equation. The notion of a
black-hole, for example, is not well-defined before one has constructed
the entire space-time. In quantum gravity, the situation gets even more
complicated. To see this, recall first that in non-relativistic quantum
mechanics particles do not have well-defined trajectories due to the uncertainty principle; time-evolution produces only a wave function $\Psi(\vec{x}, t)$
rather than a specific trajectory $\vec{x}(t)$. Similarly, in quantum gravity, even
after a complete evolution, one would *not* recover a classical space-time.

In absence of a space-time, how would one introduce familiar physical
notions such as causality, dynamics, black holes and scattering matrices?
Although these notions arise from classical considerations, presumably
they do have *some* counterparts in the quantum theory as well. I believe
that the space of connections would be the natural arena to formulate
and analyze these counterparts. Is there any reason to prefer this space
over the space of 3-metrics of geometrodynamics? The answer, I believe, is in the affirmative. We saw in Section 3 that one can indeed
single out time and re-interpret the quantum scalar constraint as the
Schrödinger equation in an approximate way on the space of connections. In geometrodynamics, on the other hand, where the wave functions are functions of 3-metrics, it has not been possible to carry out this
procedure in that (or any other) approximation. Indeed, to extract time
in the traditional variables in the weak field limit, one is forced to step
outside the space of 3-metrics and construct a "mixed" representation
in which states are functionals of conformal metrics and traces of the
extrinsic curvatures (Kuchař (1970)). (Note also that it has not been
possible to extend this mixed representation to the full theory where
it is extremely cumbersome to express the scalar constraint in terms of
these variables. The connection representation, on the other hand, was
first introduced for the full theory and the truncation was carried out
subsequently. This is possible precisely because the connection contains
information about both the 3-metric and the extrinsic curvature, in just
the right combination.)

Now, it may well be that the construction outlined in Section 3 does
not extend to the full theory. If that should happen, one would simply

say that in a generic situation one has to work entirely in the space of connections. Presumably, the solutions to constraints will exist. These will be the physical states. Physical observables will act on this space. Their algebraic properties, eigenvalues and eigenvectors would suggest their interpretation. In general, it may not be possible to speak of time and evolution at least in terms that we are familiar with. It may be that only for certain solutions of constraints can we introduce the notion of time following, e.g., the ideas of Section 3. In these cases, it would be superfluous to work in the infinite dimensional space of connections; it would be possible to construct a four dimensional space-time and, to a good approximation, describe everything that "really" takes place in the infinite dimensional space of connections in terms of structures and constructs in four dimensions. In more general situations, the four dimensional approximate structure would be unavailable and we would be forced to deal with the infinite dimensional space directly. The challenge, then, is to develop intuition for structures in this space, learn to pose physically interesting questions in terms of these structures and answer them.

I have benefited a great deal from discussions with a number of colleagues. I am especially grateful to Ted Jacobson, Chris Isham, Karel Kuchař, Lionel Mason, Ted Newman, Roger Penrose, Carlo Rovelli, Joseph Samuel and Lee Smolin for encouragement, inspiration, criticism and advice. This work was supported in part by the NSF grants PHY 86-12424 and INT88-15209 and by research funds provided by Syracuse University.

Note Added: After the Boulder conference, I received a communication from Noboru Nakanishi (Kyoto University) saying that, over the past decade, he has a constructed a satisfactory quantum theory of gravity by carrying out canonical quantization non-perturbatively, using however, an approach that is completely different from the one discussed in this chapter.

References

Ambjorn, J. (1989), Private communication.

Ashtekar, A. (1986), *Phys. Rev. Lett.* **57**, 2244.

Ashtekar, A. (1987), *Phys. Rev.* **D36**, 1587.

Ashtekar, A. (1988), in *Mathematics and General Relativity*, ed. J. Isenberg (American Mathematical Society, Providence).

Ashtekar, A. (1989), in *Conceptual Problems of Quantum Gravity*, ed. A. Ashtekar and J. Stachel (Birkhäuser, Boston).

Ashtekar, A. and Romano, J. D. (1989), SU Report.

Ashtekar, A. and Rovelli, C. (1989), preprint.

Ashtekar, A., Romano, J. D. and Tate, R. S. (1989a), *Phys. Rev.* **D40**, 2572.

Ashtekar, A., Husain, V., Rovelli, C., Samuel, J. and Smolin, L. (1989b), *Class. Quant. Grav.* **6**, L185.

Bengtsson, I. (1989), *Phys. Lett.* **220B**, 51.

Blencowe, M.P. (1989), Imperial College preprint.

Capovilla R., Jacobson, T. and Dell, J. (1989a), *Phys. Rev. Lett.* **63**, 2325.

Capovilla R., Jacobson, T. and Dell, J. (1989b), *Class. Quant. Grav.*, in press.

Dirac, P. A. M. (1987), in *Paul Adrien Maurice Dirac*, eds. B. N. Kursunoglu and E. P. Wigner (Cambridge University Press, Cambridge).

Gross, D. and Periwal, V. (1988), *Phys. Rev. Lett.* **61**, 2105.

Husain, V. (1989), *Nucl. Phys.* **B313**, 711.

Isham, C. J. (1984), in *Relativity Groups and Topology II*, eds. B. S. DeWitt and R. Stora (North-Holland, Amsterdam).

Jacobson, T. (1988), *Class. Quant. Grav.* **5**, L143.

Jacobson, T. and Smolin, L. (1987), *Phys. Lett.* **196B**, 39.

Jacobson, T. and Smolin, L. (1988), *Class. Quant. Grav.* **5**, 583.

Jacobson, T. and Smolin, L. (1988), *Nucl. Phys.* **B299**, 295.

Kodama, H. (1988), *Prog. Theor. Phys.* **80**, 1024.

Koshti, S. and Dadhich, N. (1989), *Class. Quant. Grav.*, in press.

Kuchař, K. (1970), *J. Math. Phys.* **11**, 3322.

Mason, L. and Newman, E.T. (1989), *Commun. Math. Phys.* **121**, 659.

Misner, C. W. (1989), private communication.

Newman, E. T. (1984), in *Proceedings of the 10th International Conference on General Relativity and Gravitation*, eds. B. Bertotti, F. de Felice and A. Pascolini (D. Reidel, Amsterdam).

Penrose, R. (1976), *Gen. Relat. Grav.* **7**, 31.

Rovelli, C. and Smolin., L. (1989), *Phys. Rev. Lett.* **61**, 155.

Rovelli, C. and Smolin., L. (1989), *Nucl. Phys.* **B** (in press).

Samuel, J. (1987), *Pramana J. Phys.* **28**, L429.

Samuel, J. (1988), *Class. Quant. Grav.* **5**, L123.

Smolin, L. (1989), in *Proceedings of the 1989 Johns Hopkins Workshop*, in press.

Weinberg, S. (1972), *Gravitation and Cosmology*, (John Wiley, New York).

Witten, E. (1988), *Nucl. Phys.* **B311**, 46.

15

Progress in quantum cosmology

James B. Hartle
Physics Department,
University of California,
Santa Barbara, CA 93106 USA

1. The aims of quantum cosmology

Our observations of the world give us specific facts. Here, there is a galaxy; there, there is none. Today, there is a supernova explosion; yesterday, there was a star. Here, there are fission fragments; before, there was a uranium nucleus. The task of physics is to bring order to this great mass of facts which constitutes our experience. In the language of complexity theory, the task is to compress the message which describes these facts into a shorter form — to compress it, in particular, to a form where the message consists of just a *few* observed facts together with simple universal laws of nature from which the rest can be deduced.

In the past, physics, for the most part, has concentrated on finding dynamical laws which correlate facts at different times. Such laws predict later evolution given observed initial conditions. However, there is no logical reason why we could not look for laws which correlate facts at the same time. Such laws would be, in effect, *laws* of initial conditions.

I believe it was the limited nature of our observations which led to our focus on dynamical laws. Now, however, in cosmology, in the observations of the early universe and even on familiar scales, it is possible to discern regularities of the world which may find a compressed expression in a simple, testable, theory of the initial conditions of the universe as a whole. The search for this law of the initial conditions is the subject of quantum cosmology and the subject whose recent development I shall try and review here.[1]

What are the regularities of the universe that we hope might be tied directly to a theory of its initial conditions? Below is a list of likely candidates. Interestingly, they occur on all length scales.

On the larger scales of the universe there are those observations whose explanation cosmology has traditionally assigned to initial conditions. These include:

- The approximate homogeneity and isotropy of the universe on scales above several hundred Mpc and its simple spatial topology.
- The vast age of the universe when measured in Planck time units or, what is the same thing given Einstein's equation, the approximate spatial flatness on large scales.
- The spectra of deviations from exact homogeneity and isotropy which we see today as the large-scale distribution of galaxies and clusters of galaxies.

One of the more interesting recent developments in quantum cosmology is the growing understanding that a theory of initial conditions will not only make predictions about the very large scales of traditional cosmology but also on familiar scales and on the microscopic scales of elementary particle theory. The features on common scales that we hope find explanations for quantum cosmology are so familiar that we usually take them for granted:

- The homogeneity of the thermodynamic arrow of time. Everywhere in the universe presently isolated systems evolve towards equilibrium in the same direction of time. This time asymmetry cannot arise from dynamical laws for they are approximately time reversal invariant. It can only arise from the initial conditions.
- The manifest existence of the classical world necessary for the interpretation of quantum mechanics. If the world is quantum mechanical at a fundamental level, classical behavior cannot be a general feature of it. For any system – from an atom to the spacetime of the universe itself – the number of states which imply classical behavior in any sense is but an infinitesimal fraction of the total number of states of the system. Classical behavior is a property of *particular* quantum states. The classical behavior which we find throughout the universe today can only be traceable to particular properties of its initial quantum state.

There have been many ideas for features of very small scale physics which may find their origin in a theory of initial conditions. Perhaps the most intriguing of these is the idea that initial conditions may explain:

- The present small value of the renormalized cosmological constant as measured in the natural units of elementary particle physics (10^{-118} in Planck units). We are used to thinking of the coupling constants of the elementary particle interactions as being fixed by a fundamental Lagrangian. But, the coupling constants we measure are not these, but rather the renormalized, effective coupling constants at low energies. The effective cosmological constant is measured only by the expansion of the universe — an extraordinarily low energy scale. It should, then, not be so surprising that its renormalized value may be affected by the same initial conditions which determine the large scale structure of the universe itself.

These are some facts. The enterprise in quantum cosmology is to see whether such a diverse list of facts, and others like them, can be seen to follow from one, simple, compelling law of initial conditions — a law which like other fundamental laws of physics is quantum mechanical — a prescription, in effect, for the whole quantum state of the universe.

The history of the application of quantum mechanics to cosmology goes back quite far. The quantum gravity of Dirac, Wheeler, Bergmann, DeWitt, Arnowitt, Deser and Misner (for a review, see Kuchař (1981)) was applied to cosmology in the late '60's work of DeWitt, Misner, Matzner, Ryan, MacCallum and others to construct quantum mechanical models of the dynamics of the universe. (For reviews see Misner (1972); Ryan (1972); MacCallum (1975).) Starting mostly in the '70's there are the ongoing efforts of Parker, Davies, Hu, Zel'dovich, Anderson, Ford, Starobinskiĭ, and a great many others to understand the influence of quantized matter fields on the evolution of the early universe. (For a review of some of this work see Birrell and Davies (1982).) For the most part, however, these efforts concentrated on understanding quantum dynamics — not on one state but all states.

It remained to the early '80's for the issue of quantum initial conditions to be taken seriously. There we find the influential (though entirely classical) proposals of Penrose, (Penrose (1979)) and the specifically quantum mechanical proposals of Hawking (Hawking (1982) and (1984); Hartle and Hawking (1983).) Vilenkin (Vilenkin (1982) and (1983)) and their coworkers for specific quantum states of the universe. Since that time the subject has grown greatly. Halliwell's authoritative bibliography of papers in this area (Halliwell (1990)) contains nearly three hundred entries. It will not be possible for me to review them all! Rather I shall just attempt to highlight the developments of the past three years in this subject as I see it.

2.   The quantum mechanics of cosmology

Early work on quantum cosmology tended to bypass the difficult issues of interpretation posed by the application of quantum mechanics to the universe as a whole. Recent years have seen considerable progress in our understanding. This has led to a new perspective on the long standing issues of interpretation in quantum mechanics. I would now like to describe some of these ideas.

From Bohr, to Everett, to post-Everett

The Copenhagen framework of quantum mechanics that was developed in the '30's and '40's, and that appears in most text books today, is insufficiently general on two counts to formulate a quantum mechanics of cosmology.

The first of these concerns the idea, characteristic of the various Copenhagen interpretations, that there was something external to the framework of wave function and Schrödinger equation necessary to interpret the theory. Bohr (1954) spoke of phenomena which could be alternatively described in terms of classical language. Landau and Lifshitz (1958) emphasized a preferred class of classical observables. Heisenberg (1958) stressed the essential role of an external observer for whom the wave function is but a summary of the information possessed about the observed system. All singled out the measurement process for a special role in the theory. In various ways these authors were taking the existence of the classical world which we see all about us as fundamental, and measurements as the primary focus of scientific statements.

We can have none of this in quantum cosmology. In a theory of the whole thing there can be no fundamental division into observer and observed. Measurements and observations cannot be fundamental notions in a theory which seeks to discuss the early universe when neither existed. There is no reason in general for a classical world to be fundamental or external in a basic formulation of quantum mechanics.

It was Everett (1957) who first suggested how to generalize the Copenhagen framework so as to apply to cosmology. His idea was to take quantum mechanics seriously and apply it to the universe as a whole. He started with the idea that there is one wave function for the universe always evolving by the Schrödinger equation and never reduced. Any observer would be part of the system described by quantum mechanics, not separate from it. Everett showed that, if the universe contains an observer, then its activities — measuring, recording, calculating probabilities, etc. — could be described in this generalized framework. Further, he showed how the Copenhagen structure followed from this generalization when the observer had a memory which behaved classically and the system under observation behaved quantum mechanically.[2]

Yet the Everett analysis, as I have stated it, is not complete. It neither adequately explained the origin of the classical world in quantum mechanical terms nor the meaning of the "branching" that replaces the notion of measurement. It was a theory of "many worlds", but it did not explain how these were defined nor how they arose. The subsequent, post-Everett, view on these questions is the synthesis of the work of many people — Zeh (1971), Zurek (1981, 1982) Griffiths (1984), Joos and Zeh (1985), Omnès (1988abc), Gell-Mann and the author (1989) among others. There has been considerable progress on elaborating this view in the last several years which I would now like to describe.

In familiar quantum mechanics probabilities can be assigned to quantities which are measured and in general to nothing else. This is nowhere more simply illustrated than in the two slit experiment (see Figure 1).

In an experiment where an electron is detected at the screen, suppose we ask whether it went through the upper slit or the lower slit. If we did not measure whether it went through the upper or lower slit, then we are not permitted to assign probabilities to these histories. The reason is that, if we tried, the correct probability sum rules would not be satisfied because of interference. The probability to arrive at a position y on the screen is not the sum of the probabilities to go through the upper slit or the lower slit

$$p(y) \neq p_U(y) + p_L(y), \tag{1}$$

because

$$|\psi_L(y) + \psi_U(y)|^2 \neq |\psi_L(y)|^2 + |\psi_U(y)|^2. \tag{2}$$

It is thus meaningless even to speak of the alternatives of passing through the upper or lower slit. However, if we did measure which slit the election

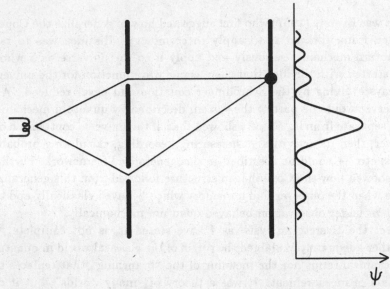

Fig. 1. The two slit experiment. An electron gun at right emits an electron traveling towards a screen with two slits, its progress in space recapitulating its evolution in time. When precise detections are made of an ensemble of such electrons at the screen it is not possible, because of interference, to assign a probability to the alternatives of whether an individual electron went through the upper slit or the lower slit. However, if the electron interacts with apparatus which measures which slit it passed through, then these alternatives decohere and probabilities can be assigned.

went through, then interference is destroyed, the sum rule obeyed, and we *can* assign meaningfully probabilities to these alternative histories. They are then said to *decohere*.

More generally, sets of alternative decohering histories of a system are defined as follows: An exhaustive set of "yes-no" alternatives at one time is represented in the Heisenberg picture by sets of projection operators $(P_1^k(t), P_2^k(t), \cdots)$. In $P_\alpha^k(t)$, k labels the set, α the alternative and t its time. An exhaustive set of exclusive alternatives satisfies

$$\sum_\alpha P_\alpha^k(t) = 1 \quad , \quad P_\alpha^k P_\beta^k = \delta_{\alpha\beta} P_\beta^k \quad . \tag{3}$$

Time sequences of such exhaustive sets of alternatives define a set of histories. A particular history is one set of alternatives $(P_{\alpha_1}^1(t_1), P_{\alpha_2}^2(t_2), \cdots \cdots P_{\alpha_n}^n(t_n))$, abbreviated $[P_\alpha]$. The decoherence functional is defined on pairs of histories by

$$D\big([P_{\alpha'}], [P_\alpha]\big) = Tr\left[P_{\alpha'_n}^n(t_n) \cdots P_{\alpha'_1}^1(t_1) \rho P_{\alpha_1}^1(t_1) \cdots P_{\alpha_n}^n(t_n) \right] \tag{4}$$

where ρ is the density matrix of the system under discussion. Probabilities can be consistently assigned to a set of alternative histories when they *decohere*, that is, when the off diagonal elements of D are sufficiently small (Griffiths (1984); Omnès 1988abc; Gell-Mann and Hartle (1989))

$$D([P_{\alpha'}], [P_\alpha]) \approx \delta_{\alpha_1' \alpha_1} \cdots \delta_{\alpha_n' \alpha_n} p(\alpha_n t_n, \cdots, \alpha_1 t,) \ . \tag{5}$$

The diagonal elements $p(\alpha_n t_n, \cdots, \alpha_1 t_1)$ are the joint probabilities of the alternatives $\alpha_1, \cdots, \alpha_n$. These probabilities are consistent because as a consequence of (3) and (5) they satisfy the necessary sum rules, for example

$$\sum_{\alpha_k} p(\alpha_n t_n, \cdots \alpha_{k+1} t_{k+1}, \alpha_k t_k, \alpha_{k-1} t_{k-1}, \cdots, \alpha_1 t_1)$$
$$\approx p(\alpha_n t_n, \cdots \alpha_{k+1} t_{k+1}, \alpha_{k-1} t_{k-1}, \cdots, \alpha_1 t_1) \ . \tag{6}$$

In quantum cosmology we apply these ideas to define probabilities for histories of the closed system which is the universe as a whole — a system which contains both the observed plus its observers if any. The ρ in Eqs (4) and (5) is the density matrix summarizing the initial conditions of the universe. Meaningful probabilities can be assigned exactly to those sets of alternative histories of the universe which decohere in the sense of (5). Decoherence thus serves the same function as "measurement" in usual quantum mechanics but in a more general, observer independent way. Sets of alternative decoherent histories of the universe are what we may discuss in the quantum mechanics of cosmology.

There are no non-trivial time sequences of exhaustive sets of projections which decohere in all circumstances. What histories decohere and what their probabilities are depends crucially on the initial conditions ρ. Mechanisms for decoherence have been discussed by many authors, notably Zurek (1981, 1982); Caldeira and Leggett (1983); and Joos and Zeh (1985). Decoherence occurs when the projections in (5) involve only a few and certain variables interacting with a much larger set of others not acted upon. In effect, the coherent phases of the followed variables are irretrievably lost to the rest by the interaction. This sort of situation is widespread in our universe due to its initial conditions. Consider, for example, a dust grain about a millimeter in size in a superposition of two positions about a millimeter apart. This superposition is decohered by the interactions with the 3° K background radiation, if by nothing else, in about a nanosecond (Joos and Zeh (1985)). So prevalent is this phenomenon that we may speak of certain variables such as the positions and momenta of sizable bodies as *habitually* decohering.

A set of alternative decohering histories may be fine grained by including more and finer projections in the defining time sequences. Progressive fine graining increases coherence. We may thus identify the *maximal sets* which are fine grained to the limit set by the standard for decoherence in (5).

Among those maximal sets some will exhibit higher levels of classicity than others, that is, in particular, some will be more correlated in time than others. Because of existence of quantum phenomena there are no *exactly* classical maximal sets, only quasiclassical ones with high levels of classicity. The maximal sets of alternative decohering histories with maximal classicity define the possible quasiclassical worlds for the universe.

It would be a striking and deeply important fact if among the maximal sets there was one with much higher classicity than all the others. That would then be *the* quasiclassical world, independent of any subjective criterion. In the post-Everett framework this is a decidable if difficult question.

To what do we owe the origin of our quasiclassical world? A quasiclassical world is not a general feature of a quantum system and neither is it a general feature of particular variables. The number of states which imply classical behavior in any sense is but an infinitesimal fraction of the total number of states the universe could have had. Rather the origin of the quasiclassical world lies in the particular state the universe does have – that is – in its quantum initial conditions. The founders of quantum mechanics were right. Something external to the framework of wave function and Schrödinger equation *is* needed to interpret the theory. But, it is not a postulated classical world to which the quantum theory does not apply. Rather, it is the initial conditions of the universe specified within quantum theory itself.

Given a theory of initial conditions we should be able to predict the decohering variables defining the classical world. In a quantum theory of spacetime the first of these must be the classical spacetime of the late universe itself. The recent period has seen a considerable clarification of how initial conditions can imply both decoherence and classicity for spacetime. Work by DeWitt (1967); Lapchinsky and Rubakov (1979); Hartle and Hawking (1983); Banks (1985); Banks, Fischler and Susskind (1985); Fischler, Ratra and Susskind (1985); Halliwell and Hawking (1985); Halliwell (1986); Hartle (1986); Joos (1986); Wada (1986); Zeh (1986); Brout, Horowitz and Weil (1987); D'Eath and Halliwell (1987); Halliwell (1987); LaFlame (1987); Vilenkin (1988); and Padmanabhan (1989), has shown how the classical evolution of geometry according to Einstein's equations can be predicted by a theory of initial conditions. Work by Zeh (1986, 1988); Kiefer (1987); Fukuyama and Morikawa (1989); Halliwell (1989); Padmanabhan (1989), and Mellor (1990) has shown how differing classical geometries are decohered by the neglect of small fluctuations of wavelength longer than horizon size. In this work one can begin to see how the quasiclassical world so familiar to us could become a genuine prediction of a theory of the origin of the universe.

Time

The second count on which the Copenhagen formulation of quantum mechanics is insufficiently general for application to cosmology concerns the preferred role of time.

Time plays a special and peculiar role in Hamiltonian quantum mechanics. All observations are assumed to be unambiguously "at one moment of time." Time is the only observable for which there are no interfering alternatives as a measurement of momentum is an interfering alternative for a measurement of position. Time is the sole observable not represented in the theory by an operator but rather enters the theory as a parameter describing evolution.

In constructing the quantum mechanics of a classical theory a first task is to identify this preferred time. Non-relativistic classical physics already possesses a preferred time which is unambiguously taken over as the preferred time of quantum mechanics. In special relativity there is already an issue but there is also a resolution. The classical spacetime of special relativity has many timelike directions. However, because of relativistic causality, the quantum mechanics constructed from any one choice is unitarily equivalent to that constructed from any other.

In each of the above examples the preferred time of quantum mechanics is given by a fixed background spacetime. In the construction of a quantum theory *of* spacetime the choice of a time variable becomes a fundamental difficulty. If the geometry of spacetime is not fixed, but fluctuating quantum mechanically and without definite value, it cannot supply a preferred time. Neither can it supply a notion of causality which would guarantee the equivalence of different choices for the preferred time. The classical theory of general relativity does not prefer one set of spacelike surfaces over all others. It is covariant. There is thus a conflict between the covariance of the classical theory and Hamiltonian quantum mechanics which needs a preferred time. This is the "problem of time."

Resolutions of this conflict are under active current investigation and no consensus can be said to have been reached. Roughly, there are three points of view:

1) The traditional resolution of canonical quantum gravity (see, e.g. Kuchař (1981) for a review) has been to keep quantum mechanics as it is but to give up on spacetime. A preferred class of spacelike surfaces is specified for the preferred time of quantum mechanics. Covariance is thus broken at the quantum level. Space and time become separate in quantum geometry with the synthesis of Einstein and Minkowski emerging only in its classical limit.

2) A second approach is to *add* distinguished variables to the theory which can serve as the distinguished time for quantum mechanics. This has been advocated by Unruh and Wald (1989); Henneaux and Teitelboim (1989); Brown and York (1989), among others. Typically these variables are "hidden" in the sense that they are not directly accessible to us in the same way that, say, the geometry of spacetime is.

3) A third approach is to keep the theory generally covariant but to generalize quantum mechanics so that it does not require a preferred time. Teitelboim (1983abc); Sorkin (1989); and the author (1986, 1988ab, 1989) have suggested, in various ways, that the sum-over-histories formulation may supply such a generalization. Of course, any such generalization of quantum mechanics has the obligation to show how the familiar Hamiltonian theory emerges in an appropriate limit. I now would like to describe schematically how this works:

In a quantum theory of cosmological spacetime the histories are four-dimensional spacetime geometries $\mathcal{G}$ with matter fields $\phi(x)$ living upon them. The sums over histories defining quantum amplitudes therefore have the form:

$$\sum_{\mathcal{G}} \sum_{\phi(x)} \mathrm{Exp}\Big(iS[\mathcal{G}, \phi(x)]\Big) \ , \tag{7}$$

where S is the action for gravity coupled to matter. The sums are restricted by the theory of initial conditions and by the observations whose probabilities one aims to compute.

The interior sum in (7),

$$\sum_{\phi(x)} \mathrm{Exp}\Big(iS[\mathcal{G}, \phi(x)]\Big) \ , \tag{8}$$

defines a quantum field theory in a temporarily fixed background space-time $\mathcal{G}$. The requirement that the fields be single valued on the fixed background $\mathcal{G}$ means that the field theory has an equivalent Hamiltonian formulation with the preferred time directions being those of the background spacetime $\mathcal{G}$.

However, when the remaining sum over $\mathcal{G}$ in (7) is carried out the equivalence with any Hamiltonian formulation disappears. There is no longer any fixed geometry, no longer any external time to define the preferred time of a Hamiltonian formulation. All that is summed over. There would be an *approximate* equivalence with a Hamiltonian formulation were the initial conditions to imply that for large scale questions in the late universe, only a single geometry $\hat{\mathcal{G}}$ (or an incoherent sum of such geometries) contributes to the sum over geometries in (7). For then,

$$\sum_{\mathcal{G}} \sum_{\phi(x)} \text{Exp}\Big(iS[\mathcal{G}, \phi(x)] \Big) \approx \sum_{\phi(x)} \text{Exp}\Big(iS[\hat{\mathcal{G}}, \phi(x)] \Big) \tag{9}$$

and the preferred time is that of the geometry $\hat{\mathcal{G}}$. It would be in this way that the familiar Hamiltonian formulation of quantum mechanics emerges as an approximation to a more general sum-over-histories framework appropriate to specific initial conditions and our special position in the universe so far from the big bang and centers of black holes.

The wave function of the universe

In the remainder of this review I would like to describe some of the proposals which have been made for theories of initial conditions and their consequences. Such proposals are in effect proposals for the quantum state of the universe, and are therefore usually stated as specifications of the wave function of the universe which describes that quantum state. While the wave function of the universe is not the only amplitude of interest, it is a useful one for extracting predictions as we shall see.

In the quantum mechanics of closed cosmologies with fixed (for simplicity) spatial topology (say that of a 3-sphere) the wave function of the universe is a function of superspace — the space of three geometries and matter field configurations on a spacelike surface. Thus, we write

$$\Psi = \Psi\left[h_{ij}(\mathbf{x}), \phi(\mathbf{x})\right] , \tag{10}$$

where $h_{ij}(\mathbf{x})$ is the metric describing the three-geometry and $\phi(\mathbf{x})$ stands for the spatial matter field configuration. There is no additional time label because the three-geometry itself carries the information about its location in spacetime (Baierlein, Sharp and Wheeler (1962)). *A law for initial conditions in quantum cosmology is a law which prescribes this wave function.*

In a theory of quantum spacetime we are not free to propose any wave function as a theory of initial conditions. It must satisfy the constraints which implement the dynamics of the particular theory of quantum gravity we have in mind. For example, in pure Einstein gravity with cosmological constant Λ, three of the constraints are

$$ iD_i \left(\frac{\delta \Psi}{\delta h_{ij}(\mathbf{x})} \right) = 0, \tag{11} $$

where D_i is the covariant derivative of the metric h_{ij}. These constraints enforce gauge invariance in the spacelike surface. The additional dynamical constraint is

$$ \mathcal{H}\Psi = \left[l^2 \nabla^2 + l^{-2} h^{1/2} \left({}^3R - 2\Lambda \right) \right] \Psi = 0, \tag{12} $$

where $l = (16\pi G)^{1/2}$ is the Planck length (in units where $\hbar = c = 1$) and ∇^2 is a wave operator in the 3-metric. Explicitly

$$ \nabla^2 = G_{ijkl} \frac{\delta^2}{\delta h_{ij}(\mathbf{x})\delta h_{kl}(\mathbf{x})} \quad + \quad \left(\begin{array}{c} \text{linear terms depending} \\ \text{on factor ordering} \end{array} \right), \tag{13a} $$

with

$$ G_{ijkl} = \frac{1}{2} h^{-1/2} \left(h_{ik}h_{jl} + h_{il}h_{jk} - h_{ij}h_{kl} \right). \tag{13b} $$

This constraint, called the Wheeler-DeWitt equation, encapsulates the dynamics of general relativity. In a different theory of quantum gravity there would be a different constraint.

One can think of the Wheeler-DeWitt equation as a kind of wave equation in superspace. Finding a law for initial conditions may be viewed as the problem of specifying the boundary conditions which select from its many solutions that one which is the wave function of our universe.

At least on large scales, most of our observations of the universe are of its classical behavior. Classical behavior is most properly understood in quantum mechanics as the prediction of a decohering string of observations which are, with high probability, correlated in time by classical laws. However, the semiclassical behavior of the wave function gives a direct and easily extractable signal of this classical behavior.

Recall how the semiclassical approximation works in ordinary quantum mechanics: Suppose $\psi(\vec{x})$ is an energy eigenfunction

$$H\psi = \left[-\frac{\hbar^2}{2M}\nabla^2 + V(\vec{x})\right]\psi = E\psi(\vec{x}). \tag{14}$$

Further, suppose it is well approximated, far from turning points, by

$$\psi(\vec{x}) \sim p(\vec{x})e^{iS(\vec{x})/\hbar}, \tag{15}$$

where $p(\vec{x})$ is a real prefactor and $S(\vec{x})$ is a classical action – a solution of the Hamilton-Jacobi equation

$$\frac{1}{2M}(\nabla S)^2 + V(\vec{x}) = E. \tag{16}$$

Then we know that successive measurements of position and momentum which are not too accurate will be correlated by the laws of classical physics with high probability. A wave function of the form (15) corresponds to an ensemble of classical trajectories differing one to the other by the initial condition required to integrate

$$M\frac{d\vec{x}}{dt} = \nabla S(\vec{x}). \tag{17}$$

When we find a wave function of the form (15) we predict the classical correlations exhibited by one of the members of this ensemble.

A semiclassical approximation to the Wheeler-DeWitt equation would be some linear combination of

$$\Psi[h_{ij}(\mathbf{x}), \phi(\mathbf{x})] \sim p[h_{ij}(\mathbf{x}), \phi(\mathbf{x})]\,\mathrm{Exp}\left(iS[h_{ij}(\mathbf{x}), \phi(\mathbf{x})]\right) \tag{18}$$

and its complex conjugate. Here, S is a classical action for general relativity coupled to matter fields. In regions of superspace where such an approximation holds, we predict that the universe will exhibit the correlations of one of the classical cosmologies in the ensemble represented by S. This will be enough for comparison of the predictions of particular theories of initial conditions with our crude astronomical observations.

3. Proposals for a law of initial conditions

Ideas for initial conditions

In recent years a number of different proposals for a law of initial conditions have been put forward. There is Roger Penrose's (1979) time asymmetric proposal that the Weyl tensor vanishes on initial singularities but not on final singularities. There is the proposal that the universe emerges as a coöperative process out of empty Minkowski space (Brout, Englert and Gunzig (1978), Gunzig and Nardone (1982)). There are the proposals that the universe nucleates "spontaneously from nothing" worked out most explicitly and clearly by Alex Vilenkin (1982, 1983). There is the proposal of Narlikar and Padmanabhan (1983) that one should quantize only the conformal factor and enforce the constraints only semiclassically. There is the proposal of Fischler, Ratra and Susskind (1985) that the wave function of the universe is singled out from other solutions of the Wheeler-DeWitt equation by the requirement that the energy in the matter field not diverge at the singularity. Frank Tipler (1986) has a proposal in which the universe "explodes from nothing." There is the proposal of A. Linde (1986, 1987) that the universe is an "eternally self-reproducing inflationary universe." There is the proposal of Suen and Young (1989) that the universe is a "leaking system." There is the "no boundary" proposal of Stephen Hawking and his collaborators (Hawking (1982) and (1984); Hartle and Hawking (1983)) that the universe is in the cosmological analog of its ground state, implemented concretely by specifying the wave function as a Euclidean sum-over-histories. There are no doubt many more.

It would be difficult to review all these proposals. I shall concentrate on just two which have received the most attention and are perhaps the most developed. These are the "no boundary" proposal of Hawking and his associates and the "spontaneous nucleation from nothing" proposal as developed by Alex Vilenkin.

The "no boundary" proposal

The "no boundary" proposal is that the quantum state of the universe is the analog for closed cosmologies of the ground state, or state of minimum excitation. To understand what this means, understand first that what is meant is not a state of minimum energy. For closed cosmologies there is no natural notion of time, therefore no natural notion of energy, therefore no natural Hamiltonian, and therefore no natural notion of a state with the lowest eigenvalue of the Hamiltonian. For systems which possess a Hamiltonian, however, finding the lowest energy eigenstate is not the only way of finding the ground state wave function. One can also calculate it directly as a Euclidean sum-over-histories. For example, for a particle in a potential $V(x)$ the ground state wave function is

$$\psi_0(x) = \sum_{\text{paths}} \text{Exp}\left(-I[x(\tau)]\right), \tag{19}$$

where $I = \int dt(m\dot{x}^2/2 + V(x))$ is the Euclidean action and the sum is over all paths, $x(\tau)$, which start at the argument of the wave function at $\tau = 0$ and proceed to a configuration of minimum action in the infinite past. This construction generalizes to the quantum mechanics of closed cosmologies.

For a closed cosmology the "no boundary" ground state wave function is given by:

$$\Psi\left[h_{ij}(\mathbf{x}), \phi(\mathbf{x})\right] = \sum_{\mathcal{G}, \phi(\mathbf{x},t)} \text{Exp}\left(-I[\mathcal{G}, \phi(\mathbf{x})]\right). \tag{20}$$

The sum over $\mathcal{G}$ is over all compact, Euclidean four-geometries which have a boundary on which the induced three-geometry is the argument of the wave function and *no other boundary*. The sum over $\phi(x)$ is over all four-dimensional field configurations which match the spatial configuration $\phi(\mathbf{x})$ on the boundary and are otherwise regular. These conditions are the analogs in the gravitational case of the conditions in the particle case that the paths start at the argument of the wave function and proceed to a configuration of minimum action in the past. Such a construction, properly implemented, automatically yields a solution of the Wheeler-DeWitt equation (Hartle and Hawking (1983); Barvinsky and Ponomariov (1986); Barvinsky (1986); Halliwell (1988); Halliwell and Hartle (1990b); and Woodard (1989)). Since the remarkable simplicity of the early universe suggests that it is in a state of low excitation, it is a natural conjecture that this "no boundary" wave function is the wave function of our universe.

An important problem for the "no boundary" proposal is the indefiniteness of the gravitational action. An integral of $\text{Exp}(-I)$ over real four geometries does not converge. Indeed it would be a disaster for the proposal if it did, for the resulting wave function would be positive, never oscillate, never behave semiclassically, and thus fail to predict one of the central facts of our experience — the classical spacetime of the late universe. Rather as proposed by Gibbons, Hawking and Perry (1978) the contour of integration must be complex. For understood examples, where the action is also indefinite, such as the theory of a free spin-2 field, such complex contours give correct results (Schleich (1987); Hartle (1984); Hartle and Schleich (1987); Arisue, Fujiwara, Kato and Ogawa

(1987)). Mazur and Mottola (1989) have also argued that a complex contour follows naturally from the measure of the sum-over-geometries. But what complex contour should be chosen? The choice is constrained by requirements that the integral in (20) converge, that the wave function satisfy the constraints (11) and (12), that it predict classical geometry when the universe is large as in the discussion of (18), and that it give the correct matter quantum field theory in the predicted classical spacetimes (Hartle (1989), Halliwell and Louko (1989) and (1990ab); Halliwell and Hartle (1990a)). Although the remaining ambiguity in the contour is unlikely to affect the classical predictions, it remains for deeper physical considerations to fix it completely. Until this is resolved, we might do better to speak of the no boundary proposals corresponding to the different possible contours.

The essence of the no boundary proposal is *topological*. It is a restriction on the *manifolds* which contribute to the sum-over-histories. It is this simple topological character which gives one hope that the idea will be generalizable wherever our search for a fundamental theory of quantum gravity may lead us.

Spontaneous nucleation from nothing

"Spontaneous nucleation from nothing" is the name associated with another proposal for initial conditions. Although the idea has been developed by many, (Tryon (1973); Pagels and Atkatz (1982); Vilenkin (1982, 1983, 1984); Linde (1984); Zel'dovich and Starobinskii (1984) and others), its clearest statement in the framework of quantum cosmology we have been discussing is due to Alex Vilenkin (e.g. in Vilenkin (1988)).

The Wheeler-DeWitt equation is a second order "functional differential" equation on superspace somewhat analogous to the Klein-Gordon equation. There is thus an associated conserved current. Superspace may be thought of as bounded by configurations with infinite three-curvature and/or scalar field. The "non-singular boundary" of superspace in Vilenkin's terminology is the part of the boundary which contains three-geometries which have only singularities attributable to singular slices of *regular* four-geometries. The singular boundary is the rest. Very large volumes, for example, would be counted as part of the singular boundary. Vilenkin's boundary condition is that on the singular boundaries of superspace Ψ contain only outgoing waves carrying flux out of superspace (in the sense of the Wheeler-DeWitt current) and be finite elsewhere. In simple minisuperspace models where the geometries are restricted to be nearly homogeneous and isotropic, these boundary conditions give a unique solution of the Wheeler-DeWitt equation which has a number of properties one would associate with a tunneling event, for example "outgoing waves" at large three geometries.

The difficult problem with this idea is to know how general its defining features are and how uniquely they fix a solution of the Wheeler-DeWitt equation when highly inhomogeneous possibilities for the geometry are considered. For example, just to define an "outgoing wave" requires a notion of time, or what is the same thing, the universal validity of the semiclassical approximation near the singular boundaries of superspace. This is true in the models, but is it true in general?

4. Predictions

In Section 1 we listed physical phenomena that might be predicted by a theory of initial conditions. I would now like to review where the two proposals described in Section 3 stand on predicting them.

Large scales

The status of the predictions of homogeneity, isotropy, and the spectrum of deviations from these exact symmetries are much the same as at our last meeting. In minisuperspace models the wave functions of both proposals are peaked about homogeneity and isotropy. The same Harrison-Zel'dovich spectrum of fluctuations is predicted by both. (See e.g. Halliwell and Hawking (1985); Vilenkin (1986); Wada (1986); D'Eath and Halliwell (1987); Vachaspati and Vilenkin (1988); Shirai and Wada (1988); and, in other theories of initial conditions, Banks, Fischler and Susskind (1985); Fischler, Ratra and Susskind (1985)). There is some evidence that, if the topology of the universe is a quantum variable, the "no boundary" proposal may favor a simple $R \times S^3$ large-scale topology of the late universe (Gibbons and Hartle (1990)).

There is considerable debate, however, on the question of whether the no boundary proposal predicts enough inflation to explain the present large size of the universe, the absence of monopoles, and the spectrum of density fluctuations that formed the galaxies. To understand this debate remember that in the semiclassical approximation the wave function does not predict a *particular* classical universe. Rather it predicts an *ensemble* of possibilities. Members of this ensemble which have a higher value of the prefactor $p[h_{ij}, \phi]$ can be said to be more likely than those with lower values.

Each proposal predicts an ensemble of universes with various values of the effective cosmological constant governing an early inflationary epoch. The question is whether the distribution determined by the prefactor favors a long period of inflation or a short one. The spontaneous nucleation from nothing proposal pretty clearly predicts a sufficiently long inflationary period in the models in which it has been analysed (Vilenkin (1988)). In simple models the "no boundary" proposal does also if the model's predictions are indicative at very large values of the inflation driving field — values so high that their energy density is above the Planck scale (Hawking and Page (1986), (1988)). The debate centers on whether it is reasonable to take such predictions seriously or whether they should be disregarded (Mijić, Morris, and Suen (1989); Vilenkin (1988)). However, even if disregarded a recent analysis by Grishchuk and Rozhansky (1988) and (1989)) argues that the no boundary prediction for inflation falls short by only a modest amount from that needed — an amount very possibly made up in a more detailed analysis of the particle physics involved.

Familiar scales

Possible features of the universe on familiar scales that are directly traceable to initial conditions are the quasiclassical world of the late universe and the thermodynamic arrow of time. The emergence of the quasiclassical world I have already discussed. The arguments for the homogeneity of thermodynamic arrow of time have not changed since our last meeting (see, e.g. Page (1985); Hawking (1985b)). The nearly exact homogeneity and isotropy predicted by these proposals at early times means that the universe is far from equilibrium in its gravitational degrees of freedom across an entire spacelike surface. The matter degrees of freedom are predicted to be far from equilibrium as well. A suitably coarse grained entropy will, therefore, increase in a homogeneous fashion.

One imagines that both a quasiclassical world and a homogeneous thermodynamic arrow of time will be predictions of many proposals for initial conditions. In this sense, these features are not likely to be of much help in discriminating among different theories. However, they so manifest that any theory which does not clearly predict them can be immediately ruled out. It is therefore well worth the effort to explore their origin in more detail.

Wormholes

I would now like to try and describe how the initial conditions for cosmology may affect physics on very small scales, and in particular, how the laws of particle physics on all accessible scales may depend on initial conditions. In order to be specific I shall focus on the "no boundary" proposal where these ideas have been most developed.

The "no boundary" wave function is defined by a sum over Euclidean geometries with no boundary other than that needed to define the argument of the wave function. In mathematics a geometry is a manifold with a metric. In a sum-over-geometries it is, therefore, as reasonable to sum-over-manifolds as it is over metrics. Such a sum is called a sum-over-topologies. An example of a possible contributing geometry with interesting topology is a wormhole geometry. A Euclidean wormhole geometry is a four-dimensional space with one or more narrow throats joining larger, flatter regions. (See Figure 2). We do not expect wormhole geometries having throats of macroscopic scales to contribute significantly to the no boundary wave function and simple dimensional estimates of their actions show that this is so. However, there is no par-

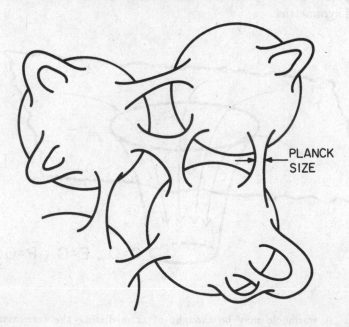

PLANCK
SIZE

Fig. 2.   Euclidean wormhole geometries have Planck scale size throats joining larger, flatter regions.

ticular reason not to expect significant contributions from wormholes of Planck-length size or less. I would now like to review the argument that Planck scale wormholes will alter the coupling constant of matter fields at accessible scales. I shall follow an argument due to Klebanov, Banks and Susskind (1989) in an exposition of Hawking (1989).

Matter fields propagating in a wormhole geometry can go down the wormhole to some region of the universe different from the large one we are in. They also can arrive through wormholes from other regions. On the scales we can observe we would see the effect of Planck-scale wormholes as local point interactions through which particles appear or disappear from our region. Such interactions could be represented in the low energy effective Lagrangian by effective interactions of the low energy fields ϕ which create and destroy particles at a point in various configurations of field. One can, in an intermediate way, think of the wormholes as carrying particles off to a separate closed universe (Figure 3). Such interactions must therefore, conserve energy, momentum and gauge-field charges for these are zero in a closed universe. Let us denote the vertices describing these interactions by terms $\theta_i(\phi)$. Roughly, we can expect to have all possible terms which are consistent with the above general symmetries.

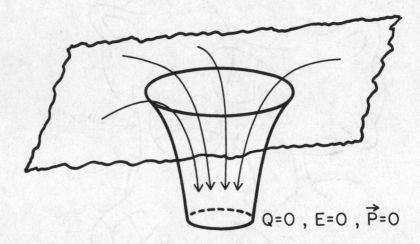

$$Q=0 \; , \; E=0 \; , \; \vec{P}=0$$

Fig. 3. A wormhole may be thought of as mediating the transition from a large region to a small closed universe. For a closed universe, energy-momentum and gauge-field charges vanish. These quantum numbers are conserved, therefore, by the effective interactions arising from wormholes.

The element of the no boundary proposal which survives these trun-cations is that every wormhole that leaves the universe must connect back to it. There are no dangling ends with extra boundaries. In the dilute wormhole approximation each wormhole is assumed to connect two large regions. The low energy effect of Planck scale wormholes may therefore be represented by the effective bilocal interaction

$$-\frac{1}{2}\sum \Delta^{ij} \int d^4 x \sqrt{g(x)}\theta_i(x) \int d^4 y \sqrt{g(y)}\theta_j(y) \ . \qquad (21)$$

This represents a wormhole, say, destroying particles at point x and creating them again at point y. The positive-definite coupling matrix Δ^{ij}, the survivor of the detailed Planck scale geometry, may be set equal to the unit matrix by suitable choice of the vertices, and we shall do so from now on.

The contribution of an effective interaction, such as (21), to the gen-erating function defining a low energy field theory would be

$$
\begin{aligned}
Z = \int \delta\phi\delta\phi \, \mathrm{Exp}[&- \int d^4 x \sqrt{g(x)}\mathcal{L}(x) \\
&+ \frac{1}{2}\sum_{ij} \int d^4 x \sqrt{g(x)}\theta_i(x) \int d^4 y \sqrt{g(y)}\theta_j(y)]
\end{aligned}
\qquad (22)
$$

where $\mathcal{L}(x)$ is the Lagrangian at point x in the absence of wormholes — a function of the local fields and their derivatives there as are the $\theta_i(x)$.

Following Klebanov, Banks and Susskind (1989) we can rewrite this expression using the identity

$$e^{\frac{1}{2}M^2} = \int_{-\infty}^{+\infty} d\alpha e^{-\frac{1}{2}\alpha^2 - \alpha M} \ . \qquad (23)$$

The generating function becomes

$$Z = \int (\Pi_i d\alpha_i) P(\alpha_j) Z(\alpha_i) \qquad (24)$$

where

$$P(\alpha_i) = e^{-\frac{1}{2}\sum \alpha_j^2} \qquad (25)$$

and

$$Z(\alpha_i) = \int \delta g \delta\phi \, \mathrm{Exp}\left[- \int d^4 x \sqrt{g(x)}[\mathcal{L}(x) + \sum_j \alpha_j \theta_j(x)]\right] \ . \qquad (26)$$

The integration is now over the geometries and fields of the large regions.

The result (26) is thought of as specifying a distribution of universes in each of which the effective low energy particle Lagrangian is

$$\mathcal{L}(x) + \sum_j \alpha_j \theta_j(x) \ . \tag{27}$$

In this Lagrangian, the α_j have become coupling constants multiplying essentially arbitrary interaction terms. Their values are not fixed by the Lagrangian $\mathcal{L}$. They are distributed with weight $P(\alpha_i)Z(\alpha_i)$ given by the initial conditions and can be determined only by observation.

How uncertain are the coupling constants? We can try and estimate $Z(\alpha_i)$ by computing the sum over geometries and fields by steepest descents. If there were just one large region, the extremum geometry would be a round sphere with a radius determined by that universe's cosmological constant having an action

$$I = -\frac{48\pi^2}{\ell^2\Lambda} + \begin{pmatrix} \text{lower order} \\ \text{in } 1/\Lambda \end{pmatrix} \ . \tag{28}$$

The leading term in the distribution in Λ, would be as Hawking (1985) originally argued,

$$e^{-I} = e^{48\pi^2/\ell^2\Lambda} \ . \tag{29}$$

This is sharply peaked about $\Lambda = 0$. However, Coleman (1988) has argued that including the effect of other large regions modifies this to

$$\text{Exp}\left(e^{48\pi^2/\ell^2\Lambda}\right) \ . \tag{30}$$

This is so sharply peaked about $\Lambda = 0$ that we may say that the theory predicts this value.

The idea that the laws of microscopic physics might not be fixed and immutable but rather find their origin in the initial conditions of the universe was first discussed, I believe, by Wheeler (1977, 1988). Here, it is implemented, in a potentially computable way, in a specific theory of initial conditions. Clearly there is much to be understood, but the change in our vision of the fundamental laws of elementary particle physics would be so far reaching that it is important to try to do so.

5.　The place of quantum cosmology in general relativity

In conclusion, it is perhaps appropriate to say a few words about the place of quantum cosmology in our field of general relativity.

The challenge of providing a theory of initial conditions for our universe seems inescapable. Most of the predictable phenomena of physics depend on these initial conditions in some way and some, as we have seen, depend on them strongly. Laboratory physics is, in effect, the effort to avoid this challenge. In cosmology, however, there is no way of unifying the diverse phenomena that I have described other than by providing such a theory. It is as fundamental a challenge as finding the basic dynamical laws.

The last several decades have seen remarkable progress in our observational understanding of the universe on the largest scales. At the same time the increased confidence of particle physicists in their understanding of matter physics at high energies has enabled us to extrapolate the history of the universe back to a very early stage. The picture of remarkable simplicity for these early stages which has emerged has, for the first time, made it seem feasible to search for a compelling and simple law of initial conditions which can be confronted with observation. The progress in quantum gravity and field theory generally has given us the tools to try to do so.

It is perhaps a bold step to extrapolate quantum mechanics to scales so far beyond its direct experimental verification. But it is a completely natural, even inevitable effort, and the early signs seem hopeful. Indeed, as I have argued, in this extrapolation to large scales we may learn more about quantum mechanics on familiar and microscopic scales.

Einstein's theory, which is the subject of our conference, is the correct dynamical law governing the structure of the late universe on the largest scales. The generalization of Einstein's geometric vision to quantum gravity could well lead us to a theory of the fundamental interactions. This, coupled with a quantum mechanical law for initial conditions, may provide a theory of the origin of the universe which will unify phenomena on scales ranging from those of elementary particle physics to the galaxies themselves.

Preparation of this review was supported in part by the National Science Foundation (PHY-85-06686) and by a John Simon Guggenheim Fellowship. The author is grateful for the hospitality of the Aspen Center for Physics where most of it was written.

Notes

1. Although there is some overlap, this review is intended to complement that of the author (Hartle (1988c)) where somewhat earlier developments of still current importance were surveyed.

2. The idea was developed by many, among them Wheeler (1957); Cooper and VanVechten (1969); DeWitt (1970); Geroch (1984); and Mukhanov (1985).

References

Arisue, H., Fujiwara, T., Kato, M. and Ogawa, K. (1987), *Phys. Rev.* **D35**, 2309.

Baierlein, R., Sharp, D. and Wheeler, J. A. (1962), *Phys. Rev.* **126**, 1864.

Banks, T. (1985), *Nucl. Phys.* **B249**, 332.

Banks, T., Fischler, W. and Susskind, L. (1985), *Nucl. Phys.* **B262**, 159.

Barvinsky, A. and Ponomariov, V. N. (1986), *Phys. Lett.* **B167**, 289.

Barvinsky A. (1986), *Phys. Lett.* **B175**, 401.

Birrell, N. D. and Davies, P. C. W. (1982), *Quantum Fields in Curved Space*, (Cambridge University Press, Cambridge).

Bohr, N. (1954), *The Unity of Knowledge*, (Doubleday and Co., New York).

Brout, R., Englert, F. and Gunzig E. (1978), *Ann. Phys. (NY)* **115**, 78.

Brout, R., Horowitz, G. and Weil, D. (1987), *Phys. Lett.* **B192**, 318.

Brown, J. and York, J. (1989), Jacobi's Action Principle and the Recovery of Time in General Relativity, (preprint).

Caldeira, A. and Leggett, A. (1983), *Physica* **121A**, 587.

Coleman, S. (1988), *Nucl. Phys.* **B307**, 643.

Cooper, L. and VanVechten, D. (1969), *Am. J. Phys.* **37**, 1212.

D'Eath, P. and Halliwell, J. (1987), *Phys. Rev.* **D35**, 1100.

DeWitt, B. (1967), *Phys. Rev.* **160**, 1113.

DeWitt, B. (1970), *Physics Today* **23**, no. 9.

Everett, H. (1957), *Rev. Mod. Phys.* **29**, 454.

Fischler, W., Ratra, B., and Susskind, L. (1985), *Nucl. Phys.* **B259**, 730.

Fukuyama, T., and Morikawa, M. (1989), *Phys. Rev.* **D39**, 462.

Gell-Mann, M. and Hartle, J. B. (1989), in *Proceedings of the Santa Fe Workshop on the Physics of Information, Entropy, and Complexity*, (to appear), or in *Proceedings of the 3rd International Symposium on the Foundations of Quantum Mechanics in the Light of New Technology*, (to appear).

Geroch, R. (1984), *Noûs* **18**, 617.

Gibbons, G. W., Hawking, S. W., and Perry, M. (1978), *Nucl. Phys.* **B138**, 141.

Gibbons, G. and Hartle, J. (1990), Real Tunneling Geometries and the Large-Scale Topology of the Universe, (preprint).

Giddings, S. and Strominger, A. (1988), *Nucl. Phys.* **B307**, 854.

Griffiths, R. (1984), *J. Stat. Phys.* **36**, 219.

Grishchuk, L. P. and Rozhansky, L. V. (1988), *Phys. Lett.* **B208**, 369.

———— (1989), Does the Hartle-Hawking Wave Function Predict the Universe We Live In? (to be published).

Gunzig, E. and Nardone, P. (1982), *Phys. Lett.* **B118**, 324.

Halliwell, J. (1986), Quantum Field Theory in Curved Spacetime as the Semi-Classical Limit of Quantum Cosmology, DAMTP preprint.

Halliwell, J. (1987), *Phys. Rev.* **D36**, 3626.

———— (1988), *Phys. Rev.* **D38**, 2468.

———— (1989), *Phys. Rev.* **D39**, 2912.

———— (1990), Bibliography of Papers on Quantum Cosmology, (to be published in *J. Mod. Phys.*).

Halliwell, J. and Hawking, S. W. (1985), *Phys. Rev.* **D31**, 1777.

Halliwell, J. and Louko, J. (1989), *Phys. Rev.* **D39**, 2206.

———— (1990a), Steepest-descent Contours in the Path-Integral Approach to Quantum Cosmology II: Microsuperspace, ITP preprint NSF-ITP-89-21, to appear in *Phys. Rev. D*.

———— (1990b), Steepest-Descent Contours in the Path-Integral Approach to Quantum Cosmology III: A General Approach with Application to some Anisotropic Models, in preparation.

Halliwell, J. and Hartle, J. (1990a), Integration Contours for the No Boundary Wave Function of the Universe, (to be published).

———— (1990b), Wave Functions Constricted from Invariant Sums-Over-Histories Satisfy Constraints, (to be published).

Hartle, J.B. (1984), *Phys. Rev.* **D29**, 2730.

———— (1986) in *Gravitation in Astrophysics* eds. Hartle, J. B. and Carter, B. (Plenum Press, New York).

———— (1988a), *Phys. Rev.* **D37**, 2818.

———— (1988b), *Phys. Rev.* **D38**, 2985.

———— (1988c), in *Highlights in Gravitation and Astronomy*, eds. B. R. Iyer, A. Kembhavi, J. V. Narlikar, and C. V. Vishveshwara (Cambridge University Press, Cambridge).

———— (1989), in *Conceptual Problems of Quantum Gravity*, eds. A. Ashtekar and J. Stachel, (Birkhauser, Boston); or in *Proceedings of the 5th Marcel Grossmann Conference on Recent Developments in Relativity*, (World Scientific, Singapore).

Hartle, J. B. and Hawking, S. W. (1983), *Phys. Rev.* **D28**, 2960.

Hartle, J. B. and Schleich, K. (1987), in *Quantum Field Theory and Quantum Statistics: Essays in Honour of the Sixtieth Birthday of E. S. Fradkin*, eds. I. A. Batalin, G. A. Vilkovisky and C. J. Isham (Hilger, Bristol).

Hawking, S. W. (1982), in *Astrophysical Cosmology*, eds. by H. A. Brück, G. V. Coyne, and M. S. Longair (Pontifica Academia Scientarium, Vatican City).

Hawking, S. W. (1984), *Nucl. Phys.* **B239**, 257.

———— (1985a), *Phys. Lett.* **B195**, 337.

———— (1985b), *Phys. Rev.* **D32**, 2989.

———— (1989), Do Wormholes Fix the Constants of Nature? (to be published).

Hawking, S. W. and Page, D. (1986), *Nucl. Phys.* **B264**, 185.

———— (1988), *Nucl. Phys.* **B298**, 789.

Heisenberg, W. (1958), *Physics and Philosophy*, (Harper Bros., New York).

Henneaux, M. and Teitelboim, C. (1989), *Phys. Lett.* **B222**, 195.

Joos, E. (1986), *Phys. Lett.* **A116**, 6.

Joos, E. and Zeh, H.D. (1985), *Zeit. Phys.* **B59**, 223.

Kiefer, C. (1987), *Class. Quant. Grav.* **4**, 1369.

———— (1988), *Phys. Lett.* **A126**, 311.

Klebanov, D., Banks, T. and Susskind, L. (1989), *Nucl. Phys.* **B317**, 665.

Kuchař, K. (1981), in *Quantum Gravity 2*, eds. Isham, C., Penrose, R. and Sciama, D. (Clarendon Press, Oxford).

Laflamme, R. (1987), *Phys. Lett.* **B198**, 156.

Landau, L. and Lifshitz, E. M. (1958), *Quantum Mechanics*, (Addison-Wesley, Reading).

Lapchinsky, V. and Rubakov, V. A. (1979), *Acta. Phys. Polonica.* **B10**, 1041.

Linde, A. (1984), [*Sov. Phys. JETP*, **60**, 211]; (1984), *Zh. Eksp. Teor. Fiz.* **87**, 369. .

———— (1986), *Mod. Phys. Lett.* **A1**, 81.

———— (1987), *Physica Scripta* **T15**, 169.

MacCallum, M. A. H. (1975), in *Quantum Gravity*, eds. C. Isham, R. Penrose, and D. Sciama (Clarendon Press, Oxford).

Mazur, P. and Mottola, E. (1990), The Gravitational Measure, Solution to the Conformal Factor Problem and Stability of the Ground State of Quantum Gravity, (preprint).

Mellor, F. (1990), Newcastle preprint.

Mijić, M. B., Morris, M. S., and W.-M. Suen (1989), *Phys. Rev.* **D39**, 1486.

Misner, C. W. (1972), in *Magic Without Magic: John Archibald Wheeler*, ed. J. R. Klauder (Freeman, San Francisco).

Mukhanov, V. F. (1985), in *Proceedings of the Third Seminar on Quantum Gravity*, eds. by M. A. Markov, V. A. Berezin, and V. P. Frolov (World Scientific, Singapore).

Narlikar, J. and Padmanabhan, T. (1983), *Physics Reports* **100**, 151.

Omnès, R. (1988a), *J. Stat. Phys.* **53**, 893.

———— (1988b), *J. Stat. Phys.* **53**, 933.

———— (1988c), *J. Stat. Phys.* **53**, 957.

Padmanabhan, T. (1989), *Phys. Rev.* **D39**, 2924.

Pagels, H. and Atkatz, D. (1982), *Phys. Rev.* **D25**, 2065.

Page, D. (1985), *Phys. Rev.* **D32**, 2496.

Penrose, R. (1979), in *General Relativity: An Einstein Centenary Survey*, eds. by Hawking, S. W. and Israel, W. (Cambridge University Press, Cambridge).

Ryan, M. (1972), *Hamiltonian Cosmologies*, (Springer, New York).

Schleich, K. (1987), *Phys. Rev.* **D36**, 2342.

Shirai, I. and Wada, S. (1988), *Nucl. Phys.* **B303**, 728.

Sorkin, R. (1989), in *History of Modern Gauge Theories*, eds. M. Dresden and A. Rosenblum (Plenum Press, New York).

Suen, W.-M. and Young, K. (1989), *Phys. Rev.* **D39**, 2205.

Teitelboim, C. (1983a), *Phys. Rev.* **D25**, 3159.

———— (1983b), *Phys. Rev.* **D28**, 297.

———— (1983c), *Phys. Rev.* **D28**, 310.

Tryon, E. P. (1973), *Nature* **246**, 396.

Tipler, F. (1986), *Physics Reports* **137**, 231.

Unruh, W. and Wald, R. (1989), *Phys. Rev.* **D40**, 2598.

Vachaspati, T. and Vilenkin, A. (1988), *Phys. Rev.* **D37**, 898.

Vilenkin, A. (1982), *Phys. Lett.* **B117**, 25.

———— (1983), *Phys. Rev.* **D27**, 2848.

———— (1984), *Phys. Rev.* **D30**, 509.

_____ (1986), *Phys. Rev.* **D33**, 3560.
_____ (1988), *Phys. Rev.* **D37**, 888.
Wada, S. (1986), *Nucl. Phys.* **B276**, 729.
Wheeler, J. A. (1957), *Rev. Mod. Phys.* **29**, 463.
_____ (1977), in *Foundational Problems in the Special Sciences*, eds. R. E. Butts and K. J. Hintikka (D. Reidel, Dordrecht).
_____ (1988), *IBM Jour. Res. Dev.* **32**, 4.
Woodard, R. (1989), Enforcing the Wheeler-DeWitt Constraint the Easy Way, (to be published).
Zeh, H.D. (1971), *Found. Phys.* **1**, 69.
_____ (1986), *Phys. Lett.* **A116**, 9.
_____ (1988), *Phys. Lett.* **A126**, 311.
Zel'dovich, Y. B. and Starobinskiĭ, A. A. (1984), *Pis'ma Astron. Zh.* **10**, 323; [(1984), *Sov. Astron. Lett.* **10**, 135.]
Zurek, W. (1981), *Phys. Rev.* **D24**, 1516.
_____ (1982), *Phys. Rev.* **D26**, 1862.

16

String theory as a quantum theory of gravity

Gary T. Horowitz
Department of Physics,
University of California,
Santa Barbara, CA 93106 USA

1. Introduction

Although the roots of string theory go back to the late 1960's, the first connection between string theory and gravity was noticed in 1974 independently by Yoneya (1974) and by Scherk and Schwarz (1974). By the early 1980's it became clear that the recently developed superstring theory was an excellent candidate for our first perturbatively finite quantum theory of gravity. One loop calculations were shown to be finite and general arguments suggested that this should hold to all orders. (For a review of what was known at that time see Schwarz (1982).) Since then, an enormous amount of work has been done and progress made in our understanding of string perturbation theory. The evidence for finiteness has grown stronger and stronger (see e.g. D'Hoker and Phong (1988); Atick, Moore and Sen (1988); La and Nelson (1989), and references therein). Although there is always a chance of some unexpected results, no one who works on this subject doubts that it is true.

Conspicuously absent from this brief history is the remarkable explosion of interest in string theory beginning in the fall of 1984 and the almost equally remarkable drop in interest in the past year or so. This mood swing had nothing to do with string theory providing a consistent quantum theory of gravity. Rather, it resulted from the hope that the "uniqueness" of string theory would lead to definite low energy predic-

419

tions in a simple way. Unfortunately, this hope has not been fulfilled. In retrospect, it was perhaps too optimistic to expect that a theory whose fundamental scale is the Planck scale would have a simple extrapolation over 17 orders of magnitude to observable energies. After the difficulty of relating string theory to experiment became clear, many researchers lost interest. In my opinion, while it is certainly necessary for any physical theory to make contact with observations eventually, the difficulty of uniting gravity and quantum mechanics is so great that any theory which claims to do so deserves to be extensively studied.

My plan for this talk is to first discuss the connection between string theory and gravity. At first sight, a theory of strings seems to have nothing to do with gravity. I will try to argue that there is in fact an intimate connection. Next, I will discuss the quantum perturbation expansion. Third, I will consider string theory as a classical theory of gravity. There has been some recent work on whether string theory has spacetime singularities which I will review. Finally, I will discuss some recent speculation about a phase of string theory which is independent of a spacetime metric. In most talks one discusses the classical theory before the quantum theory. I have chosen this admittedly unconventional approach for the following reason. String theory is an extension of general relativity in which one adds to the observed fields of nature an infinite tower of new fields with increasing mass and spin. Although these new fields and their interactions follow simply from the assumption that particles are one dimensional extended objects which interact by splitting and joining, it cannot be denied that the result is a more complicated (a string enthusiast would say "richer") theory than that proposed by Einstein. The motivation for considering this jump in complexity lies mainly in the fact that the quantum theory is much better behaved. So I will discuss the quantum theory first. But after reviewing the status of the quantum theory, I will return to some interesting issues in the classical theory. I do this despite the title since it is clear that we have little hope to understand the quantum theory properly without a better understanding of the classical theory.

2. Connection between string theory and general relativity

We still do not have a complete description of string theory. We have a fairly good understanding of one sector of this theory which can be characterized by the fact that classical solutions have a well defined spacetime metric and quantum effects are calculated perturbatively about them. I will refer to this as the metric phase of string theory. Even in this phase string theory exhibits some qualitatively new and surprising features. However there is growing evidence that string theory also includes a nonmetric phase with even more bizarre properties. We will return to this point later.

In the metric phase, we have a rule for selecting classical solutions and a rule for calculating scattering amplitudes perturbatively about these solutions. Previous experience would suggest that the rule for selecting classical solutions should be to extremize an action which is a functional of the basic fields. Although some progress has been made in formulating string theory this way (for reviews see Horowitz (1987); Siegel (1988)), the most common approach is quite different. Classical solutions are selected by the condition of two dimensional conformal invariance. This means the following. Consider a curved spacetime. A string propagating in this background is described by the two dimensional action

$$S[q_{ab}, X^\mu] = \frac{1}{4\pi\alpha'} \int d^2\sigma \sqrt{q} q^{ab} \partial_a X^\mu \partial_b X^\nu g_{\mu\nu}, \qquad (1)$$

where q_{ab} is a metric on the worldsheet and X^μ describes its location in spacetime. α' is a dimensionful constant related to the string tension, and is believed to be of order the Planck scale. S also depends on the spacetime metric $g_{\mu\nu}$ but at this stage, $g_{\mu\nu}$ is not a dynamical field but rather a fixed background field for the motion of the string. Classically S is conformally invariant

$$S[\Omega^2 q_{ab}, X^\mu] = S[q_{ab}, X^\mu]. \qquad (2)$$

(This is true only in two dimensions, and is one of the reasons why a straightforward generalization of string theory to higher dimensional objects does not exist.)

In general, the quantum theory will not be conformally invariant and the theory is said to have a conformal anomaly. However for certain values of the spacetime metric this anomaly vanishes. The condition for the vanishing of the anomaly can be written

$$\mathcal{T}_a^a = 0 \qquad (3)$$

where $\mathcal{T}_{ab}$ is the quantum stress tensor for the two dimensional theory. This equation implies an equation for the spacetime metric which takes the form (Friedan (1980)):

$$0 = R_{\mu\nu} + \frac{1}{2}\alpha' R_{\mu\rho\omega\tau} R_\nu^{\rho\omega\tau} + \cdots \tag{4}$$

where the dots denote higher order terms involving derivatives and powers of the curvature. This can be viewed as a classical field equation for the spacetime metric and is the string generalization of Einstein's equation. Note that whenever the curvature is small compared to the Planck curvature, the first term dominates the others and this equation reduces to Einstein's equation. Thus all vacuum solutions in general relativity with curvature less than the Planck curvature are approximate solutions to string theory. I want to emphasize that one is not restricted to small perturbations about flat spacetime. (Eq. (4) is indeed obtained by doing a perturbative calculation but its based on a Riemann normal coordinate expansion in a neighborhood of a point $g_{\mu\nu} = \eta_{\mu\nu} - \frac{1}{3}R_{\mu\alpha\nu\beta}x^\alpha x^\beta + \cdots$ rather than small deviations from a flat spacetime.) Since the curvature at the event horizon of a black hole with mass larger than the Planck mass is small compared to the Planck curvature, one knows immediately that string theory has black hole solutions. Of course near the singularity the solution will deviate significantly from that of general relativity. It remains an important open problem to find the exact solution for a black hole in string theory.

What about nonvacuum solutions? In the above discussion we considered a string coupled only to the spacetime metric $g_{\mu\nu}$. One can also introduce other tensor fields on the spacetime and couple the string to them. In this case one finds that the condition for conformal invariance Eq. (3) now implies that these fields satisfy certain classical field equations as well (Callan et. al. (1985); Das and Sathiapalan (1986)). In addition, the equation for the metric is modified to $G_{\mu\nu} = T_{\mu\nu} +$ higher order terms, where $T_{\mu\nu}$ is a stress energy tensor for the background fields. Thus $T_a^a = 0$ can be viewed as the classical field equation for string theory.* It is perhaps worth emphasizing that the *classical* equation for spacetime fields comes from a condition on the two dimensional *quantum* theory. An exact solution to this equation is called a conformal field theory. General relativity is obtained in the low energy (i.e. curvature $<<$ Planck curvature) limit. Most of the additional fields turn out to satisfy massive field equations. With closed string boundary

---

*The superstring also couples to spacetime fermions. In this case, one demands a supersymmetric generalization of conformal invariance - superconformal invariance - to derive the field equations for the fermions.

conditions, only two of these fields are massless. They are a scalar field called the dilaton and an antisymmetric second rank tensor field. With open string boundary conditions, one also obtains a Maxwell field (or Yang-Mills field).

Conformal invariance also imposes constraints on the dimension of spacetime. In the simplest case of a spacetime metric but no other background fields this leads to the critical dimension of 26 for the bosonic string and 10 for the superstring. It would thus seem that any hope of making contact with observations would require some form of Kaluza-Klein compactification. However it has been shown that the addition of other background fields can change the critical dimension (Nemeschansky and Yankielowicz (1985); Myers (1987)). In particular, we will see an example of an exact solution with a nonzero dilaton describing an expanding universe in which one can change the critical dimension by changing the expansion rate. Thus the number of dimensions of string theory seems to become a free parameter (Antoniadis et. al. (1988)).

Two further arguments indicate that the critical dimension is not fundamental. First, the superstring is described by adding fermionic fields to the two dimensional worldsheet in addition to the X^μ. The dimension of spacetime is usually taken to be the number of scalar fields X^μ. But this is not a well defined quantity.

Since the worldsheet is two dimensional, under certain conditions one can replace some of the bosonic fields with fermionic ones and vice versa without changing the theory. Second, if one decomposes $X^\mu(\sigma, \tau)$ into left moving and right moving fields one finds that they are independent. So one can compactify them on different internal spaces and produce string theories which in no sense have more than four spacetime dimensions (Narain et. al. (1987); Kawai et. al. (1987)).

What is so special about conformal invariance? Why should it be elevated to a principle which underlies string theory? We do not yet have a completely satisfactory answers to these questions. It is sometimes claimed that string theory is inconsistent if conformal invariance is lost. This is far from clear. In the absence of conformal invariance, the quantum string acquires another dynamical field. The two dimensional conformal factor is on the same footing as the X^μ. It is clear that string theory is inconsistent if it is not conformally invariant and one does not include this extra dynamical field. Including the conformal factor is technically more difficult to analyze and the consistency has not yet been shown. One possible viewpoint is that the principle of conformal invariance is analogous to the principle of general covariance in general relativity. It is known that any law of physics, e.g. Newtonian gravity, can be written in a covariant form by introducing extra structure such as preferred spacelike surfaces. General relativity is based on the idea that this should not be needed. It is a minimal theory. String theory is also a minimal theory in the sense that the conformal factor is not dynamical.

One might worry that the string field equation (3) is derived by considering the quantum theory of a single string in a curved spacetime. Yet it is well known that if one tries to consider a single particle in a time dependent spacetime, particle creation effects require a second quantized description. Similarly, one expects "string creation" in time dependent backgrounds. How can one derive a consistent field equation from the single string theory? The answer is that the difficulty that requires a second quantized description is the lack of a unique inner product on the space of solutions to the constraints. This inner product is needed to convert the solution space into a one particle Hilbert space. (Note that the problem is uniqueness not existence. There always exists a large class of conserved, positive definite inner products on the space of solutions to the curved space wave equation.*) This inner product is of course crucial for the probability interpretation of the theory. However the string field equation is derived by considering a two dimensional functional integral and requiring that it be conformally invariant. No probability interpretation is needed at this stage.

3. Quantum perturbation expansion

To begin this discussion, one must choose a solution to perturb. One usually takes this to be a static configuration so that it represents a "vacuum" state. By far the majority of work in this area has focused on the simplest possibility of 10 (or 26) dimensional flat spacetime. If one tried to treat string theory as an ordinary field theory, one would consider small perturbations of all the fields (infinite in number) and start computing Feynman diagrams. Not only would this be extremely complicated, it would yield the wrong answer. It is much simpler and correct to work directly with strings. The basic starting point is that strings interact by splitting and joining. The full scattering amplitude between a

---

*R. Wald (private communication).

set of initial and final string states involves a sum over worldsheets of all possible topologies which connect the initial and final states (see Figure 1). The worldsheet without holes describes the classical scattering, the higher genus surfaces describe quantum corrections. Notice that unlike ordinary field theories, there is just one diagram for each order in the quantum perturbation expansion.

The amplitude is computed by doing the following functional integral for each worldsheet topology and summing over topologies:

$$\int DX^\mu \frac{Dq_{ab}}{V_{\text{gauge}}} \text{ (asymptotic states) } e^{-S}. \tag{5}$$

(Both the worldsheet and spacetime metrics are taken to be positive definite to ensure convergence of this functional integral.) To avoid overcounting we have divided the measure by the volume of the group of gauge transformations.

Since we have chosen our background fields to be a classical solution, the action S is invariant under conformal rescalings. It is also invariant under diffeomorphisms or coordinate transformations on the worldsheet. Thus we should not integrate over all Riemannian metrics but only those which cannot be related by conformal rescalings and diffeomorphisms. The space of inequivalent metrics is finite dimensional and called the *moduli space*.

There is an important consistency check on the rules I have described. On the one hand, we have exact classical field equations coming from the condition that the quantum string is conformally invariant. We can certainly expand these equations about a particular static solution and discuss classical scattering. On the other, I have just described an apparently different procedure for computing these same amplitudes

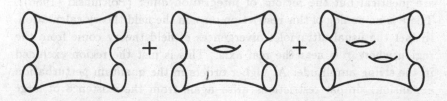

Fig. 1. The string scattering amplitude involves a sum over worldsheets of different topology.

by considering functional integrals over surfaces without holes. It has
been shown by explicit calculation that to low orders in the expansion
these methods agree (Metsaev and Tseytlin (1987); Jack et. al. (1989)).
General arguments have been given that the agreement should hold to
all orders (Tseytlin (1988); Hughes et. al. (1989)). However some issues
still remain to be understood (Jevicki (1989)).

What is the basic reason why superstring amplitudes are finite? One's
first thought is probably that it is based on the usual cancellation of infinities between bosons and fermions in a supersymmetric theory. Of
course it's known that supersymmetry by itself is not sufficient to guarantee finiteness to all orders, e.g. supergravity is not finite, but perhaps
by adding a few extra fields or extra dimensions it can be made to work.
This is not the answer. The basic reason that superstring amplitudes are
finite is that strings are extended objects and cannot probe arbitrarily
short distances. Mathematically, this idea is realized by the existence
of large diffeomorphisms on the worldsheet. That is, diffeomorphisms
which cannot be connected to the identity.

Let me illustrate this with a simple example. Consider the two dimensional torus with an arbitrary metric q_{ab}. One can always choose
a conformal factor such that the rescaled metric is flat. The resulting
flat torus is diffeomorphic to what one gets by considering the complex
plane and identifying the sides of the parallelogram for some τ shown in
Figure 2(a). But τ is not uniquely determined. It can be changed by a
diffeomorphism. If we cut the torus, rotate one end by $2\pi n$ and glue it
back then we have the same torus but have changed τ. So the moduli
space is not the entire upper half plane but rather a subset illustrated in
Figure 2(b). Each point in the moduli space denotes a genuinely different metric on the torus. To compare with ordinary field theory, one can
view the string as an infinite collection of fields and compute the one
loop amplitude using standard methods. One finds that the integrands
are identical but the regions of integration differ (Polchinski (1986)).
There is no analog of this restriction on τ in the field theory calculation.
In fact the usual ultraviolet divergences of field theory come from the
region where τ is near the real axis. This is just the region excluded
in the string amplitude. At higher orders in the quantum perturbation
expansion, similar restrictions arise again from the existence of large
diffeomorphisms on higher genus Riemann surfaces.

Since this discussion applies equally well to the bosonic string, one
might wonder why it does not have finite scattering amplitudes. The
answer is believed to be that we are not starting with the correct vacuum.

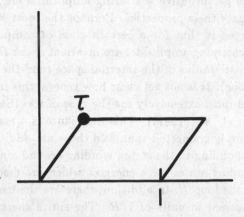

Fig. 2(a). Every flat torus can be obtained by identifying the sides of the parallelogram shown from some τ.

All known solutions for the bosonic string have a tachyon which is an indication that the solution is unstable. It is quite possible that if a stable solution for the bosonic string is found, it will again have finite scattering amplitudes.

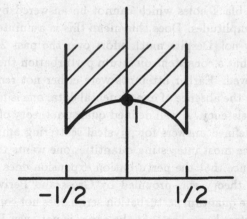

Fig. 2(b). The tori are inequivalent only for τ in the shaded region.

Given that the perturbative scattering amplitudes are finite, one can begin to investigate their properties. Perhaps the most striking feature which has emerged is that for a certain class of compactifications of spacetime, the scattering amplitudes are invariant under $R \to 1/R$ where R is a characteristic radius of the internal space (and the Planck length has been set to one). It is not yet clear how general this invariance is. It has been studied most extensively for the case of tori (Sakai and Senda (1986); and Nair et. al. (1987)). This invariance is a result of the fact that on a nonsimply connected manifold there are additional states of the string corresponding to the string winding around a noncontractible loop. These winding states have energies which are discrete multiples of the radius of the loop R. In addition, there are the usual states with momentum quantized in units of $1/R$. The entire spectrum turns out to be invariant under $R \to 1/R$. Furthermore, all scattering amplitudes are invariant under this inversion. Thus using strings as probes, there is simply no way to distinguish the two cases. It is as if there were two kinds of meter sticks in the world, one made of winding states and the other made of momentum states. Given a noncontractible loop, one measures R and the other measures $1/R$. One can view this result as one way string theory incorporates the idea that lengths should not be well defined below the Planck scale.

Is a finite quantum perturbation theory enough to answer all interesting questions about quantum gravity? Clearly not. Most of the important questions in quantum gravity concern the big bang or the evaporation of black holes which cannot be answered by a calculation of scattering amplitudes. Does this mean this is a uninteresting result? Equally clearly no! Despite much effort over the past 25 years, this is the first time that a consistent quantum perturbation theory for gravity has been achieved. Earlier attempts were either not renormalizable or not unitary. In the absence of experimental data, one is forced to rely on theoretical consistency. A well defined quantum theory of gravity should produce well defined answers for physical scattering amplitudes even if they are not the most interesting quantities one wants to calculate.

Direct evidence that the perturbation expansion does not completely determine the theory was provided by Gross and Periwal (1988) who showed that the quantum perturbation series does not converge. The nth term grows faster than $n!$ so that the series is not even Borel summable without extra nonperturbative information such as a choice of contour in the complex coupling constant plane.* This shows that non-perturbative effects are important. The situation is perhaps analogous to a double well potential in quantum mechanics. One cannot obtain the correct ground state by perturbing about one of the classical minima. One must include nonperturbative instanton effects to arrive at the true ground state.

4. Classical string theory

As we discussed in Sec. 2, classical solutions in string theory correspond to conformal field theories in two dimensions. The direct approach to finding solutions is to solve the infinite order equations (4). Since this is prohibitively difficult, a variety of tricks and approximation methods have been developed to obtain solutions. One approach is simply to start with a solution to Einstein's equation with curvature everywhere much less than the Planck curvature. This is an approximate solution to (4). Since the higher order terms are small, one can take them into account by considering small perturbations of the original metric. The problem with this approach is that by construction the effects of the new terms are small. One does not obtain any qualitatively new results this way. Note that truncating the field equation at say order R^2 and finding exact solutions to the $R + R^2$ equations is of little interest for string theory. The R^2 term is negligible until the curvature is of order the Planck scale. At this point *all* the higher order terms are equally important.

One of the main tricks for obtaining exact solutions is to use the fact that the worldsheet supersymmetry of the superstring causes many of the higher order terms in (4) to vanish. The number of terms which vanish depends on the number of independent supersymmetry transformations.

The superstring always has at least one such transformation. If there is precisely one worldsheet supersymmetry, it turns out that there are no R^2 or R^3 terms in the field equation (Alvarez-Gaume (1981); Ketov (1988)), but non-trivial terms arise starting at order R^4 (Grisaru et. al. (1986); Gross and Witten (1986)). (By "nontrivial" I mean terms which

---

*Strictly speaking, they considered the bosonic string in 26 dimensional flat spacetime for which the individual terms in the perturbation expansion diverge. However they removed these divergences with an appropriate regulator to obtain their result. They argue that the superstring perturbation expansion should behave the same way.

cannot be absorbed by field redefinitions.) It is rather surprising that the only difference between the way general relativity and superstring theory describes gravity is the existence of these R^4 (and higher) terms in the field equation. If there are two supersymmetry transformations, *there are no nontrivial higher order terms in the field equations at all* (Witten (1986); Nemeschansky and Sen (1986)). Thus exact solutions can be found by solving the lowest order equations. There exist two supersymmetry transformations only if the background fields satisfy certain extra conditions such as the spacetime must be complex and the metric must be Kahler (Zumino (1979)). (This turns out to be equivalent to spacetime supersymmetry.) Using this fact a large class of solutions have been found which are the product of a four dimensional Minkowski spacetime and a six dimensional compact internal space with $R_{MN} = 0$ (Candelas et. al. (1985)). These solutions may seem a little strange since there are nonzero fields on the internal space which do not cause any gravitational effects in the four macroscopic dimensions. Indeed, they have no analog in four dimensional general relativity since in three dimensions or fewer, a Ricci flat space is flat. But they are quite common in Kaluza-Klein theories. In fact these spacetimes are solutions of Einstein's vacuum equation in ten dimensions. Unfortunately, although these solutions tell us a great deal about the particle spectrum and interactions of the low energy four dimensional field theory, they do not tell us much about gravity in the strong field region.

One trick for finding exact solutions without using supersymmetry is to consider strings moving on a group manifold. Witten (1984) has shown that by taking the metric to be the Cartan-Killing metric constructed from the structure constants, and including a certain nonzero value for the antisymmetric tensor field $B_{\mu\nu}$, this theory is conformally invariant. Since the underlying manifold of $SU(2)$ is a three sphere, and the translation group is just R, one example of this construction is an exact solution on $R \times S^3$ which is isometric to the Einstein static universe.

Strings moving on a group manifold have a much larger symmetry group (generated by a Kac-Moody algebra) than strings on an arbitrary spacetime. But this is not the main reason that the theory is conformally invariant. When $B_{\mu\nu}$ is nonzero, the equation for the metric includes terms constructed from the curl of $B_{\mu\nu}$. Mukhi (1985) has shown that if one introduces a connection with torsion $\partial_{[\mu}B_{\nu\rho]}$, then every term in the equation for the metric involves at least one power of the curvature with torsion. Thus, if this curvature vanishes, one has an exact conformal field theory. In the case of a Lie group with structure constants $C^{\nu}_{\rho\sigma}$ and invariant metric $g_{\mu\nu}$, $g_{\mu\nu}C^{\nu}_{\rho\sigma}$ is a closed three form. One can thus define $B_{\mu\nu}$ (at least locally) so that its curl is proportional to this three form. It is not difficult to verify that the curvature of a connection with torsion proportional to this three form vanishes. A nonperturbative topological argument (Rohm and Witten (1986); Teitelboim (1986)) shows that in some cases, the coefficient of this torsion is quantized. For the above example of the Einstein static universe, this implies that radius of the three sphere can only take a discrete set of values.

In the case of the Einstein static universe, the timelike direction was added "by hand" by taking the product with $\mathbb{R}$. This is not always necessary. Some Lie groups are naturally spacetimes in that the Cartan-Killing metric has Lorentz signature. One example is the three dimensional Lorentz group $SO(2,1)$. This is a three dimensional group and the metric is Lorentzian. It is easy to check that this space has constant negative curvature and is thus three dimensional anti-de Sitter spacetime. To avoid closed timelike curves it is natural to work on the universal covering space. By adding $B_{\mu\nu}$ in the above way one obtains an exact solution to string theory (Balog et. al. (1988)). Like the earlier examples, anti-de Sitter space is globally static, but its causal structure is more interesting since there are no Cauchy surfaces. In other words, given initial data on a spacelike surface one cannot evolve everywhere in the spacetime. This shows that Cauchy horizons exist in classical string theory even for large curvature.

It is tempting to try to find nonstatic exact solutions to string theory by using the following observation. We have just seen that the reason group manifolds lead to exact conformal field theories is that there exists a flat metric compatible connection whose torsion (after lowering the index) is a closed three form. But it is known that for *every* metric (on a parallelizable manifold) there exists a metric compatible flat connection with torsion. Thus the only condition that has to be imposed in order to obtain new exact solutions is that $g_{\mu\nu}T^{\nu}_{\rho\sigma}$ be a closed three form where $T^{\nu}_{\rho\sigma}$ is the torsion which flattens the curvature of the metric $g_{\mu\nu}$. This condition turns out to surprisingly restrictive. In fact, it may only be satisfied by Lie groups.

Of course, the most interesting questions about classical string theory are those which deal with the high curvature regime. In particular: Are there singular solutions in string theory? Are generic solutions singular? In other words, is there some analog of the singularity theorems in this theory? Realistically, one expects quantum effects to be important in any regions of large curvature. But it would certainly be interesting to know whether the ubiquitous singularities of general relativity are removed in string theory even at the classical level.

Until recently all known exact solutions were static, so there was no evidence one way or the other. Recently, however a simple time dependent model cosmology has been found (Antoniadis et. al. (1988)). This solution includes a nonzero value of the dilaton Φ as well as the metric. It is given by:

$$ds = -dt^2 + t^2 dx_i dx^i \,,$$
$$\Phi = \log t \,.$$

(6)

This is a spatially flat Robertson Walker metric with a linearly growing scale factor. One can show that this is an exact solution for all $t > 0$. Actually, the four dimensional spacetime (6) by itself is not a solution since it does not have the right critical dimension. However, one can always take the product of this spacetime with an appropriate time independent "internal" conformal field theory to obtain a valid solution which does have the right dimension.

There is a simple generalization of the solution (6) in which one rescales the expansion rate by an arbitrary factor Q: $ds^2 = -dt^2 + Q^2 t^2 dx_i dx^i$ and $\Phi = \log Qt$. This solution is almost identical to the one above with the important exception that the critical dimension depends on Q! Unfortunately, in the present case the critical dimension is increased rather than decreased: $D = 26 + 3Q^2$. Nevertheless, this example illustrates the general principle that solutions to bosonic string theory do not have to have 26 dimensions. (Despite appearances, Q need not even be an integer. There are some conformal field theories which have "effective dimensions" which are not integral.) For simplicity, we will set $Q = 1$, and ignore the time independent internal part of the solution.

The scalar curvature of the linearly expanding Robertson-Walker metric behaves like $R = 6/t^2$, so $t = 0$ looks like a standard curvature singularity. However, before one jumps to the conclusion that classical string theory has singular solutions, one must consider the following. If we introduce a new time coordinate $t = e^\eta$ then

$$ds^2 = e^{2\eta}[-d\eta^2 + dx^i dx_i] \,,$$
$$\Phi = \eta \,.$$

(7)

So the conformally rescaled metric $e^{-2\Phi} \tilde{g}_{\mu\nu}$ is flat and clearly nonsingular. But in a theory with a metric and a scalar field, this is just a field redefinition which should not change the physical properties of a solution.

In general relativity the issue of field redefinitions is resolved by considering the motion of test particles. These particles follow geodesics with respect to one particular metric $g_{\mu\nu}$. In addition, ideal clocks moving along a geodesic will measure the proper time as calculated from $g_{\mu\nu}$. These properties justify the interpretation of $g_{\mu\nu}$ as the "physical" metric. In the late 1960's there was much discussion over the proper definition of a singular spacetime in general relativity. Naive notions of the curvature components becoming infinite were plagued by ambiguities in the asymptotic behavior of the coordinates. The final consensus (Geroch (1968)) was that if the physical metric is geodesically incomplete then the spacetime should be called singular. This seems reasonable since in such a spacetime, certain test particles could only exist for a finite time. However in principle any (invertible) combination of the matter fields and the physical metric can be used as the metric in general relativity. The motion of test particles and the calculation of elapsed time will be more complicated, but one will obtain the same physical answers.

In string theory, the situation is not so simple. To illustrate the problem let me compare the low energy limit of string theory with the Brans-Dicke theory . The action for the Brans-Dicke theory is

$$S_{BD} = \int [e^{2\Phi}(R - 4\omega(\nabla\Phi)^2) + L_m]\sqrt{g}\ d^4x\,, \tag{8}$$

where ω is the Brans-Dicke coupling constant and L_m is the matter Lagrangian. Since Φ does not couple directly to the matter, test particles will again follow geodesics of the metric $g_{\mu\nu}$. The low energy effective action from string theory is

$$S_{ST} = \int [e^{2\Phi}(R + 4(\nabla\Phi)^2) + e^{2\Phi}L_m]\sqrt{g}\ d^4x\,. \tag{9}$$

One difference is that the Brans-Dicke coupling constant is -1. But more importantly, there is direct coupling between the matter and Φ. This means that string theory violates the equivalence principle. Particles of different composition will follow different trajectories. (These trajectories can be found by rescaling $g_{\mu\nu}$ by the appropriate powers of e^{Φ} so that the Lagrangian for the particular matter field of interest appears with no explicit Φ coupling. Particles of this field will then follow geodesics of this rescaled metric.) Furthermore, clocks of different composition will measure different times even when they are following the same trajectory! These "predictions" clearly violate experiment, so if string theory is to describe our world the dilaton must acquire a mass through quantum corrections. (For the superstring, the dilaton is in the same supermultiplet as the graviton. So it cannot obtain a mass until supersymmetry is broken. Unfortunately, this is a nonperturbative effect and beyond our present ability to calculate reliably.) But it is reasonable to assume that near the singularity the mass of the dilaton will be negligible and we must face the question of how we are going to define a singularity in string theory. Which metric should one look at to examine geodesic incompleteness?

One possibility is to consider a purely gravitational test particle. That is, a small black hole. The motion of these objects will be approximate geodesics of the metric with the standard Einstein action. This is related to the above metric by $\tilde{g}_{\mu\nu} = e^{2\Phi} g_{\mu\nu}$. Now for the cosmological solution (6) the metric $\tilde{g}_{\mu\nu}$ describes the expanding universe and $g_{\mu\nu} = e^{-2\Phi} \tilde{g}_{\mu\nu}$ is flat. This suggests that test black holes will live only a finite time as measured say by their normal modes of oscillation. Hence one might be tempted to declare the solution singular. But it is far from clear whether there are objects analogous to black holes with $M < M_{Planck}$ in string theory. If not, we cannot use them as test particles near a singularity. A better approach seems to be to consider "test strings" rather than test particles. The metric that the string couples directly to turns out to be the flat metric. (The fact that $g_{\mu\nu}$ is flat and Φ is a linear function of one coordinate is the reason that this configuration could be shown to be an exact conformal field theory.) From this standpoint, the solution is nonsingular. But even this is not completely satisfactory. It seems more natural to define a singularity in string theory not as a property of one field (geodesic incompleteness of some metric) but rather some condition involving all fields. An appropriate and convenient formulation of such a condition is needed before a deeper understanding of singularities in string theory is possible.

One can construct additional time dependent solutions to string theory by starting with essentially any static solution with constant dilaton, and replacing the dilaton with a function which is linear in time. But these solutions are still too special to shed light on the dynamical effects of string theory at high curvature. Can one find other solutions? There must exist a huge class of time dependent exact solutions since there are an infinite number of dynamical fields. The fact that we know essentially none of them is a remarkable statement of our ignorance of string theory. Just as our knowledge of general relativity was enormously enhanced by a few key exact solutions (Schwarzschild, Friedmann, etc.), string theory will take a giant step forward when we find the string analog of these solutions.

5. Speculations

In the previous section we considered the classical limit of string theory and saw that general relativity can be recovered in the limit when the curvature is small compared to the Planck scale. We also discussed some strong curvature solutions. But all these solutions have a spacetime metric. The other fields in the theory are often taken to be zero but not the metric. Why should the metric be singled out in this way? String theory is supposed to be a unified theory of gravity with the other fields. (Although this unification is admittedly obscured in the formulation described here.) If the other fields can be zero, we should also allow the metric to be zero. Note that this does not mean just zero curvature, but actually zero metric. Of course the key question is whether the equations of string theory can be formulated in a way which do not depend on an invertible metric. There is strong evidence that this is the case (Horowitz et. al. (1986); Hata et. al. (1986); Strominger (1987)). In this alternative formulation, zero metric solutions not only exist but are preferred in the sense that they have maximal symmetry. This strongly suggests that there is a nonmetric phase of string theory in which physics is quite different from anything we have seen before.

Over the past year, several examples of theories without a spacetime metric have been studied (Witten (1988a, 1989)). Theories of this type, which are known as "topological field theories," contain no local dynamics. Nevertheless, the theories are not entirely trivial. There is typically a finite dimensional space of classical solutions which are related to the topology of the underlying manifold or field configurations. In many cases the quantum theory is exactly soluble and all physical quantities are topological invariants.

Perhaps the simplest example is the following (Horowitz (1989)). Let M be a four dimensional manifold without boundary. Let A_μ be a vector field and $B_{\mu\nu}$ be a two form on M. The action $S = \int_M B_{[\mu\nu} \, p_\rho A_{\sigma]}$ is well defined without any further structure such as a spacetime metric. The classical field equations are simply $\partial_{[\mu} A_{\nu]} = 0$ and $\partial_{[\mu} B_{\nu\rho]} = 0$.

The action is invariant under diffeomorphisms, and the gauge transformations $\delta A_\mu = \partial_\mu \lambda$ and $\delta B_{\mu\nu} = \partial_{[\mu} v_{\nu]}$. So the space of gauge inequivalent solutions is the space of closed forms modulo the exact forms. In other words it is the de Rham cohomology groups $H^1(M) \times H^2(M)$. The quantum theory can be constructed via path integrals (or canonical quantization if $M = \mathbb{R} \times \Sigma$ for some three manifold Σ). One can show (Horowitz and Srednicki (1989)) that if U is a closed curve and V is a closed two surface in M which do not intersect and are homologically trivial then

$$L(U, V) = \int [dA][dB] \int_U A \int_V B \; e^{iS} \tag{10}$$

is the linking number of the surfaces U and V. It is interesting that in order to define the measure in the functional integral one does have to introduce a metric, but the final answer turns out to be independent of this choice.

A second example is gravity in three dimensions. The standard formulation of three dimensional gravity involves a metric but no local dynamics since the field equation implies that the spacetime is flat. However in first order form, the Einstein action is

$$S = \int e^a_{[\mu} R^{bc}_{\nu\rho]} \epsilon_{abc} \tag{11}$$

where e^a_μ denotes a linearly independent triad of vectors and $R^{bc}_{\nu\rho}$ is the curvature two form of an $SO(2,1)$ connection. The metric is obtained by requiring that e^a_μ be an orthonormal basis. However since the inverse of the triad does not appear in this action one can generalize to the case where e^a_μ are now an arbitrary collection of three vectors. In fact this is very natural in the following respect. In addition to the usual symmetries of Lorentz rotations and diffeomorphisms, the action (11) is invariant under the transformation $e^a_\mu = D_\mu \lambda^a$ where D_μ denotes the covariant derivative with the Lorentz connection. This is because one can integrate by parts and use the Bianchi identity for the curvature. However this transformation does not preserve the linear independence of e^a_μ. In fact $e^a_\mu = 0$ is gauge equivalent to an invertible triad under this transformation. Thus three dimensional gravity resembles a topological field theory. Its quantization can be carried out exactly (Witten (1988b)).

Even in four dimensions, the Einstein action in first order form is simply

$$S = \int e^a_\mu e^b_\nu R^{cd}_{\rho\sigma} \epsilon_{abcd}. \tag{12}$$

This action involves the tetrad e^a_μ but not its inverse. Thus the field equations are perfectly well defined if we drop the usual condition that the tetrad be linearly independent and consider e.g. the solution $e^a_\mu = 0$ (Tseytlin (1982)). Although unlike the three dimensional case, there is no analog of the inhomogeneous gauge transformations which relate invertible and noninvertible tetrads. So it is more natural to restrict to invertible ones.

Topological field theories have provided a new approach for studying a variety of mathematical structures. However their physical interpretation remains obscure. The nonmetric phase of string theory will probably be richer than the topological field theories that have been studied so far. An appropriate physical interpretation may only arise after we have a better understanding of how it is related to the more familiar metric phase of the theory.

To conclude, it should be clear that string theory is still in its infancy. There are many areas in which more insight is needed. One is the issue of classical dynamics in a region of large curvature. This includes a better description of singularities in classical string theory and some results about how frequently they exist. Even a solution to the simplest case of spherically symmetric gravitational collapse would be enormously helpful. Another area in which progress is needed is the search for a more fundamental formulation of the theory. The rules I have described for selecting classical solutions and computing quantum scattering amplitudes perturbatively seem somewhat ad-hoc. What is the underlying principle behind string theory? The answer to this question will undoubtedly provide a framework for discussing nonperturbative effects as well. Finally we must strive to find experimentally testable predictions. This will be difficult but not impossible. In particular it need not require a Planck scale particle accelerator. What is needed is simply a new effect which–although small–can be seen by examining large quantities of matter (in the spirit of the proton decay experiments). I hope that at the GR-13 meeting in 1992, progress will be reported on one or more of these issues.

Note added: After this lecture was given, it was realized that all solutions to Einstein's equation with a covariantly constant null vector (pp waves) are exact solutions to string theory. The higher order terms in the equation of motion all vanish since the curvature is null. This includes a large class of spacetimes with curvature singularities. Furthermore, the dilaton can be taken to be constant so the ambiguity discussed in Sec. 4 does not arise. Thus classical string theory does not remove the singularities of general relativity. For more details, see Horowitz and Steif (1989).

It is a pleasure to thank Soo-Jong Rey, Mark Srednicki, Alan Steif, and Andy Strominger for discussions. I also wish to thank the organizing committee of GR12 for their invitation to lecture. This work was supported in part by NSF Contract No. PHY85-06686.

References

Alvarez-Gaume, L. (1981), *Nucl. Phys.* **B184**, 180.
Antoniadis, I., Bachas, C., Ellis, J. and Nanopoulis, D. (1988), *Phys. Lett.* **B211**, 393; (1989), An expanding universe in string theory, CERN preprint CERN-TH.5231/89.
Atick, J., Moore, G. and Sen, A. (1988), *Nucl. Phys.* **B307**, 221.
Balog, J., O'Raifeartaigh, L., Forgacs, P. and Wipf, A. (1988), preprint MPI-PAE/PTh66/88.
Callan, C., Friedan, D., Martinec, E. and Perry, M. (1985), *Nucl. Phys.* **B262**, 593.
Candelas, P., Horowitz, G., Strominger, A. and Witten, E. (1985), *Nucl. Phys.* **B258**, 46.
Das, S. and Sathiapalan, B. (1986), *Phys. Rev. Lett.* **56**, 2664.
D'Hoker, E. and Phong, D. (1988), *Rev. Mod. Phys.* **60**, 917; Superstrings, super Riemann surfaces, and supermoduli space, to appear in the *Proceedings of the String Theory Conference*, Rome (1988).
Friedan, D. (1980), *Phys. Rev. Lett.* **45**, 1057.
Geroch, R. (1968), *Ann. Phys.* **48**, 526.
Grisaru, M., van de Ven, A. and Zanon, D. (1986), *Nucl. Phys.* **B277**, 409.
Gross, D. and Periwal, V. (1988), *Phys. Rev. Lett.* **60**, 2105; **61**, 1517.
Gross, D. and Witten, E. (1986), *Nucl. Phys.* **B277**, 1.
Hata, H., Itoh, K., Kugo, T., Kunitomo, H. and Ogawa, K. (1986), *Phys. Lett.* **B175**, 138.
Horowitz, G., Lykken J., Rohm, R. and Strominger, A. (1986). *Phys. Rev. Lett.* **57**, 283.
Horowitz, G. (1987), Introduction to String Field Theory, in *Superstrings '87*, eds. Alvarez Gaume et. al. (World Scientific, Singapore).
Horowitz, G. (1989), Exactly soluble diffeomorphism invariant theories, *Commun. Math. Phys.*, to appear.
Horowitz, G. and Srednicki, M. (1989), Quantum field theoretic description of linking numbers and their generalization, *Commun. Math. Phys.*, to appear.
Horowitz, G. and Steif, A. (1989), Singularities in String Theory, Santa Barbara preprint UCSB-TH-89-38.
Hughes, J., Liu, J. and Polchinski, J. (1989), *Nucl. Phys.* **B316**, 15.
Jack, I., Jones, D. and Mohammedi, N. (1989), *Nucl. Phys.* **B322**, 431.
Jevicki, A. (1989), Renormalization group and string dynamics, to appear in the *Proceedings of the XXV Winter School of Theoretical Physics, Karpacz*, Brown preprint HET-713.
Kawai, H., Lewellen, D. and Tye, S. (1987), *Nucl. Phys.* **B288**, 1.
Ketov, S. (1988), *Phys. Lett.* **B207**, 140.
La, H. and Nelson, P. (1989), *Phys. Rev. Lett.* **63**, 24.

Metsaev, R. and Tseytlin, A. (1987), *Phys. Lett.* **B191**, 354.

Mukhi, S. (1985), *Phys. Lett.* **B162**, 345.

Myers, R. (1987), *Phys. Lett.* **B199**, 371.

Nair, V., Shapere, A., Strominger, A. and Wilczek, F. (1987), *Nucl. Phys.* **B287**, 402.

Narain, K., Sarmadi, M. and Vafa, C. (1987), *Nucl. Phys.* **B288**, 551.

Nemeschansky, D. and Sen, A. (1986), *Phys. Lett.* **B178**, 365.

Nemeschansky, D. and Yankielowicz, S. (1985), *Phys. Lett.* **54**, 620.

Polchinski, J. (1986), *Commun. Math. Phys.* **104**, 37.

Rohm, R. and Witten, E. (1986), *Ann. Phys. (NY)* **170**, 454.

Sakai, N. and Senda, I. (1986), *Prog. Theor. Phys.* **75**, 692.

Scherk, J. and Schwarz, J. (1974), *Nucl. Phys.* **B81**, 118.

Schwarz, J. (1982), *Phys. Rept.* **89**, 223.

Siegel, W. (1988), *Introduction to String Field Theory*, (World Scientific, Singapore).

Strominger, A. (1987), Lectures on closed string field theory, in *Superstrings '87*, eds. Alvarez-Gaume et. al. (World Scientific, Singapore).

Teitelboim, C. (1986), *Phys. Lett.* **B167**, 69.

Tseytlin, A. (1982), *J. Phys.* **15**, L105.

Tseytlin, A. (1988), *Phys. Lett.* **B208**, 221.

Witten, E. (1984), *Commun. Math. Phys.* **92**, 455.

Witten, E. (1986), *Nucl. Phys.* **B268**, 79.

Witten, E. (1988a), *Commun. Math. Phys.* **117**, 353; **118**, 411.

Witten, E. (1988b), *Nucl. Phys.* **B311**, 46.

Witten, E. (1989), *Commun. Math. Phys.* **121**, 351.

Yoneya, T. (1974), *Prog. Theor. Phys.* **51**, 1907.

Zumino, B. (1979), *Phys. Lett.* **B87**, 203.

D1

Theories of quantum gravity
I: superstring theory

C. Aragone

Dpto. de Fisica, Universidad S. Bolivar,
Apartado 89000 - Caracas 1080 A, Venezuela

Superstrings continue to be a source of inspiration for the basic understanding of quantum gravity. They seem to provide a more fundamental arena than quantum field theory. Even though we still do not have a theory of everything, string concepts bring a new theoretical richness to research in quantum and classical gravity.[1]

Based in previous work on general gauge theories, S. Ichinose analyzed in this session 2D conformal gravity centering his attention on gauge fixing, physical quantities, the energy momentum tensor, and renormalization. C. G. Torre in his talk presented a new formulation of Hamiltonian 2D-gravity which is covariant under all the relevant groups: the spacetime diffeomorphism group, the slice diffeomorphism, and the group of conformal isometries. The key ingredients that allow covariance with respect to the above groups are the enlargement of the phase space by the inclusion of the cotangent bundle over the space of embeddings of a Cauchy surface into the spacetime and the extensive use of conformal 2D isometries.

I. Bakas showed that the Sugawara formalism used in 2D conformal field theory to construct the stress-energy tensor of some non-linear σ-models may have a natural geometric interpretation as a gauge fixing mechanism for current algebras. In the simplest case ($SL(2, C)$) this procedure provides a realization of the diffeomorphism group of the circle in terms of a 2-D non linear σ-model. Its connection with the choice of the variables proposed by Isham, Klauder and others in the context of canonical quantum gravity was discussed.

For the other cases $(SL(n,c); n > 2)$ the basic coordinate functions were given a physical meaning using symmetries generated by higher spin fields.

Then H. Suzuki addressed the audience on the Becchi-Rouet-Stora (BRS) formulation of string theory and the properties of non-critical strings,[2] problems which are currently being extensively investigated by Fujikawa's group. He reviewed their recent study of a construction of covariant tensorial (BRS, energy-momentum, and ghost number) operators for string theory in the conformal approach, based on the covariant path integral measure. By incorporating the 2D world sheet metric freedom explicitly in the theory, it was shown that these covariances are maintained for an arbitrary space-time dimension D. Then, the covariant quantization of the metric freedom was examined, showing an interesting solution at $D <$ (critical dimension) which conserves both the above covariances and the closure of the BRS symmetry. These results, it was pointed out, suggest the validity of the covariant path integral approach of string theory and might lay the formulation of a possible quantization of non critical strings.

S. V. Ketov and O. A. Soloviev presented their results on the anomalies of the heterotic σ-model in the curved superspace of 2D-(1,0) supergravity. They developed the covariant (in space-time) background field method for the computation of the supergravitational or superconformal anomaly for the heterotic string propagating on the background of its massless modes.

This theory can be interpreted just as the non-linear heterotic σ-model defined in a curved superspace of the 2D-(1,0) supergravity. The conformal anomaly (or the central charge coefficient of the superVirasoro algebra) is calculated in the five loop model approximation. It was shown to agree with the known results for the low energy heterotic string effective action calculated in the tree-string S-matrix amplitude.[3] Absence of dilaton dependent contributions at two and three loops was checked explicitly. In an additional contribution A. A. Deriglasov and S. V. Ketov presented partial results concerning the calculation of the four loop β-function for the 2D-supersymmetric non linear σ-model with torsion and the $\zeta(3)$-correction for the low energy effective action of type II superstrings.

A. Das, J. Maharana and S. Roy studied the evolution of a type II closed superstring in a curved target manifold with torsion. The classical BRST charge they obtained is nilpotent. The quantum BRST-charge gives rise to the equations of motion for the background fields in order to be nilpotent and to obtain the quantum superconformal algebra.

Gravity from strings appears also in the contribution of V. N. Shchyotochkin and V. N. Yefremov. They presented an algorithm for computation of the low energy coupling constants based on the numerical characteristics of the $E8 \times E8$ superstring model. Through the fact that the exponents and the Coxeter number of the Weyl group $W(E8)$ are related to topological invariants of Coxeter type orbifolds, which provide a unifying framework for the construction of 4D string theories, this work establishes a connection with this new area of research in string theory.[4]

The method gives a consistent set of the dimensionless strong, weak, and gravitational coupling constants with precisions corresponding to that of the fine structure constant's latest determination. Their time variation is also predicted. A number theoretical formalism for computing the energy dependent gauge couplings and the hierarchy of energy scales for the orbifold-heterotic string was also presented.

D. C. Dunbar discussed the consistency conditions for the existence of type I superstring theories in $D < 10$. Also, in these dimensions the possibility of a cancellation mechanism between oriented and unoriented surfaces for open string loop amplitude was considered. Due to its close connection with finiteness, the vanishing of Yang-Mills and gravitational anomalies in $D < 10$ was analyzed.

The work of J. Gegenberg, G. Kunstatter and R. B. Mann is on the quantum properties of algebraically extended bosonic σ-models. The torsion term arises naturally as a consequence of the extended geometry. The quantum theory is developed "a la Polyakov." The Weyl anomaly is calculated and does not cancel in any spacetime dimension even though the Liouville factors cancel in $D = 26$.

Supermembranes might constitute an alternative to string theory as a fundamental model. Its development is too recent and incipient in order to be able to guess today the basic role, if any, they will play in a fundamental theory of gravity.

R. Güven reported on a new family of supermembranes possessing spacetime supersymmetry and Siegel symmetry. This family consists of bosonic solutions of supermembrane field equations and describes static, toroidal membranes in $D = 11$ black-hole spacetimes. These black holes are obtained by embedding the $D = 4$ Reissner-Nordstrom solutions into $D = 11$, $N = 1$ supergravity. It is shown in this work that supermembranes pick, as their backgrounds, only the extreme Reissner-Nordstrom black holes and require the $D = 4$ magnetic charge not to vanish. These peculiarities suggest that the quantization about them is by itself an interesting problem. It is worth noticing that membranes are intrinsically non-linear objects and finding exact solutions is not a trivial endeavour.

In the more classical context of Kaluza-Klein theories, M. I. Beciu addressed the question of why the extra dimensions are not perceived. He showed that the extra dimensions are unobservables if our 4D spacetime is an asymptotically null membrane, i.e., a timelike membrane which for late times becomes infinitely close to the null cone, evolving in an embedding $D > 4$ world. The idea is illustrated by a 4D membrane moving in a $D = 5$ world having a de Sitter and a Schwarzschild phase.

Beside the Nambu-Goto strings, the simplest relativistic strings one can imagine, there is the rigid string which contains a term depending upon the extrinsic curvature of the world sheet.

The idea can be pursued also in the context of relativistic points. M. Pav̌sič presented what happens if one considers a point particle with extrinsic curvature and its quantization. Essentially, after quantization one obtains the Dirac equation. Classical angular frequency of the helical motion turns out to be equal to the frequency of the quantum Zitterbewegung.

At a purely geometrical level, going in the direction of Dirac's suggestion of making our equations invariant under wider groups, D. Pandres proposed a group which is so large that superpoints do not have an invariant meaning. The most primitive entities which have invariant meaning are superpaths. In this scheme a particle localized at a superpath is a superstring.

On the analytical side, U. Bruzzo contributed by proposing suitable definitions of relative Picard group, relative Picard variety and relative Chern class of a line bundle over a family of graded Riemann surfaces. As the Polyakov approach to string theories leads to expressions of the string amplitudes as integrals over the moduli space of Riemann surfaces, it is usually thought that a similar analysis can be performed working in super Riemann surfaces and over their moduli space.

V. A. Kostelecky and S. Samuel reported on their extensive results of gravitational phenomenology in higher dimensional theories and strings. They investigate gravitational phenomenology in compactified higher dimensional theories, with particular emphasis on the consequences of the mechanism arising from string theory for tensor-induced spontaneous Lorentz symmetry breaking.

The role played by this mechanism in causing a gravitational version of the Higgs effect and in compactification of the extra dimensions was explored.

A general argument based on the specific behavior of a tensor field acquiring an expectation value was given, showing that no mass terms may be generated through this mechanism. They also show for this string inspired model that the compactified manifold has to be a fiber bundle over a circle. The experimental consequences of compactified higher dimensional theories were examined by looking at generalized Schwarzschild solutions and at different types of results from cosmology in higher dimensions. Their results provide relevant constraints to higher dimensional theories. Concerning the cosmological aspects, the higher dimensional models satisfying the standard cosmology and the available observational constraints are found to have the fine tuning problem. It is pointed out in this work that an increase in the precision of the measurements of the mass density of the universe, the Hubble constant and the acceleration parameter may destroy their validity.

Their conclusions are very interesting: i) the perturbative sector of any higher dimensional theory is excluded on the basis of current experimental observations; and ii) the phenomenological difficulties may be overcome if a mechanism is found providing a mass to the massless modes of the higher dimensional components of the metric tensor.

In view that many attractive models for unification and quantum gravity contain dimensions higher than 4, this set of results constitute a serious challenge to them, especially because there does not seem to be at hand the existence of an essentially non covariant mechanism for giving selectively masses in the higher dimensions.

Anomaly cancellations in $D = 10$ supergravity require the presence of Lorentz Chern-Simons terms in the initially supersymmetric action. If these terms are left alone local supersymmetry is broken.

This action therefore can not be the low energy effective action of realistic superstring models like the heterotic string.

The very deep and brilliant talk of M. de Roo described the recent results obtained (with E. A. Bergshoeff) concerning the detailed structure of the quartic part of the effective action for the heterotic string containing the Lorentz Chern-Simons terms, as well as the supersymmetric transformation rules of the supergravity multiplet up to cubic order in the two fundamental constants, appearing in the action: the inverse of the string tension and the inverse of the square of the gauge coupling constant.

They use the Noether method in the component approach and the main idea is to treat in a symmetric way the supergravity multiplet and the Yang-Mills one. More precisely, they represent the Lorentz affinity and its corresponding Riemann tensor in a form that can be identified with an $SO(9, 1)$ Yang-Mills multiplet.

They show that this symmetry persists at least to the fourth order action $\sim R^4$. The construction of the supersymmetric $\sim R^2$ action is rather simple, the $\sim R^3$ part does not contain bosonic contributions, in agreement with the result of string calculations, and finally the term $\sim \alpha(\alpha R^2 + \beta F^2)^2$ was given.

The geometric origin and calculation of the other constituent of the quartic action, which at tree level involves $\zeta(3)$ and is known to belong to an independent invariant having the form $R + R^4$, remains an open question in the component approach.

String effective actions thus allow the analysis of the phenomenological 4D compactified models.

The presence in the bosonic sector of an additional contribution to the Einstein scalar, proportional to the square of the Riemann tensor was clear from the very beginning of the 1984 revival of string theory. However an arbitrary term of this type gives rise to ghosts which strings do not have. The way out found by Zwiebach[6] was to postulate that the right structure which has to appear in $D = 10$ effective actions was the Gauss-Bonnet combination. More generally, Zumino[7] rediscovered in a very elegant geometrical formulation the generalization of the Gauss-Bonnet actions found some time ago by Lovelock.[8]

Y. Choquet-Bruhat in her talk pointed out the interest of analyzing them, both from the mathematical and physical point of view. They are the only ones leading to second order differential equations for the metric in $D > 4$. The corresponding field equations are fully nonlinear. She presented fundamental properties of high frequency wave and two cases were considered: i) when the background metric has constant curvature; and ii) when the coupling constant of the Gauss-Bonnet term is small. Taking the example of a high frequency perturbation of a plane wave there was shown to exist a distortion of the shape for the disturbance which does not appear in the pure Einstein case.

Gravity as a gauge theory seems to be better represented in its vielbein form than in a metric way. Supergravities also need gravity through its D-bein representation, due to the presence of fermions. The mathematical structure of D-bein gravity is largely different from the metric Palatini-type of action.

C. Aragone in his talk presented how vielbein gravity looks in the light-front gauge. In this work (with A. Khoudeir), D-dimensional gravity is analyzed in the light front gauge when it is given in its first order vielbein formulation. The unconstrained action coincides (as it must) with previous results using the metric formulation. It turns out that, in $D > 4$ dimensions the expression for the light front energy does not appear to be explicitly non-negative due to the presence of the D-2 transverse scalar curvature. It is shown that however, in the linearized approximation, the light front energy is non-negative, being the transverse metric energy of a pure helicity two excitation.

Summing up, the session provided a wide spectrum of the many ways in which string ideas enrich the current understanding of quantum gravity.

References

1. A. M. Polyakov (1987), *Mod. Phys. Lett.* **A2**, 893; V. G. Knizhnik, A. M. Polyakov and A. B. Zamolodchikov, (1988), *Mod. Phys. Lett.* **A3**, 819.
2. K. Fujikawa, N. Nakazawa and H. Suzuki (1989), *Phys. Lett.* **221B**, 289; K. Fujikawa, T. Inakagi and H Suzuki (1989), Ritp Hiroshima Univ. preprint RK 89 - 17.
3. D. J. Gross and J. H. Sloan (1987), *Nucl. Phys.* **B91**, 49.
4. A. N. Schellekens and N. P. Warner (1988), *Nucl. Phys.* **B308**, 397.
5. M. B. Green and J. H. Schwartz (1984), *Phys. Lett.* **B149**, 117.
6. B. Zwiebach (1985), *Phys. Lett.* **B156**, 315.
7. B. Zumino (1986), *Phys. Rep.* **137**, 109.
8. D. Lovelock (1971), *J. Math. Phys.* **12**, 498.

D2

New Hamiltonian variables

Lee Smolin

Department of Physics, Syracuse University,
Syracuse, NY 13244-1130 USA

The present status of the new variables program for canonical quantum gravity is discussed. A summary is given of the papers which were presented at the New Variables Workshop at GR-12, and particular attention is given to those issues which are crucial at the present stage of development of this program. Chief among these issues is whether a theory can be quantized nonperturbatively in the absence of any information about the physical observables algebra of the corresponding classical theory. Finally, a wild speculation about the relationship between nonperturbative quantum general relativity and perturbative string theory is made.

Introduction

Abhay Ashtekar introduced his new variables in the Fall of 1985. In the intervening four years this new formalism has been developed and applied to both classical and quantum general relativity. On both sides significant new results have been achieved, and these developments are being actively pursued by a growing number of people. It was thus appropriate to have a workshop at GR-12 dedicated to this subject. In this, a summary of that workshop, I will try to describe briefly several of the directions that this work has taken, with an emphasis on the present status and open problems. In doing so I will touch on each of the six presentations that were given in the workshop, but I will not strictly follow the format of the workshop itself.

449

In this report I will assume that the reader has read the article of Ashtekar in this volume.[1] The basic definitions of the variables and constraints of the new variables formalism are given there and in Ref. 2.

Before beginning I would like to emphasize what I think is an obvious, but important, point. Ashtekar's new variables have turned out to be useful for a variety of problems because they simplify the structure of Einstein's equations, both in the canonical and in the four dimensional formalism. As such, it is likely that they will be useful for a great many problems in gravitational physics, many of which have yet to be touched on. Up to now, a rather small group of people have been actively using them, and their interests have certainly flavored the subject as it has so far developed. However, I do not believe that one has to share the interests and prejudices of this small group to find the new variables a useful tool.

I will discuss first some applications of the new variables to problems in classical general relativity. After this I will make some remarks about the role of classical general relativity in the quantum theory, as a prelude to a discussion of the recent progress in quantum general relativity. I close by taking advantage of the unrefereed format to record an implausible but, to me, intriguing speculation about the relationship between nonperturbative quantum general relativity and perturbative string theory.

Developments in the classical theory

In a very general sense, the idea behind the new variables is to express the theory in a form which is "close to" a "solvable" sector of the theory. In perturbatively sensible theories, this solvable sector is the free theory. In general relativity, there is a large nonlinear sector which is still, in some sense "solvable." This is the self-dual sector of the theory. We know from the developments of twistor theory, and the work of Newman and collaborators, that the self-dual sector of Einstein's equations are solvable in the sense that the general solution can be written in closed form in terms of certain complex manifolds. (Whether the self-dual sector is actually integrable, in the sense of there being an infinite number of conserved quantities which are involutions is an important unsolved problem.) The Ashtekar formalism[2,3] was then partly motivated by the desire to express the canonical theory in terms of variables in which the simplifications which occur, when the theory is restricted to the selfdual sector, are manifest. The miracle was that it then turned out to be the case that not only the Hamiltonian formalism, but also the Lagrangian and the equations of motion, for the *full theory* simplify significantly, when expressed in terms of variables which are natural for the description of the self-dual sector.

This fact is brought out very strongly in some results which Samuel presented during the workshop. I will list these results here, the proofs are completely straightforward, and can be found in Ref. 4.

Let $^4F_{ab}^i$ be the (four dimensional) self-dual curvature defined after Eq. (4) of Ashtekar's contribution. Let the self-dual connection A_a^i be related to a spacetime metric g_{ab} through Ashtekar's Eq. (3). Then,

I) $^4F_{ab}^i$ is selfdual with respect to g_{ab} if and only if g_{ab} is an Einstein metric satisfying $R_{ab} = \lambda g_{ab}$ for some value of the cosmological constant λ. Here selfdual means $F_{ab}^i = \frac{i}{2}\epsilon_{ab}^{cd}F_{cd}^i$. It is intriguing that the condition that $^4F_{ab}^i$ be self-dual restricts the metric to being Einstein, without determining the value of the cosmological constant.

Samuel also told us two additional results which hold when the cosmological constant is non-vanishing.

II) Let $\epsilon^{abc}F_{bc}^i = \lambda \tilde{E}_i^a$, where here F_{ab}^i is the three dimensional self-dual curvature, λ is the cosmological constant, ϵ^{abc} is the *three dimensional* Levi-Civita tensor and $\tilde{E}_i^a$ is the triad field defined in Ashtekar's contribution. Then: 1) Einstein's equations are satisfied, with λ the cosmological constant; and 2) the Weyl tensor is anti-selfdual.

III) Every solution to Einstein's equations with non-vanishing cosmological constant and antiself-dual Weyl tensor is of the form $\epsilon^{abc}F_{bc}^i = \lambda \tilde{E}_i^a$. To this last result can be added a result which was found by Capovilla, Dell and Jacobson just after the Boulder conference. In a paper with several fundamental and new results[5] they show that for the case of vanishing cosmological constant the *general solution to the vector and scalar constraints* (the second and third equation of Ashtekar's Eqs. (4)) is given by $\tilde{E}_i^a = M_j^i \epsilon^{abc} F_{bc}^j$, where M_j^i is expressed in terms of a tracefree symmetric tensor field, ϕ_j^i as $M_j^i = \phi_k^i \phi_j^k - \frac{1}{2}Tr\phi^2 \delta_j^i$.

If one contemplates these results one cannot, I think, avoid the impression that they express something profound about the structure of the Einstein's equations. The problem, and I think it is one of the most important open problems at present, is to understand exactly what that something is.

Another aspect of this whole situation is that the self-dual Einstein equations can themselves be written in a very simple form when expressed in terms of Ashtekar's variables.[6] This form is so simple that it can be expressed in a few lines. Let us consider a four dimensional manifold with topology $\Sigma \times I\!R$, where Σ is a three manifold, which we will assume is endowed with a fixed, non-dynamical volume element ϵ. (The choice of this is pure gauge.) On $\Sigma \times I\!R$ let us have three vector fields $V_i^a(t)$, where a is a spatial index, i labels the three vector fields and t is a coordinate on $I\!R$. Then the self-dual Einstein equations are equivalent to three constraint equations and an evolution equation on the $V_i^a(t)$. The constraint equations are simply that the $V_i^a(t)$ be divergence free with respect to ϵ,

$$\partial_a \epsilon V_i^a = 0.$$

The evolution equation is,

$$\frac{\partial V_i^a}{\partial t} = \epsilon^{ijk}[V_j, V_k]^a,$$

where the bracket is just the Lie bracket.

Recently, this result has been greatly expanded and exploited by Mason and Newman[7] in work that Ted Newman presented in the workshop. They showed how both the self-dual and the full Einstein equations can be expressed as a Yang-Mills theory in which the usual compact Lie algebra is replaced by the infinite dimensional algebra of vector fields on a four manifold. That is, they consider a gauge theory on a fixed manifold in which the gauge group (actually the Lie algebra) is the diffeomorphism group of another four dimensional manifold. Again, one has the feeling with this result that something important is being revealed about the structure of the Einstein equations, the question, again, is what?

The role of classical general relativity in the quantum theory

As our understanding of both classical and quantum general relativity grows, I think it is important to keep in mind something that was stressed by Professor Ferrari during her talk at the conference.[8] This is that the Einstein equations are not just some generic non-linear field theory. General relativity is not hydrodynamics. A great many special solutions are known, the study of which teaches us much about the characteristic behavior of the full, non-linear, gravitational field that we could not learn from perturbation expansions. Furthermore, the theory contains large non-linear sectors which are integrable. For example, besides the possibility I mentioned that the self-dual sector is integrable, the two Killing field sector is known to be integrable. What the Ashtekar variables are telling us is that, when expressed in terms of the right variables, the non-linearities have a much simpler structure than previously thought. It also seems to me very possible that the Ashtekar variables are not the last word, that with further work we will find further simplifications of the Einstein equations.

Now, I think that the same lesson may hold for the quantum theory. That is, it seems to me likely that we will only make progress on the construction of quantum general relativity to the extent that we use nontrivial information about the dynamics of general relativity. To make this point, I would like for the purposes of this discussion to divide methods of quantization of field theories into two classes, generic and special. Generic methods of quantization are those methods in which no special information about the dynamics of the theory are used. The two most important examples are perturbation theory and Monte-Carlo simulations. When these methods succeed, they succeed without using any special knowledge about the structure of solutions or the algebra of observables of the classical theory.

The generic methods have turned out to be very useful for a variety of theories, but I would like to suggest here that there may be theories for which a quantization exists but for which the non-linearities of the classical theory are important enough that the generic methods will not succeed. The best examples of these are the integrable field theories. In an integrable field theory there exists on the physical phase space, or space of solutions, a physical observables algebra which is complete— that is, a set of physical observables whose Poisson algebra closes is known which together give a good set of coordinates on the space of solutions. When one has this one can quantize the theory directly by constructing a representation (up to deformations) of this algebra of observables. This is an example of what I would like to call a special method of quantization, that is a quantization method that uses detailed information about the structure of the solutions of the full, nonlinear, classical theory.

Now, we know that perturbation theory does not work for general relativity and, so far, it is an open question whether brute force numerical simulations will lead to progress (although this is, I believe, an important thing to pursue.) At the same time general relativity is almost certainly not an integrable theory. (Think of the problem of three black holes.) However, I would like to suggest that the only real hope that general relativity exists as a quantum theory lies in the possibility that its behavior is close enough to one of its integrable subsystems that a special method of quantization can be devised for it. That is, we know that a great many of the solutions of classical general relativity are not close to solutions of the linearized theory. For the compact case we expect there to be no exact solutions which remain close to flat spacetime for all time while, for the asymptotically flat case, we expect that such solutions exist only for sufficiently weak initial data. However, what is relevant for the question of the existence of perturbation theory is the high energy behavior, and we expect that at high energies singularities and horizons which have no counterpart in the linearized theory are important. For this reason, it is reasonable (at least many people think so[9]) to hope that the failure of the perturbation theory is not a sign that quantum general relativity does not exist.

On the other hand, it is not easy to make progress with the quantization of a theory which is not an integrable system without the use of perturbation theory. An indication of the magnitude of the difficulty we are facing is the following fact: for classical general relativity, with compact boundary conditions and without matter, *not a single physical observable of the theory is known explicitly*. By a physical observable I mean here a function on the phase space which has (weakly) vanishing Poisson brackets with the constraints, so that it is invariant under four dimensional diffeomorphisms. Now, this is not a surprising fact, for such an observable must be a function of the variables on a three slice of a spatially compact spacetime that is independent of which spatial slice is used. That is, because the gauge group includes the four dimensional diffeomorphism group any gauge invariant observable is also a constant of motion. But it seems a very difficult problem to construct a constant of motion for the Einstein equations expressed as an explicit function of the phase space variables.

However, while our ignorance of the gauge invariant observables of general relativity is not surprising, it is also, I would like to claim, a very serious problem for any attempt at a non-perturbative quantization of the theory. The reason is the following. Suppose someone arrived from Mars claiming to have a non-perturbative quantization of general relativity, without matter, for the case of spatially compact topology. What would we want to check to verify their claim? What they would be handing us would be a Hilbert space, together with an algebra of operators, which satisfied certain Hermiticity conditions with respect to the inner product of the Hilbert space. Since there is no Hamiltonian we cannot check anything about evolution. Furthermore, what they give us is only the space of physical states and the algebra of physical operators; on both of these the constraints have a trivial action, so there is nothing concerning the constraints we could check. To check if the alleged quantization is correct we need a correspondence between at least a subalgebra of their physical operator algebra and a subalgebra of the classical physical observables algebra. If we have such a correspondence we can check two things: that the commutator algebra of the quantum observables is a deformation of the Poisson algebra of the classical observables and that the Hermiticity conditions satisfied by the quantum operators reflect the reality conditions satisfied by the corresponding classical observables.

This is then the point: without any information about at least a subalgebra of the physical observables algebra of the classical theory we cannot check either of these. Thus, without this information we would be unable to verify our Martian friend's claim!

I emphasize this point because this is not far from the situation we are actually in, with respect to recent results in quantum general relativity.

Developments in quantum general relativity

The use of the new variables has led to progress in quantum general relativity, because it is now possible to find exact solutions to the quantum constraint equations. Such exact solutions have been found, and are being studied, both in the full theory[10,11] and in certain models which result from truncating the degrees of freedom of general relativity. In the workshop we had several presentations describing these developments. I will summarize here only the main points of these presentations, and refer the reader to the original papers for the details.

In the full theory the approach which has, so far, been the most fruitful is the loop representation approach. In the workshop this was described by Rovelli, however as Ashtekar has already described the basic ideas of this approach in his contribution to the conference, I will limit myself here to a statement of exactly what has and what has not, as of this time, been achieved with this approach.[11]

What has been done: 1) The loop representation gives a representation of the unconstrained observable algebra, which is fully regulated. Most importantly, the regulated operator algebra is spatially-diffeomorphism covariant, it carries a representation of the exact spatial diffeomorphism group. 2) The spatial diffeomorphism constraints are well defined when acting on the representation space of the regulated theory, and the general solution to the diffeomorphism constraints can be found in terms of an explicit countable basis. This basis is in one to one correspondence with the diffeomorphism equivalence classes of sets of piecewise smooth loops in the spatial manifold. These equivalence classes are called generalized link classes. 3) The Hamiltonian constraint can be regulated and ordered in such a way that an infinite dimensional space of solutions is found to the regularized constraints. This means that they are solutions to the equation,

$$\lim_{\epsilon \to 0} (\mathcal{C}^{\epsilon} |\Psi >) = 0,$$

where $\mathcal{C}^{\epsilon}$ stands for the regulated Hamiltonian constraint with regulated ϵ. Note that no inner product is involved in this equation; this is because the physical inner product in Dirac quantization is imposed only on the space of solutions to the constraints. 4) A simultaneous set of exact solutions to the Hamiltonian and diffeomorphism constraints can be found. There exists, first of all, a space of such solutions which has a countable basis which is in one to one correspondence with the ordinary link classes of the three manifold. (These are the diffeomorphism equivalence classes of smooth, nonintersecting loops.) Some additional sets of solutions involving intersections are known. These include the cases in which two and three loops intersect at a point.

What is not known: 1) The general solution to the Hamiltonian constraint is not known. Certainly there is a large phenomenology of solutions involving multiple loops and multiple intersections; this has only begun to be explored, and it seems likely to be important for the eventual physical interpretation of the solutions. However, it is also not known to what extent the solutions which can be found exactly using the techniques so far developed span the full space of solutions. 2) While an infinite set of physical operators can easily be constructed, because we have a subspace of the physical state space in terms of an explicit basis, we do not know what classical observables these correspond to. This is because, as I stressed above, we do not have any information about the classical physical observables. 3) Because we know nothing about the correspondence between the quantum operators and the physical classical observables, we do not have any information concerning the reality conditions which would allow us to choose the inner product for the theory. Thus, we have no way of selecting out of the infinite number of possible inner products on our space of solutions the correct physical inner product.

Thus, it should be clear why I stressed the issues concerning our ignorance of the classical physical observables. Without them it is not clear how to proceed to construct a quantum theory of gravity from the solutions to the constraints we have found. The main challenge faced by this approach is then to find some way to give a physical interpretation to some operators, on the space of solutions to the constraints which have been found. Work aimed at doing this is proceeding in two directions. 1) Extend the formalism to the asymptotically flat case, in which some physical observables are known such as energy, momentum and angular momentum.[12] 2) Couple the theory to matter, which, again, allows us to construct some physical observables.[13] There is also a third direction, which seems very intriguing, but on which no progress has been made. Knot theory occurs ubiquitously in the mathematics of conformal field theories. It is very interesting to speculate that the occurrence of knot theory in quantum general relativity, as well, is not a coincidence, and that conformal fields could somehow be used to help construct a physical interpretation for the quantum states of the gravitational field.

I now turn to a discussion of the results which have been achieved with respect to the quantization of truncations of general relativity. At the workshop, H. Kodama presented a paper in which he showed that the dynamics of the Bianchi models take a very simple form when they are expressed in terms of the new variables.[14] He also was able to find some exact solutions to the quantum constraint of the Bianchi IX model. This work has stimulated interest in using the new variables to study both the classical and quantum mechanics of the Bianchi models. Some further work about the Bianchi models is discussed in Ashtekar's contribution.[15]

As I mentioned above, the two Killing field reductions of general relativity are a non-linear, but integrable, sector of the theory. This sector of the theory has been studied from many points of view, both classically and quantum mechanically.[16] In the workshop, V. Husain presented work in which cosmological models with two spatial Killing fields were expressed in terms of the new variables.[17] The results in this case, so far, are essentially the same as the results I summarized above, for the full theory. However, it is interesting to note that in this case the algebra of the physical operators on the space of solutions found using the loop representation form a $GL(2)$ Kac-Moody algebra. This is very intriguing and given this result, the integrability of the theory, and the many results known concerning the classical theory, it seems likely that much more can be accomplished. It seems, in fact, possible to hope that in this case the physical observables and inner product can be found, thus completing the program of canonical quantization. If progress of this sort can be achieved we will have model quantum cosmologies which have infinite numbers of degrees of freedom, but are still tractable. This could provide nontrivial tests of many of the ideas which have been proposed concerning quantum gravity and quantum cosmology.

Another development which was not described in the workshop, but which has been very useful for our understanding of the whole new variable program has been the study of quantum gravity in $2+1$ dimensions. Although general relativity in $2+1$ dimensions has been studied for many years,[18] it was only realized recently that the quantum theory is, in this case, exactly solvable. This was realized for the case of the toroidal topology by Martinec,[19] and then in complete generality by Witten.[20] Witten writes the dynamics of $2+1$ general relativity in a form which is closely analogous to the Ashtekar variables for the $3+1$ theory, and then shows that the quantum mechanics for any spatial topology is exactly solvable using a representation in which the states are functions of the spacetime connection. This has stimulated a number of interesting works, but there is a great deal more that remains to be done with quantum gravity in $2+1$ dimensions.

The last presentation of the workshop, by Torre, differed from the others in that the emphasis was on approximate rather than exact solutions to quantum gravity using the new variables. In particular, Torre described work which concerned the strong coupling expansion of the theory, using the new variables.[21] The strong coupling *limit*, which is the limit in which G_{Newton} is taken to infinity, has been studied in quantum gravity from a number of different points of view[22] including that of the new variables.[23] The work that Torre described was, however, aimed at producing a strong coupling *expansion* for the quantum theory. Torre showed in his talk how to develop such an expansion by making an expansion in powers of $1/G_{Newton}$ of the phase space BRST path integral. It is interesting to note that, because of the way that the gravitational constant occurs in the definition of the Ashtekar connection A_a^i, the strong coupling expansion in terms of the new variables is different than the expansion in the metric representation.

A wild speculation

I would like to close this summary, as I closed my remarks introducing the workshop at Boulder, with a remark to which I find my own thoughts repeatedly returning, in spite of its apparent implausibility. Several people who are familiar with both string theory and the recent work involving the new variables and the loop representation have remarked that there is more than one possible point of contact between the two developments. These are: 1) The fundamental role that observables parametrized in terms of loops rather than points play in both formalisms; 2) the fact that the quantum states of linearized Maxwell theory and linearized general relativity are products of functions which are identical in form to the vertex operators that create linear spin one and spin two excitations in perturbative string theory;[24] 3) The fact that knot theory appears as a basic tool to classify the physical states of both theories. This last point might be said more definitely as follows: string theory vacuum states are described by conformal field theories. There is a large class of conformal theories which are known to be associated in a one to one fashion with link invariants.[25] Given any link invariant one may construct an exact solution to the quantum constraints of general relativity. Thus there seems to be a mapping between the space of exact physical states of general relativity and vacuum states of string theory (or, what is the same thing, perturbative string theories.)

Reflecting on these points of contact, it is tempting to conjecture that at the non-perturbative level, general relativity and string theory may be more closely related than one would guess from the perturbative formalisms. This impression is strengthened if one recalls that, as we were taught several years ago by Friedman and Sorkin,[26] there are quantum states of general relativity in the asymptotically flat case that have either integral or half-integral spin. Suppose now that it is the case that, for reasons of consistency, any sensible perturbative description of fields whose classical limit includes perturbative classical general relativity must be a string theory. Suppose also that non-perturbative quantum general relativity exists in the asymptotically flat case, and has a spectrum of states that includes excitations of various spins and masses (as we may expect to be the case given both the work of Friedman and Sorkin and the general expectation that there will be quantum states of non-perturbative quantum gravity representing excitations whose classical description would be black holes of various masses, spins and charges.) If both of these suppositions are true then any perturbative description of the scattering of these excitations will be a string theory. One may note that the obvious objection to this view, that general relativity and string theory describe dynamics in spaces of different dimensionality, is now moot, given the discovery of consistent string theories which exist in four dimensions. These theories have a very complicated spectrum of states, but then, we expect that the spectrum of states in quantum general relativity will also be very complicated. Thus, to put it another way, one of the basic questions in string theory is to find a non-perturbative dynamics for the theory. Is it possible that the non-perturbative dynamics that underlies at least some perturbative string theories is general relativity itself?

My understanding of this area has been shaped by discussions with many people, including Abhay Ashtekar, Ted Jacobson, Carlo Rovelli and the contributors to both the workshop and the associated poster session.

References

1. A. Ashtekar (1990), Chapter 14 in this volume.
2. A. Ashtekar (1988), *New Perspectives in Canonical Gravity*, with invited contributions by T. Jacobson, P. Renteln, D. C. Robinson, C. Rovelli, L. Smolin and C. Torre (Bibliopolis, Naples).
3. A. Ashtekar (1986), *Phys. Rev. Lett.* **57**, 2244 and (1987), *Phys. Rev.* **D36**, 1587.
4. A. Ashtekar and J. Samuel, in preparation. Related results are also described in *New Perspectives... op. cit.*; P. Renteln (1988), Harvard Ph. D. thesis, and R. Capovilla, J. Dell and T. Jacobson (1990), *Class. Quant. Grav.* **7** L1. These results are also described by S. Koshti and N. Dadhich (1990), *Class. Quant. Grav.* **7**, L5.
5. R. Capovilla, J. Dell and T. Jacobson (1989), *Phys. Rev. Lett.* **63**, 2325.
6. A. Ashtekar, T. Jacobson and L. Smolin (1988), A new characterization of half-flat solutions to Einstein's equations, *Commun. Math. Phys.* **115**, 631-648.
7. L. Mason and E. T. Newman (1989), *Commun. Math. Phys.* **121**, 659.
8. V. Ferrari (1990), Chapter 1 of this volume.
9. O. Klein (1955), in *Niels Bohr and the Development of Physics*, eds. W. Pauli, L. Rosenfeld and V. Weisskopf (Pergamon, Oxford, 1955); L. Landau, in *Niels Bohr and the Development of Physics, op. cit.*; W. Pauli (1956), *Helv. Phys. Acta Suppl.* **4**, 69; S. Deser (1957), *Rev. Mod. Phys.* **29**, 417; R. Arnowitt, S. Deser and C. W. Misner (1960), *Phys. Rev.* **120**, 313; B. S. DeWitt, *Phys. Rev. Lett.* **13**, 114; C. J. Isham, A. Salam and J. Strathdee (1971), *Phys. Rev.* **D3**, 1805 and (1972), *Phys. Rev.* **D5**, 2548.
10. T. Jacobson and L. Smolin (1988), *Nucl. Phys.* **B299**, 583.
11. C. Rovelli and L. Smolin (1988), *Phys. Rev. Lett.* **61**, 1155; preprint (1988), to appear in *Nucl. Phys. B*.
12. C. Rovelli and L. Smolin, in progress.
13. C. Rovelli (1989), *What is observable in classical and quantum gravity?*, Trieste and Pittsburgh preprint, 1989.
14. H. Kodama (1988), *Prog. Theor. Phys.* **80**, 1024.

15. A. Ashtekar and J. Pullin (1989), Syracuse University preprint.
16. A very incomplete selection of works on various aspects of two Killing field space times is: R. Geroch (1971), *J. Math. Phys.* **12**, 918 and (1972), *J. Math. Phys.* **13**, 394; I. Hauser and F. J. Ernst (1980), *J. Math. Phys.* **21**, 1126 and (1981), *J. Math. Phys.* **22**, 1051; Y. S. Wu and M. L. Ge (1983), *J. Math. Phys.* **24**, 1187; Y. S. Wu (1983), *Phys. Lett.* **A96**, 179; K. Kuchař (1971), *Phys. Rev.* **D4**, 955; B. K. Berger (1974), *Ann. Phys. (N.Y.)* **83**, 458 and (1984), *Ann. Phys. (N.Y.)* **156**, 155; V. Moncrief (1981), *Phys. Rev.* **D23**, 312 and (1981) *Ann. Phys. (N.Y.)* **132**, 87; S. Chandrasekhar and V. Ferrari (1984), *Proc. Roy. Soc. Lond.* **A396**, 55.
17. V. Husain and L. Smolin (1989), Exactly solvable quantum cosmologies from two Killing field reductions of general relativity, *Nucl. Phys.* **B327**, 205-238.
18. S. Deser, R. Jackiw and G. 't Hooft (1984), *Ann. Phys. (N.Y.)* **152**, 220; G. 't Hooft (1988), *Commun. Math. Phys.* **117**, 685; S. Deser and R. Jackiw (1989), *Commun. Math. Phys.* **118**, 495; P. de Sousa Gerbert and R. Jackiw, MIT preprint CTP **1594**; J. Nelson and T. Regge (1989), *Nucl. Phys.* **B328**, 190.
19. E. Martinec (1984), *Phys. Rev.* **D30**, 1198.
20. E. Witten (1988), *Nucl. Phys.* **B311**, 46.
21. C. Torre (1988), *Class. Quant. Grav.* **5** L63.
22. D. Eardley, E. Liang and R. Sachs (1972), *J. Math. Phys.* **13**, 99; C. J. Isham (1984), in *Relativity, Groups and Topology II*, eds. B. S. DeWitt and R. Stora (North Holland, Amsterdam); M. Pilati, (1982) *Phys. Rev.* **D26**, 2645 and (1983), *Phys. Rev.* **D28**, 729; C. Rovelli (1987), *Phys. Rev.* **D35**, 2987; J. N. Goldberg (1988), *Gen. Relat. Grav.* **20**, 881.
23. A. Ashtekar, in *New Perspectives, op. cit.*; V. Husain (1988), *Class. Quant. Grav.* **5**, 575.
24. A. Ashtekar and C. Rovelli, *Quantum Faraday lines: Loop quantization of the Maxwell field*, preprint in preparation; A. Ashtekar, C. Rovelli and L. Smolin, *Loop representation for linearized quantum gravity*, preprint in preparation.
25. E. Witten (1989), *Commun. Math. Phys.* **121**, 351; M. F. Atiyah (1988), in *The Mathematical Heritage of Herman Weyl*, Proc. Symp. Pure Math. **48**, ed. R. Wells (Amer. Math. Soc.).
26. J. Friedman and R. Sorkin (1980), *Phys. Rev. Lett.* **44**, 1100; (1980), *Phys. Rev. Lett.* **45**, 148; (1982), *Gen. Relat. Grav.* **14**, 615.

D3

Quantum cosmology
and baby universes

Leonid P. Grishchuk
Sternberg Astronomy Institute,
Moscow University,
119899 Moscow V-234, USSR

The contributed papers presented to the GR-12 workshop on "Quantum Cosmology and Baby Universes" have demonstrated the great interest in, and rapid development of, the field of quantum cosmology. In my view, there are at least three areas of active research at present. The first area can be defined as that of practical calculations. Here researchers are dealing with the basic quantum cosmological equation, which is the Wheeler-DeWitt equation. They try to classify all possible solutions to the Wheeler-DeWitt equation or seek a specific integration contour in order to select one particular wave function or generalize the simple minisuperspace models to more complicted cases, including various inhomogeneities, anisotropies, etc. The second area of research deals with the interpretational issues of quantum cosmology. There are still many questions about how to extract the observational consequences from a given cosmological wave function, the role of time in quantum cosmology, and how to reformulate the rules of quantum mechanics in such a way that they could be applicable to the single system which is our Universe. The third area of research is concerned with the so-called "third quantization" of gravity. In this approach a wave function satisfying the Wheeler-DeWitt equation becomes an operator acting on a Wave Function of the many-universes system. Within this approach one operates with Euclidean worm-holes joining different Lorentzian universes. This is, perhaps, one of the most fascinating, although not entirely clear, subjects considered recently.

The papers presented orally at the workshop covered all of these areas. The talk given by J.J. Halliwell described the path-integral representation of the wave function of the universe. The objective was to find the complex contours satisfying a number of reasonable criteria. It was shown that the chosen criteria are still not sufficient to fix the contour (and the wave function) uniquely. In particular, it was emphasized that there are many possible contours corresponding to the "no boundary" proposal of Hartle and Hawking.

The talk by M. Morikawa was devoted to the "quantum field theory of the Universe." The transition from the "first-quantized" to the "second-quantized" level of a theory formulated in a fixed space-time was, by analogy, extrapolated to the quantum cosmology. In this way the speaker reached the conclusion that the multiple production of baby universes is an inevitable consequence of such a "field" theory of the Universe. He went so far as to introduce a detector for the particles-universes in analogy with the Unruh detector for particles which could be detected in accelerated frames.

A.D. Linde spoke about the life after inflation and the cosmological constant problem. He argued that if the present-day vacuum energy density exceeded some extremely small critical value, the lifetime of any civilization should be finite despite the endless existence of the Universe as a whole. The possible importance of the one or another form of the anthropic principle was discussed.

The paper presented by S. Sinha dealt with the validity of the minisuperspace approximation in quantum cosmology. It was shown that under some conditions the minisuperspace cosmology (where all degrees of freedom are neglected except a few) can be regarded as the infrared limit of the full scale quantum gravity theory (where all degrees of freedom are treated on an equal footing.)

K. Maeda described work on quantum cosmological generality of inflation in the Bianchi type IX spacetimes. In other words, the anisotropies were included in the set of superspace variables. The conclusion of this work was that the Hartle-Hawking prescription for constructing the wave function of the Universe predicts mostly the isotropic universes while the Vilenkin prescription is likely to assign the largest probability to anisotropic universes.

W.-M. Suen presented a paper where the Universe was treated as a "leaking" system. In many respects the proposed wave function is similar to the so-called "tunneling" wave functions, though it seems to be differently defined. The author uses the definition according to which the integration is performed over all nonsingular Lorentzian 4-

geometries which include the arguments of the wave function on one of their boundaries.

The talk by C. Kiefer was devoted to the intrinsic time and the emergence of classical properties in quantum gravity. The main emphasis was on the notion of how classical properties emerge through the interaction of a quantum system with its environment. The author argued that in quantum gravity the influence of the "environment" (inhomogeneous and anisotropic degrees of freedom) on minisuperspace variables can be seen in the form of the density matrix $\rho(a, a')$ resulting from averaging over the additional degrees of freedom. The argument is that the interference between different scale factors a and a' is suppressed and that this suppression becomes more effective with increasing a.

The importance of the unitarity requirement was stressed by T.A. Jacobson in his talk. An example was suggested in which nonunitary evolution, although confined to a spatially localized region remote from the region where measurements are being made, can lead to predictions which are ambiguous or acausal. It was concluded that to abandon unitarity in attempting to interpret quantum gravity is likely to lead to ambiguity or causality violation at the macroscopic, semi-classical level.

The worm-hole solutions for Yang-Mills fields coupled to gravity were discussed by Y. Verbin. The lower and upper limits for the numerical values of the action were considered as well as possible applications of the solutions to the problem of tunneling between various vacua.

In conclusion, I would like to say that although there is still a lack of understanding of some conceptual issues of quantum cosmology as well as its relationship with the "less" and "more" quantized neighboring theories, it seems to me that a coherent picture is gradually emerging and quite possibly may be presented by someone at the next GR meeting.

D4

Quantum field theory
in curved space-time

Jürgen Audretsch
Universität Konstanz,
Fakultät für Physik,
D-7750 Konstanz 1, Federal Republic of Germany

In order to enable serious discussions, to allow the speakers to describe their subject in detail and give self-contained reviews, only three talks were scheduled for the workshop. The other submitted papers which partly contained substantial contributions to quantum field theory in curved space-time and its applications, have been presented to the participants through the abstracts and posters.

Coarse-grained effective action

In many studies of quantum fields in dynamic space-times one often needs to treat the high frequency and the low frequency normal modes differently. One familiar example in early universe quantum processes is in the cosmological particle creation backreaction problem (Lukash and Starobinsky (1974); Hu and Parker (1978)), where a division is made at the non-adiabatic limit for each mode to distinguish (quantum) particle creation from (classical) red-shifting effects. Another is the recently proposed stochastic inflation model where a cut-off of the fluctuation field momenta at the horizon introduces a Markovian noise source (Starobinsky (1986); Rey (1987)). However, these simple cut-off procedures cannot account for the mixing of high and low frequency modes due to non-linear interaction or nonadiabatic effects.

In his talk B. L. Hu proposed a coarse-grained effective action for these problems. Most work on coarse-grained free energy is done in critical phenomena physics carried out by condensed matter physicists.

467

After briefly covering these approaches, B. L. Hu turned to his work done in collaboration with Yuhong Zhang. Their coarse-grained effective action is based on the iteration of a momentum space cut-off followed by a rescaling. It incorporates the backreaction of the high frequency modes on the low frequency modes in a natural way (as in the averaging of high-frequency gravitational waves done e.g., by Isaacson (1968) or MacCallum and Taub (1973)). The same procedure can also be used to describe the effect of fast variables on the slow variables of a nonlinear system (as in time-dependent Hartree-Fock calculations, see e.g., Balian and Veneroni (1981).

For the inflationary cosmology, B. L. Hu showed that the universe in the inflation regime manifests scaling behavior. Using the differential renormalization group equations constructed from each iteration, he derived the time variation of the field parameters as the system evolves to the infrared regime. This method can also be used to discuss entropy generation and quantum decoherence problems for dynamical fields.

Quantum violations of energy conditions

That quantized fields can violate classical energy conditions such as the weak energy condition was one of the motivations to study quantum field theory in curved space-time. The Casimir effect demonstrates that these violations arise even for free fields in flat space-time. Analogous vacuum energy effects in curved space-time are known.

It is perhaps less well known that negative energy densities can arise through quantum coherence effects, i.e., effects due to the choice of a quantum state that is not an eigenstate of the particle number. L. Ford reviewed this latter type of energy condition violation effect, gave a summary of earlier work on it and discussed some simple examples. Some years ago, it was argued that there should be inequalities which constrain the magnitude of any negative energies due to quantum coherence effects (see Ford 1978). In the particular case of a flux of negative energy in two-dimensional flat spacetime, it was conjectured that the mean magnitude of the flux, F, and its duration in time, t, obey an inequality of the form $|F| < t^{-2}$. This inequality was shown to be satisfied for particular choices of the quantum state. It appears to constrain the negative energy sufficiently to avoid macroscopic violations of the second law of thermodynamics. In the subsequent years, there has been a considerable amount of discussion in the literature about the significance and range of validity of this inequality. Summing up this discussion, Ford stressed that at present no general proof of the universality of the inequality exists. However he presented a heuristic argument supporting the notion that there should be inequalities of this form, and which may serve as the basis for future attempts to find a general proof.

Renormalized stress tensors and black holes

A. C. Ottewill began his talk in mentioning, that Candelas (1980) was the first to calculate the value of the renormalized stress tensor for a conformally invariant scalar field on the horizon of a Schwarzschild black hole. Candelas' argument was later repeated by Elster (1984) for the case of the electromagnetic field in a Schwarzschild spacetime, and Elster's result was subsequently confirmed by Frolov and Zel'nikov (1985) as the non-rotating limit of their expression for the renormalized electromagnetic stress tensor at the pole of a Kerr black hole.

To investigate the related approximations Jensen and Ottewill (1989) performed the numerical evaluation of the electromagnetic stress tensor in the Hartle-Hawking vacuum following the methods of Howard (1984). Their numerical results led Jensen, McLaughlin and Ottewill (1987) to discover a deviation from the calculations of Elster, Frolov and Zel'nikov. The latter authors neglected a linearly divergent term in the point-separation renormalization they used, which Christensen (1978) had directed users to eliminate by averaging. A. C. Ottewill claimed that the correct result for the renormalized energy density of the horizon is $-19\pi^2/7680M^4$. The negativity of this quantity invalidates the blackhole evaporation halting mechanism of Balbinot and Barletta (1988). A natural analytic approximation was found which agreed with the above result extremely well but was none of the previously suggested approximations.

A. C. Ottewill continued his talk in pointing out that any discussion of back reaction in quantum field theory in curved space-time should also include the effect of massive fields and the effect of linear gravitons, which contribute to the one-loop effective action a term of the same order as those from ordinary matter fields. Anderson (1989) has calculated $< \Phi^2 >$ for massive fields and used this to calculate $< T_\mu^\mu >$. The formalism for the calculation of the graviton stress tensor was given in Allen, Folacci and Ottewill (1988). The back-reaction problem would also require an evaluation of the renormalized stress tensor inside the black hole, as discussed by Poisson and Israel (1988). The value of $< \Phi^2 >$ inside the black hole was calculated by Candelas and Jensen (1986).

Preliminary work towards evaluating the renormalized stress tensor in a rotating black-hole space-time is contained in Frolov and Thorne (1989). However, the physical significance of these results is not yet clear given the theorems of Kay and Wald (1989) which suggest that there is no non-singular analogue of the Hartle-Hawking state in this case.

An intensive discussion incorporating other aspects of quantum field theory in curved space-time concluded the workshop.

References

Allen, B., Folacci A. and Ottewill, A. C. (1988), *Phys. Rev.* **D38**, 1069.

Anderson, P. R. (1989), $< \Phi^2 >$ for Massive Fields in Schwarzschild Spacetime, Montana State Univ. preprint, to appear in *Physical Review* **D**.

Balbinot, R. and Barletta A. (1988), *Class. Quant. Grav.* **5**, L11.

Balian, R. and Veneroni, M. (1981), *Annals of Physics* **135**, 270.

Candelas, P. (1980), *Phys. Rev.* **D21**, 2185.

Candelas, P. and Jensen, B. P. (1986), *Phys. Rev.* **D33**, 1596.

Christensen, S. M. (1978), *Phys. Rev.* **D17**, 946.

Elster, T. (1984), *Class. Quant. Grav.* **1**, 43.

Ford, L. H. (1978), *Proc. Roy. Soc. Lond.* **A364**, 227.

Frolov, V. P. and Zel'nikov, A. I. (1985), *Phys. Rev.* **D32**, 3150.

Frolov, V. P. and Thorne, K. S. (1989), *Phys. Rev.* **D39**, 2125.

Howard, K. W. (1984), *Phys. Rev.* **D30**, 2532.

Hu, B. L. and Parker, L. (1978), *Phys. Rev.* **D17**, 933.

Isaacson, R. (1968), *Phys. Rev.* **166**, 1263, 1272.

Jensen, B., McLaughlin, J. and Ottewill, A. C. (1988), *Class. Quant. Grav.* **5**, L187.

Jensen, B. P. and Ottewill, A. C. (1989), *Phys. Rev.* **D39**, 1130.

Kay, B. and Wald, R. M. (1989), Uniqueness of stationary non- singular quasi-free states in spacetimes with a bifurcate Killing horizon, *Class. Quant. Grav.*, to appear.

Lukash, V. N. and Starobinsky, A. A. (1974), *Sov. Phys. JETP* **32**, 742.

MacCallum, M. A. and Taub, A. H. (1973), *Comm. Math. Phys.* **30**, 153.

Poisson, E. and Israel, W. (1988), *Class. Quant. Grav.* **5**, L201.

Rey, S. J. (1987), *Nucl. Phys* **B284**, 706.

Starobinsky, A. A. (1986), in *Field Theory, Quantum Gravity and Strings*, eds. H. J. de Vega and N. Sanchez (Springer-Verlag, Berlin), **246**, 107.

D5

Theories of Quantum Gravity II (not superstring theory)

Arthur Komar

Department of Physics, Yeshiva University,
500 W. 185th Street,
New York, NY 10033 USA

Introduction

Quantum theory and relativity theory, the two great revolutions of twentieth century physics, do not seem to mesh very well. Indeed the only fully consistent relativistic quantum theories seem to be linear free fields. It is not clear whether the difficulty of combining the two theories is one of principle or merely one of practice. On the most pedestrian level, the Hamiltonian formalism of classical mechanics, which is most suitable for quantizing a classical theory, requires an explicit choice of a time-like coordinate and dynamical variable adapted to this choice; relativistic theories, in contrast, treat space and time variables on essentially equal footing and are expressed most naturally in the Lagrangian formalism. Of course, for classical physical theories the two formalisms are equivalent in the sense that one can map from one to the other by means of a Legendre transformation. Thus one can, for example, follow the quantization procedure of a classical theory in the Lagrangian formalism, or demonstrate the relativistic covariance of a theory in the Hamiltonian formalism. But the exercises are decidedly awkward, and, given the ensuing difficulties which occur for non-trivial theories, one does have cause for puzzlement.

On a more fundamental level, the measurement theory for the standard interpretation of the quantum mechanical formalism requires that if an observation of a complete commuting set of local observables is performed by an observer in some finite neighborhood of a space-time

471

point, the state of the system is reduced everywhere, including in regions well outside the forward light cone of that neighborhood. It is hard to see how such a reduction can be understood by another observer in inertial motion relative to the first one if one is obliged by relativity theory to consider the two observers as equally privileged. For linear, Lorentz-covariant theories, one can, and generally does, Fourier analyze the local field variables before quantizing the theory. Thus, the set of variables preferred for quantization is decidedly non-local, which of course muddies the analysis of the relativistic reduction of the wave function. For a field theory of vanishing rest mass, such as electromagnetism, there is no proper position representation having a standard probability interpretation. The non-local Fourier transformed fields are thus particularly adapted for quantization. But although suggestive, it is not clear whether the employment of such non-local field variables is an essential requirement for the construction of a consistent relativistic quantum theory.

When we turn to the consideration of the general relativistic theory of gravitation the difficulties of quantization are further compounded. The theory is essentially non-linear thereby foreclosing the possibility of Fourier analyzing the local field variables in a relativistically sensible fashion. The relativity group of the theory is no longer a finite dimensional Lie group, but rather a gauge-like non-Abelian function group. This has as a consequence that the field variables must satisfy non-linear constraint relations. These constraint relations in turn satisfy a rather intricate closed algebra which, moreover, differs significantly from a Lie algebra in that the analogue of the structure constants depend materially on the field variables. Furthermore it has thus far proven impossible to exhibit any functions of the classical field variables which are invariant under the gauge group of the theory and could therefore qualify as candidates for quantum observables.

But perhaps the most radical conceptual difficulty entailed by a quantum theory of gravity is that the arena of the physics is no longer a space-time. A pure state of such a theory could not contain the full complementary information required to specify a classical Riemannian manifold, whereas a wave packet is unlikely to approximate an acceptable Riemannian manifold in the mean, as there is no way to assure that the mean values of the dynamical variable in the mixed state will satisfy the classical constraint relations even approximately. In other words it is hard to see how such a quantum theory will satisfy an Ehrenfest type correspondence principle. (For a more complete discussion of this difficult point see Komar (1979).)

Recent developments

In view of the many intertwining complications besetting the quantization program for gravitation for more than half century (the first paper on the subject was Rosenfeld (1930)), it is perhaps not surprising that many physicists have by now despaired of finding a quantized version of the traditional Einstein's general relativity. Among the many modifications that have been proposed are increasing the number of dimensions, employing additional field variables, altering the action principle by the inclusion of terms of higher differential order, introducing a non-dynamical classical background metric, abandoning the point-set continuum. Few however question the need for a quantum theory of gravitation, as it is rather evident that the gravitational field has independent degrees of freedom which can transfer energy, momentum and angular momentum to and from material bodies, much as does the Maxwell field.

Of the 34 papers submitted to this session of the GR12 conference, at most only 3 or 4 papers could be recognized as being concerned with the traditional canonical quantization program of general relativity. That is, the standard classical dynamical variables of the Einstein theory are re-expressed in terms of canonically conjugate pairs in phase space, and the dynamics of the theory is contained in the Poisson bracket algebra of four local constraints which are related to the diffeomorphism symmetry of the theory (Dirac (1958)). A key problem in the canonical quantization program is the realization of this algebra of constraints by means of Hermitian operators on a linear vector space. This is necessary to assure that the resulting quantum theory remains covariant under diffeomorphisms, that is, that the quantum theory remain general relativistic. As the Dirac constraints are non-linear–indeed non-polynomial–functions of the canonical variables, the problem of a consistent factor ordering of the quantum dynamical variables occurring in the constraint equations is decidedly non-trivial.

Should this factor ordering be solved, it still remains to find a measure on the linear vector space which will allow a probabilistic interpretation of the physical state vectors. Indeed, since the arena of the resulting quantum theory would presumably be an abstract function space rather

than space-time, it is difficult to envision the guise or form under which
such an interpretation would appear. A question also remains concerning
the uniqueness of the quantization procedure - that is, whether alternate
or modified approaches yield equivalent quantum theories.

Reported results

A paper presented by Kuchař (Utah) reports a consistent factor ordering
for the Dirac constraints. A curious feature of the factor ordering is that
it depends critically on the choice of the space-like foliating surfaces
on which the canonically conjugate dynamical variables are defined. A
change in the choice of foliation causes the constraint algebra to grow
Schwinger anomaly terms. However Kuchař reports that these terms can
be canceled by subtracting a potential from the constraints, an operation
which cannot be employed in the classical theory precisely because it
would destroy the algebra of the Dirac constraints.

Rather than realizing the constraint algebra quantum mechanically,
an alternate quantization procedure would be to solve the classical constraints, thereby obtaining the true independent degrees of freedom of
the gravitational field on a reduced phase space which could then be
quantized in a straight forward fashion. J. Ramano and R. Tate (Syracuse) reported comparing these two methods of quantization for a simple
finite dimensional model. They found that the two different methods of
quantization led to materially different spectra. For the particular case
of the Coulomb potential they found that Dirac procedure gave the
correct hydrogen spectra. Of course the spectra obtained by both procedures agreed in the limit of large quantum numbers, indicating that
both procedures are consistent with the correspondence principle.

A paper by C. Rovelli (Rome) was concerned with the attempt to
recover from a quantized gravitation theory a variable which could be
interpreted as a geometric surrogate for time. He reported that he was
only able to accomplish that by "modifying," that is, abandoning, the
Heisenberg equations of motion for the dynamical variables. But of
course the group theoretic meaning of the Hamiltonian as the generator
of the time translations, is thereby undetermined. This is such a central
requirement for all of physics we must unequivocally reject Rovelli's
conclusion, and thereby his geometric interpretation of time.

H.-H. Von Borzeszkowksi (Potsdam, DDR) reported work of H.-J.
Treder and himself on the measurement problem for the gravitational

field. Their conclusion is that for a meaningful interpretation of quantum gravity it is essential to have a classical background metric against which to refer the measurements. They therefore claim that the only sensible way to construct a theory of quantum gravity is to consider as quantizable dynamic variables only perturbations of the background metric, sometimes called the "covariant quantization procedure." A conspicuous weakness of any measurement theory argument is that one must first know the correct quantum theory before one can deduce the limitations imposed on measurements by the commutation relations. It is not even clear at this time which variables are gauge invariant and hence subject to measurement.

Conclusion

As already noted, the remaining papers presented at this session proposed much more speculative modifications of the Einstein theory of gravitation. We can conclude that the search for a quantum version of the Einstein theory of gravitation is alive if not exactly well. After more than half century of herculean efforts by many of the best scientific minds of our time the problem has resisted solution. One begins to have an uncomfortable feeling that it may not be merely that the technical details of the program are cumbersome and difficult (which they most certainly are), but that perhaps there may be a fundamental question of principle which is being overlooked. Could it be that quantum theory and relativity theory are not really compatible? After all, the only known consistent relativistic quantum field theories are linear free fields.

This work was supported in part by the National Science Foundation under grant number PHY-8813423.

References

P. A. M. Dirac (1958), *Proc. Roy. Soc. Lond.* **A333**.
A. Komar (1979), *Phys. Rev.* **D19**, 2908.
L. Rosenfeld (1930), *Ann. Phys. Lpz.* **5**, 113.

Part E.

Overviews—past, present, and future

17

Views from a distant past

E. L. Schucking
Institute of Theoretical Physics,
University of New York,
29 Washington St. West,
New York, NY 10011, USA

I apologize for uttering this talk in English if that is not your native tongue. Perhaps, at future conferences you will be wearing little earphones and can listen to the translation of such a speech into Hindustani, Japanese, American and other idioms — or, even better, switch to Chopin or Bach.

The first of these conferences, GR-0, took place in Berne three months after Einstein's death, thus probably not a contributing factor. The talks were in English, German, and French with Pauli's Schlusswort in German.[1] This was once the language of relativity, Einstein's language. When asked in old age how his English was, he answered: "Immer besser, niemals gut."

The declining knowledge of German has had the lamentable effect that among the least read authors in relativity is Saint Albert. His works are being published now and you can read the young man's love letters.[2] Optimistically, I expect that by 2155–remember, they are still working on Euler—you might be able to enjoy reading his thoughts on gravitation. But by then the English edition is possibly no longer the appropriate medium when billions of Chinese are steeped in lerativity.

What we need, within our lifetime, is an edition of Einstein's scientific papers translated without comment. It could even be a best seller among physicists who'd shelve it in their study next to the Einstein icon. This is something we relativists owe the man who put us into business.

The next GR, I don't mean the ominous thirteen, took place in Chapel Hill in January 1957.[3] A somewhat improbable historian of science attended this conference and this is how he got there:[4]

I don't know why, but I'm always very careless, when I go on a trip, about the address or telephone number or anything of the people who invited me. I figure I'll be met, or somebody else will know where we're going; it'll get straightened out somehow.

One time, in 1957, I went to a gravity conference at the University of North Carolina. I was supposed to be an expert in a different field who looks at gravity.

I landed at the airport a day late for the conference (I couldn't make it the first day), and I went out to where the taxis were. I said to the dispatcher, "I'd like to go to the University of North Carolina."

"Which do you mean," he said, "the State University of North Carolina at Raleigh, or the University of North Carolina at Chapel Hill?"

Needless to say, I hadn't the slightest idea. "Where are they?" I asked, figuring that one must be near the other.

"One's north of here, and the other is south of here, about the same distance."

I had nothing with me that showed which one it was, and there was nobody else going to the conference a day late like I was.

That gave me an idea. "Listen," I said to the dispatcher. "The main meeting began yesterday, so there were a whole lot of guys going to the meeting who must have come through here yesterday. Let me describe them to you: They would have their heads kind of in the air, and they would be talking to each other, not paying attention to where they were going, saying things to each other, like 'G-mu-nu. G-mu-nu.'"

His face lit up. "Ah, yes," he said. "You mean Chapel Hill!" He called the next taxi waiting in line. "Take this man to the University at Chapel Hill."

"Thank you," I said, and I went to the conference.

André Lichnerowicz arranged the second in the beautiful setting of the Ile de France at the Abbey Royaumont.[5] Here are some images that now overlap in the memory banks: the Gallic charm of our host who always put his arm round your shoulders when making his cryptic remarks, between sucking his pipe, in what was English but sounded like French. The Bel-Robinson tensor unveiled—in Deser's words,[6] "the cure for a disease that needs to be discovered." The dramatic return of two-component spinors launched by Papa Witten and this imaginative fellow from St. John's, Cambridge. Stan Deser introducing us to eating snails and the young American from the Middle West who didn't believe what bidets were supposed to be for. Honest gravitational waves were discovered and rediscovered from exact solutions. The beautiful null structures and congruences appearing on the blackboards wielded by Sachs, Ehlers, Robinson, Pirani and many more.

And all these lively discussions! Here is one saved from oblivion: Robinson finding himself trapped with the laconic Dirac is desperately trying to break the engulfing silence. He picks up a reprint of physicist Y's and shows it to PAM with the question: "Would you agree, that this is the second most stupidest man at this conference?" After a long pause, Dirac says: "Who do you think is the most stupidest man at this conference?" Robinson shoots back: "X." Dirac, deeply in thought apparently going through the roster of the participants, finally says: "I don't agree with you." More silence.

Dirac's talk on quantization seemed to upset a fidgeting Pirani. He finally could no longer remain quiet: "Professor Dirac," he said, "this is not covariant" (Yiddish translation: not koscher). Dirac's answer: "But it is useful," left Felix Pirani speechless. I overheard Pirani borrowing money from a friend: "Let me give you minus ten-thousand Francs." I suspect that the ingenous Felix—"I'm not anti-American, I don't hate America more than necessary."—was in the funny proceedings the author of the skit that roasted Wheelers geometrodynamics:[7]

This paper is entitled "Space-time without space-time." It forms Part 23 of a critique of classical field theory. Parts 24 to 30 have been published already. Parts 1 to 22 will appear during the next few decades.

I shall speak about the already-unified field theory. I should like to begin with a quotation from Sir Isaac Newton, who has reminded us so often and with such importance that "force is proportional to acceleration."

In the already-unified field theory, we have a set of field equations, some of the 2nd differential order, some to the 4th order, in the metric tensor. The already-unified theory provides simultaneously a description of the gravitational and electromagnetic fields.

Recently, a young student at Cambridge has proposed a new description of the theory which seems to me to represent a major contribution to the field. One may understand it in this way:

In order to reduce the order of the equations from 4 to 2 so that one may apply the well-known techniques of the theory of hyperbolic differential equations, and also formulate a satisfactory variational principle, one introduces an auxiliary anti-symmetric tensor field F_{ik}, which we propose to name the electromagnetic field tensor, since by its use one may in a very elegant and beautiful way separate in the already-unified theory the gravitational field. One might describe the form of this new theory as "unification without unification." It turns out that the new electromagnetic field tensor satisfies two sets of equations of the first order, which are very simple and which one may hope to be able to solve explicitly in some cases. At the same time, the electromagnetic field, through a symmetric tensor, which we intend to refer to as the Maxwell energy tensor, after the student who proposed this formalism. A very promising student, I might say: my best since Misner. The Maxwell energy tensor has the same canonical structure as the energy-momentum tensors in Lorentz-invariant field theories, a remarkable fact of which one has not yet a complete understanding. One hopes that the exploration of these ideas will make it possible to develop the great richness of general relativity up to the hilt, in the exciting arena of geometrodynamics. And the chorus sang:

> Its a long way to quantization,
> Its a long way to go.
> Its a long way to quantization
> As Bergmann ought to know.
> Good-bye, Palatini,
> So-long, Inverse Square
> Its a long way to quantization,
> But my heart's right there.

Thirty years later we can intone now: "As Abhay ought to know." Alex Harvey remembers: "We were bused on the excursion to see one of the magnificent gothic cathedrals. Ray Sachs who had just discovered the beautiful peeling-off theorems for the asymptotic behavior of the Weyl tensor describing gravitational radiation explained his ideas to a colleague on the bus. They continued their discussion when they got off to go with the others into the cathedral. They went into the cathedral talking intensely, they left the cathedral with the others, got back on the bus, still going like Feynman said: ge-mu-nu, ge-mu-nu." They never saw the jewel of flamboyant gothic Notré-Dame de l'Epine in Champagne—but they did alright.

Jabłonna was next—Infeld's conference, masterfully orchestrated by the Trautmans and their colleagues. They all kept a low profile. Though they had much to say about relativity, they let their guests do the talking.[8]

For many from the West it's the first time behind the curtain. We are housed in the Grand Hotel. The Americans compete in spy stories and search their rooms for mikes. The Russian colossus Fock is steered around by his daughter. When his Soviet colleague Ivanenko begins to talk Fock puts his hearing aid away. Fock delivers in Russian, but then he provides his own translation.

The charming Ginzburg—for the first time allowed out of the USSR into the Polish demi-West starts his talk in Russian, argues with his translator and continues his survey of gravitational tests in English. Wide acclaim, but the Russian delegation freezes into hostility.

During a break outside Ivanenko explains to Feynman that Ginzburg is immature and speaks patronizingly of Ginzburg's physics. Feynman does not let him get away with that: "What have you ever done in physics, Ivanenko?"

"I've written a book with Sokolov."

"How do I know what you contributed to it. Ivanenko, what is the integral of e-to-the-minus-x-squared from minus to plus infinity?" No answer.

"Ivanenko, what is one and one?"

And he turns away before Ivanenko can solve this problem. All these relativists have rubbed Feynman the wrong way. On Sunday, before his talk tomorrow he writes to his wife in Pasadena:[9]

I am not getting anything out of the meeting. I am learning nothing. Because there are no experiments this field is not an active one, so few of the best men are doing work in it. The result is that there are hosts of dopes here (126) and it is not good for my blood pressure: such inane things are said and seriously discussed that I get into arguments outside the formal sessions (say, at lunch) whenever anyone asks me a question or starts to tell me about his "work." The "work" is always: (1) completely un-understandable,

(2) vague and indefinite, (3) something correct that is obvious and self-evident, but worked out by a long and difficult analysis, and presented as an important discovery, or (4) a claim based on the stupidity of the author that some obvious and correct fact, accepted and checked for years, is, in fact, false (these are the worst: no argument will convince the idiot), (5) an attempt to do something probably impossible, but certainly of no utility, which, it is finally revealed at the end, fails (dessert arrives and is eaten), or (6) just plain wrong. There is a great deal of "activity in the field" these days, but this "activity" is mainly in showing that the previous "activity" of somebody else resulted in an error or in nothing useful or in something promising. It is like a lot of worms trying to get out of a bottle by crawling all over each other. It is not that the subject is hard; it is that the good men are occupied elsewhere. Remind me not to come to any more gravity conferences!

He didn't. But it is Feynman's talk here, to be published in Acta Physica Polonica,[10] that will later inspire Faddeev to his seminal contributions in the quantization of non-Abelian gauge theories.

An improbable second lieutenant of the US Army from Fort Monmouth, New Jersey, the absent minded Ray Sachs, wearing a crew-cut, reports about the exciting developments connected with Bondi's news function. He prefaces his talk with a sober assessment of relativity that reflects the spirit of many of its younger practitioners:[11] "Since 1916 we have had a slow, rather painful accumulation of minute technical improvements which have advanced our understandinng of the mathematical content of this theory and the physics of gravity. I think that the attempt to continue obtaining such minute improvements constitutes a legitimate and fascinating part of mathematical physics. If something really exciting turns up, fine; in any case, routine improvements will certainly be obtained and that, for me, is exciting enough. Of course it may happen that all our rather sophisticated attempts will be swept into obsolescence by some simple wholly new idea or experiment; but it may also be that the only real way to understand the nature of space, time, and gravitation is to continue a careful and impartial analysis of the present theory."

Schild, Robinson and I reported about the conference for *Physics Today*.[12] That issue had the Jablonna picture of Dirac with Feynman on the cover that was recently reproduced in the Feynman issue.[13] I opened the article with our approach to Warsaw in the Penrose VW to lead up to the Polish joke about the Palace of Culture and Science (according to Feynman "the craziest monstrosity on land") because you see it already from far away. It's the most beautiful place in Warsaw because that's the only spot in Warsaw from where you can't see it. We fought over this article. Alfred was adamant: "The VW has got to go." "Why?" I asked. "Cause," Alfred said, "I'm a Jew and I'm not going to write any commercial for a damn German car." The Penrose VW became the Penrose Peugeot. I still think it was wrong.

The strongest impression we take with us from our stay is the terrible destruction of the city and its ghetto wrought by the Germans.

We also had some skits. Feynman's impersonation of Mme. Tonnelat had to be seen to be believed. I defended a new and precise formulation of Mach's principle:[14] "Space-time is flat in the absence of physicists." This theorem still characterizes some recent theories of gravitation — like Logunov's. After Jabłonna, I suppose, the fun went out of the GR conferences.

Now, I didn't tell you about the Penrose diagram,[15] and all the other things. It's all too one-sided and personal, but, perhaps, if it weren't–it would be even more boring.

On to London, England, 1965, organized by Bondi et al., the als being Bonnor, Kilmister, Newman and Whitrow.[16] My recollection of this meeting is largely one of embarrassment.

The meeting opened at Imperial. I sat next to Chandra and I opened my folder. And here is what my friend Bill Bonnor of Queen Elizabeth College did to me: since I had happily lived for years in St. Pauli, the red light district of Hamburg, I had taken Bill and other visiting scholars on a guided tour of my neighborhood. Some savants find this kind of research in the social sciences just as much fun as pushing indices. In my folder were the usual things: pen, pad, program, map, etc., and a large envelope, neatly and discreetly imprinted, the way the British do it, "with the compliments of Queen Elizabeth College"–official stationery.

Out fell, fell to the floor, because I hadn't believed my eyes, the most explicit collection of pornographic pictures I had ever seen. If our president, I mean Ted Newman, had sat next to me, he might have said: "Gee, that's a new one, even to me!" and helped me to pick them up. Not so Chandra. He showed no visible reaction. I realized that any explanation could only have added the word liar to pervert in his mind. Ever since Chandra has treated me in a correct but slightly distant way– the way the civilized treat the unfortunately afflicted. At GR-11, Bill Bonnor was rumored to be out of the reach of justice in South Africa.

In 1965 General Relativity, the favorite daughter of Einstein's imagination, was 50. Her engagement to the universe, though widely hailed by a band of mavens, had not resulted–in most physicist's views–in significant issue. But in December 1963 she had got married to her life-long suitor in Dallas. The wedding was announced as the Texas Symposium on Relativistic Astrophysics.[17] And she became the queen of heaven. Roy Kerr wrote her wedding vows to produce a race of beautiful black holes. Bridesmaid Wheeler guided her along the aisle. Robert Oppeneimer, the father of the black hole, administered the rites while the universe was represented by the quasar 3C273 and a galaxy of astronomers. All was made final the following year when Penrose pronounced that the final issue was inescapable.[18] They had to get married. A torrid affair began. John Wheeler and his acolytes turned his Dallas talk on the "issue of the final state" into the full-size book: "Gravitation Theory and Gravitational Collapse."[19]

As Tommy Gold toasted in Dallas:[20]

It was, I believe, chiefly Hoyle's genius which produced the extremely attractive idea that here we have a case that allowed one to suggest that the relativists with their sophisticated work were not only magnificent cultural ornaments but might actually be useful to science! Everyone is pleased: the relativists who feel they are being appreciated, who are suddenly experts in a field they hardly knew existed; the astrophysicists for having enlarged their domain, their empire, by the annexation of another subject–general relativity. It is all very pleasing, so let us all hope that it is right. What a shame it would be if we had to go and dismiss all the relativists again.

An extraordinary transformation had taken place. General Relativity, the almost menopausal lady, was rejuvenated, engaged in happy intercourse with her lawful husband and was going to become pregnant with the final issues of the universe. The London conference reflected this revolution only marginally in a talk by Novikov who did not yet embrace the inevitable collapse of large masses.[21] The traditional view of relativity as a geometrical model for a physical theory prevailed. Some of the relativists saw their major challenge in the quantization of this model while others searched for a deeper understanding of its geometrical content. The stress was on issues of principle, not on a lively give-and-take between theory and observation or experiment.

This situation had set relativists apart from the cohorts of theoretical physicists who worked in that down to earth environment. When Einstein tried to understand electromagnetism in a geometrical way, he was a rather lonely man. "In Princeton," he said, "they regard me as an old fool (ein alter Trottel)."[22] A belief that seems to have persisted.

In an interview with *Physics Today*, published last month, Marvin Goldberger, Head of the Institute for Advanced Study, mused:[23] "I think that Einstein would probably have been better off if he had gone to Caltech, where he had a competing offer–to some place where he would not be treated like a god and someone would have spoken sharply to him about the things he was doing."

Not untypical for the attitude of such particular theoreticians towards Einstein's relativity is the following quote from the introduction to a textbook on gravitation written in 1971:*

> But now the passage of time has taught us not to expect that the strong, weak and electromagnetic interactions can be understood in geometrical terms, and too great an emphasis on geometry can only obscure the deep connections between gravitation and the rest of physics.

The full perversity of this statement can only be appreciated if one knows that its author wasn't a California S-matrix fanatic but got his Nobel later for a gauge theory.

Relativists are happy to see now that their life-long faith in geometry has improved connections with their colleagues who work in associated bundles. The often criticized penchant of ours for fundamental speculation untarnished by experiment or observation has now become the standard model for the guiding spirits of physical theory. A mind-boggling cottage industry has sprung up with products that bring to mind the Goethe words:[25] "Getretner Quark wird breit, nicht stark" (tread upon cottage cheese becomes wide, not strong). Maybe, some future historian will see the simultaneous rise of superstrings, inflation, quantum cosmology and Islamic fundamentalism as correlated pieces in the wave function of the universe. Anyhow, I, for one, would like to see more experiment and observation than wormhole metaphysics–but I'm oldfashioned. Still, I realize that theophysics is enjoyable. I've spent a lot of time thinking about the good old days before the big bang.

I'm probably speaking for all of you if I am thanking Neil Ashby and his helpers for having organized a very enjoyable and successful meeting.

---

*S. Weinberg (1972), *Gravitation and Cosmology*, (Wiley, New York).

They must have worked behind the scenes like Maxwell's demons to keep us comfortable in a state of low entropy and maximal information.

Let us just hope that everything we've heard is right, and if it isn't it reminds me of the Hasidic tale where the rabbi in Cracow has a vision seeing rabbi Cohen in Lemberg on his deathbed. Weeks later, a traveler returning from Lemberg informs the Cracow rabbi that Cohen is alive and sending greetings: "Your vision from Cracow to Lemberg, rabbi, was mistaken." "Still," the rabbi marveled, "consider the distance, what vision."

I am grateful to J. Ehlers, A. Harvey, A. Komar, and Ivor Robinson for sharing fond memories, correcting my lapses thereof and suppressing some tasteless remarks. A. Harvey kindly supplied me with a copy of the funny proceedings at Royaumont from which I quoted. I also thank O. Greengard for advice.

References

1. *Fünfzig Jahre Relativistätstheorie*, Berlin, 11.-16. July 1955. Verhandlungen herausgegeben von André Mercier et Michel Kervaire (1956), *Helvetica Physica Acta*, Supplement **IV**. (Birkhäuser Verlag, Basel).
2. *The Collected Papers of Albert Einstein. Vol. 1. The Early Years, 1879-1902* (1987), ed. John Stachel (Princeton University Press, Princeton).
3. B. DeWitt et al. (1957), *Rev. Mod. Phys.* **29**, 351.
4. R. P. Feynman (1989), *"Surely, You're Joking Mr. Feynman!"* (Bantam Books, New York), page 235.
5. *Les Théories Relativistes de la Gravitation* (Royaumont, 21-27 Juin 1959), Colloques Internationaux du Centre National de Recherche Scientifique (Paris, 1962).
6. *Gravitation and Geometry.*, eds. W. Rindler and A. Trautman (Bibliopolis, Napoli 1987), page 116.
7. From a copy provided by Professor A. Harvey, Queen's College, NY.
8. *Proceedings of Theory of Gravitation Conference in Warszawa and Jabłonna, 25-31 July, 1962*. ed. L. Infeld, Gauthier-Villars Éditeur. 55, quai des Grands Augustins, Paris VIe. (Villars, PWN-Éditions Scientifiques de Pologne, Warszawa 1964).
9. R. P. Feynman (1988), *What do You Care what other People Think?* (W. W. Norton, New York), page 91.
10. R. P. Feynman (1963), *Acta Physica Polonica* **24**, 697.
11. Ref. 8, page 93.
12. *Physics Today*, August (1963), page 17.
13. *Physics Today*, February 1989.
14. Ref. 8, page 375.
15. Ref. 8, page 369.
16. *International Conference on Relativistic Theories of Gravitation*, 2 vols. London, July 1965. King's College London.

17. *Quasi-Stellar Sources and Gravitational Collapse*, eds. Ivor Robinson, Alfred Schild and E. L. Schucking, (The University of Chicago Press, Chicago 1965).
18. R. Penrose (1965), *Phys. Rev. Lett.* **14**, 57.
19.. B. K. Harrison, K. S. Thorne, M. Wakano, J. A. Wheeler (1965), *Gravitational Theory and Gravitational Collapse*, (The University of Chicago Press, Chicago).
20. Ref. 17, page 470.
21. I. D. Novikov, Ja. B. Zeldovich (1965), Physics of Relativistic Collapse, Ref. 16, pp. 0-25.
22. Ref. 8, page XV.
23. *Physics Today*, June 1989, page 66.
24. J. W. v. Goethe, Divan, Sprüche 49. Werke 2, 58. (Christian Wegner Verlag, Hamburg 1949).

18

Conference summary

Jürgen Ehlers
Max-Planck-Institut für Astrophysik,
Karl-Schwarzschild-Str. 1,
8046 Garching b. München, FRG

Dear colleagues and friends, Having to summarize this meeting with its many contributions to a great variety of topics is a questionable honour. Looking through my notes I realize all too well that my wish to understand details and at the same time not to get lost in them, is larger than my ability to do so. I shall nevertheless try and give you a kind of overview of our field as it came to light–or remained in partial darkness–during this meeting.

The development of general relativity and, more generally, the physics of gravitation as it was and is reflected in the GR-meetings, starting with the Bern conference in 1955 as recalled by Engelbert Schucking in his splendid talk Thursday night, has some similarity with that of the universe, or at least with the standard model of it: a rather smooth, uniform beginning, then the evolution of more and more structure, and now a field that looks rather inhomogeneous, expanding here, contracting there, transparant in some regions, opaque (to me) in others, and partly chaotic. Important aspects of the present state are the interconnections of general relativity with gauge theories, particle physics and astrophysics. The most conspicuous and important feature of this evolution, however, is that experimental and observational research on the properties of gravity on the laboratory, terrestrial, solar system, galactic and cosmological scale has grown considerably from very small beginnings and occupies now a sizeable part of the plenary talks and, in particular, of the workshops. At long last, lively interaction between observation, experiment and theory, the essential mark of an active area of science, is increasingly taking place although, at present, theory is still dominating, at least quantitatively.

I should like to arrange my remarks in the following sequence:

1. Einstein's classical theory
   Observations and experiments
   "Pure" theory
   Numerical relativity
   Astrophysical applications and approximation methods

2. Cosmology
   Observations
   Theory

3. Quantum gravity
   Quantum cosmology
   New canonical variables
   (Super-)string theory
   Alternative theories

It goes without saying that under each heading I can mention a few items only, chosen according to my limited viewpoint and degree of understanding. Rather than referring to particular speakers and (unavoidably) leaving out others, I shall mention no persons, but address topics.

1.   Einstein's classical theory

Observations and experiments

Four by now well-known *standard effects* which serve to test Einstein's gravitational field equation in the solar system at the first post- Newtonian level of accuracy – light deflection, signal retardation, "anomalous" perihelion shift, (absence of) Nordtvedt-effect – now have been shown to agree with the theoretical predictions within fractional uncertainties of a few parts in 10^{-3}. These results imply that the GR-values of the Parametrized Post-Newtonian (PPN) parameters β and γ are correct to within standard deviations $\sigma(\beta) = 0.001$, $\sigma(\gamma) = 0.002$. Due to helioseismological observations the uncertainty concerning the contribution of the Sun's quadrupole moment to Mercury's perihelion shift seems to have been removed; $J_2 \lesssim 10^{-7}$. Another good news to relativists: The groups at JPL and at CfA have been able to extract from lunar laser ranging data another 1PN GR-effect, the *geodetic precession* of the moon's orbit predicted by de Sitter already in 1916. Although the effect amounts to a mere $2''$/century it has been verified with an accuracy of 4%. The same groups have also obtained tight upper bounds, $\lesssim 10^{-11}$/year, on a possible fractional rate of change of G, the ratio of active gravitational to inertial mass. Further development of VLBI, spacecraft technology, improvement of long term clock stability by perhaps two orders of magnitude, and other experimental advances, may lead to 2PN measurements of redshift and light deflection and, at long last, to tests of gravitomagnetic effects ("dragging of local inertial frames") – but that is music of the (distant?) future.

In spite of several experimental efforts the disturbing question whether deviations from the *inverse square law* for the gravitational acceleration exist on the laboratory or geophysical scale has not been settled yet, but contradictory results claiming deviations seem to be decreasing. A composition dependence (on baryonic or leptonic charge) of g as has been hypothesized seems to be ruled out, however. Thus Newton may be right after all.

The rate of decrease of the orbit period of "the" *binary pulsar* ascribed to the emission of gravitational waves, a measure of the highest-order GR-effect observed so far, has now been determined so accurately as to confirm the GR-prediction to within 1%. Direct information about the *gravitational waves* surrounding and pervading us is still meager. An upper bound of 4×10^{-14} on the gravitational wave strain amplitude in the period range 200–2000s from pioneer 10 Doppler data was reported here, and the effective mass density of the stochastic gravitational wave background with periods of the order of weeks seems to be below the cosmic closure density.

Many efforts to improve *gravitational wave detectors* are being made. Resonant bar antennas have reached a strain sensitivity $h \sim 5 \times 10^{-19}$ (at 2 K), intermediate-size laser interferometer antennas lag behind by about one order of magnitude, but in the long run the latter may have the larger potential for improvement. The most ambitious project is a huge laser interferometer (10^6km) in space, with dragfree satellites instead of mirrors. Very fortunately, no principal obstacles on the way towards obtaining sensitivities of the order of 10^{-21} or even better (10^{-23}?) have turned up so far, so we may hope that these efforts will bear fruit, perhaps before the turn of the century.

In view of the now quite manifold and accurate empirical evidence it seems, then, that there is no reason, at least in the macro-domain, to look for an alternative to Einstein's theory.

I have to say that, again I was very impressed by the ingenuity and persistence of our experimenting colleagues. Being confronted with their attention to noise sources and error assessments, I could not help thinking of W. Thirring's remark (in the preface to Volume 3 of his unique textbooks series on mathematical physics) that it was his "principal aim to substitute for the usual calculations of uncertain accuracy ones with error bounds, in order to replace the rough manners of theoretical physics by the more cultivated customs of experimental physics." Perhaps we should take this to our hearts when neglecting small terms in our equations and piling one "approximation" on top of another one.

"Pure" theory

In the area of "pure", *classical general relativity* significant new results about the existence of some *global*, respectively semi-global, *vacuum* and Einstein-Yang-Mill's *spacetimes* have been obtained. They confirm to a large extent, though not yet completely, the picture about the structure of outgoing gravitational (and/or Y.-M.) radiation far from sources which emerged almost 30 years ago from the work of Bondi, Penrose et al. Not long ago it would have appeared rather hopeless to justify rigorously Bondi's expansions and to show that Penrose's definition of asymptotic flatness is compatible with Einstein's equation for generic, radiative spacetimes. Now this has been achieved to a great extent. Ironically, an example of a global, asymptotically simple, non-stationary Einstein-Maxwell spacetime without material sources has been "constructed", and the existence of a large class of past asymptotically simple, non-stationary Einstein-Yang-Mill's spacetimes with positive cosmological constant has been established, but corresponding results for pure gravitational fields, or for ones with $\Lambda = 0$, are yet lacking. Perhaps even more important than the results are the *methods* used: The regular conformal field equations which permit to study the asymptotic behaviour of fields in terms of local properties, gauge-source functions to allow arbitrary coordinate and tetrad conditions, spinor variables to achieve symmetric hyperbolic partial differential equations as evolution equations in the study of the Cauchy initial value problem. The main global problem of classical general relativity is still whether some version of the *cosmic censorship* hypothesis is true. Another important remaining question, hopefully accessible to these methods, concerns the structure of the gravitational field with mass *and* radiation near spatial infinity.

Another problem, posed almost 30 years ago and variously attacked, has only recently been solved satisfactorily: The *characteristic initial value problem* for vacuum or perfect fluid spacetimes, with data on two null-hypersurfaces intersecting in a two-surface; a promising attack on the corresponding problem for an initial null cone, of interest both for gravitational radiation problems and cosmology, is being made.

Very unfortunately from the point of view of mathematical physics, the inclusion of spatially compact bodies, for example balls of perfect fluids, in studies of this kind still seems to be out of reach of rigorous methods. Even in Newtonian gravitation theory, the Kepler problem for two separate fluid balls is unsolved except for the case of circular orbits.

The problem of *colliding plane gravitational waves* has attracted a lot of attention. Are the (curvature or topological) singularities behind the collision surface which occur in various explicit examples, removable by more "generic" models? New methods, some involving twistors, may throw light on this still open question. In those – at first sight puzzling – models where colliding plane electromagnetic and gravitational waves appear to transform into different kinds of (possibly massive) matter, it has been argued that once the physical assumptions about the sources are specified in advance, ambiguities and spontaneous matter creation will no longer occur. This sounds convincing to me and should be clarified.

Only in passing I mention that (i) twistor geometry has been used to understand particular non-linear PDE's (e.g., the Korteweg-de Vries (KdV) equation); (ii) a "dense" set of self-dual YM's solutions could be constructed explicitly; (iii) new insights into the role of the Witten spinor and the Sparling 3-form in proofs of the positive mass theorem have been obtained.

Numerical relativity

Thanks to the construction of ever larger and faster computers and the skill of some of our colleagues in developing codes, *numerical relativity* has now reached the stage where processes involving in an essential way all *3 dimensions* of space can, and in some cases have been treated. As we have heard, the ambitious project to handle numerically the general initial value problem for Einstein's equation with, e.g., perfect fluid sources, can now in principle be realized. This opens up the possibility of treating quantitatively essentially relativistic, as opposed to nearly Newtonian dynamic processes accompanied by gravitational wave emis-

sion such as triaxial gravitational collapse, fission of a rapidly rotating
star, coalescence of binary neutron stars. The results of such calculations
are expected to provide detailed *predictions* hopefully *testable by gravitational wave detectors*. Newtonian simulations already indicate that
gravitational radiation efficencies $\Delta E_{\mathrm{grav.waves}}/Mc^2$ may be as large as
a few percent. Undoubtedly this field of research will extend rapidly, at
least at places which can afford buying supercomputers. Perhaps and
hopefully, interaction between numerical studies and analytic work will,
in the long run, also throw new light on such basic issues of pure GR as
cosmic censorship and the nature of singularities.

Astrophysical applications and approximation methods

Besides being still a challenging arena for mathematically inclined theorists, *general relativity* has become *a tool* – if only a rarely used one
– *for astrophysicists*. One of its applications has been the remarkably
precise "measurement" of the masses of the binary pulsar and its companion (to within 2×10^{-3}) by combining GR-formulae with pulse arrival
time data. Another impressive example of "applied GR" was presented
to us here: Comparison of numerically computed *GR-models of rapidly
rotating neutron stars* based on a large collection of equations of state
with observed pulsar rotation rates (0.5 msec pulsar) and masses shows
that only very few equations of state (including that of a quark-pion
condensate) survive. This conclusion is supported by stability analyses
of Newtonian models of compact stars. Thus astronomical observations
combined with general relativity may be used to test nuclear many-body
calculations. Other, future astrophysical applications of GR have been,
respectively will be mentioned in other parts of this summary.

Applications to real objects or processes usually require *approximation methods*. Only rather few contributions to this meeting have been
concerned with the difficult and important task of narrowing the gap between the "exact" theory and approximation methods. As an example,
a fairly complete theory of the convergence of linearized Regge calculus has been developed, with indications of extension to the non-linear
case. As a second example I mention that a post-Minkowskian formal
power series expansion, recently designed to represent the general radiative vacuum solution generated by a bounded source, has been shown to
provide, in fact, an asymptotic approximation to an exact solution.

Perhaps too little attention is paid to such activities. It might be useful to have at the next GR-meeting, a plenary talk and/or a workshop

specifically concerned with the question of (more or less rigourous) justifications of approximation methods, and to point out open problems and avenues of approach in this area which is needed to link theory and reality.

2. Cosmology

Observations

Not so long ago *cosmology* was characterized as the search for two numbers which would pick out of the collection of Friedmann-Lemaître models the "true," smoothed-out background universe. Now the search is directed more towards the real, complicated, *inhomogeneous matter distribution* and the deviations from the Hubble flow caused by "local" gravitational potential wells. In particular one is looking for *temperature anisotropies and deviations* from the Planck-distribution in the microwave background radiation, to obtain clues for the formation of clusters, superclusters, voids etc. No new results of this kind have been reported, however; but theoretical expectations and implications of expected anisotropies have been discussed.

A relatively new, very actively pursued means of obtaining information on large scale mass distributions (including dark matter) is the *gravitational lens phenomenon.* Various theoretically possible aspects have now been observed: Einstein rings and luminous arcs, images of different parities, macrolensing due to clusters and galaxies as well as brightness variations due to microlensing caused by individual stars. The time delay between the images of the double quasar has been determined to be 415 ± 20 days, but unfortunately the use of this figure to obtain the Hubble constant is hampered because of uncertainties as to the correct lens model. One particularly interesting question is whether cosmic strings can be identified through their lensing effects; so far there does not seem to be compelling evidence for their existence.

Theory

Relativistic geometrical optics, i.e. the study of systems of null geodesics on Lorentz manifolds, has been revitalized in *gravitational lens theory* as a tool to interpret, inter alia, shapes and intensity variations of quasar images. In order to survey the qualitative properties of the mapping which, for an assumed deflector mass distribution, assigns to an image in the sky the corresponding source position, one has to determine the singularities of that lens mapping. Here the first mathematical result of what is now called catastrophe theory, which says that such mappings can have but two types of stable singularities, folds and cusps (Whitney (1954)), now finds application in astrophysics. It is expected that the combination of space telescope observations with a range of theoretical models of deflector mass distributions will provide more detailed information on the structures of quasars, galaxies and clusters.

A fundamental problem of theoretical cosmology which presumably can be tackled successfully only with numerical methods, is the construction of models of *statistically homogeneous and isotropic*, but *locally inhomogeneous*, relativistic or Newtonian *universe models with large spatial density variations*, and the study of light propagation through such a clumpy medium. Only by means of such models can the relation between the various fine-scale descriptions and the smoothed-out Friedmann-Lemître background model, the "fitting problem", be clarified. Several promising first steps in this direction have been made, but it appears that much remains to be done.

As to *inflation*, we have heard a vivid, persuasive description of the picture of a universe consisting of many parts whose dynamics are determined by different vacua and symmetry groups, a universe whose early evolution is governed by a scalar field acting as the dominant source in a nearly homogeneous part of the universe. According to this view, what we can observe is only a tiny part of "the universe", a part which owes its structure to an inflationary phase. To which extent this, on the whole untestable, grand world-view can be related to experience remains to be seen when more details have been worked out quantitatively. Perhaps the main question is: *Do scalar fields actually exist*, or are they theoretical *dei ex machinae* helping us to give masses to gauge particles and to overcome alleged shortcomings of the old, standard hot big bang model? The prediction $\Omega = 1$ neither follows from the theory, except as an order-of-magnitude statement, nor is it observationally established. To my information, observation only says that Ω, λ (the dimensionless version of the cosmological constant), κ (the dimensionless version of the curvature parameter) are all about 1, with comparable uncertainties.

According to the chaotic inflationists view, inhomogeneities of matter originated as quantum fluctuations of a scalar field which were stretched by inflation and then "froze". An (according to my understanding) more detailed picture of structure formation employs free as well as interacting *cosmic strings*, left-overs (topological defects) from phase-transitions in the early universe. The moving strings may attract and focus dark or ordinary matter, and generate shocks, a process which may have produced galaxies and formed a foam-like structure such as seems to be indicated by observations.

3. Quantum gravity

Quantum cosmology

The ambitious program of *quantum cosmology* was outlined to us as being much more than a cosmology in the usual, astrophysical sense. It is an attempt to create a new kind of *quantum theory for closed systems* which, in contrast to standard (Copenhagen) quantum mechanics, is intended to be such that, (i) the physical interpretation of its formalism does not require the choice of a "cut" between "quantum object" and "apparatus", with the latter described classically in its role as measuring device – and which therefore has no place for a reduction of wave packets, (ii) it contains *laws for initial conditions* of the universe (fixing that state uniquely?), (iii) it assigns a preferred role to "time" in its formalism and interpretation. By working out probabilities of possible histories, quantum cosmologists then aim at predicting large scale (homogeneity, isotropy, fluctuation spectrum) as well as small scale (coupling constants, masses, ...) properties of the universe and its ingredients, including its classical features due to "decohering" observables.

The topics which attracted most attention in the workshop were baby universes, third quantization, prescriptions concerning possible wave functions of the universe and probability measures on the space of those states, questions of interpretation, and the derivation of observational consequences.

I feel unable to comment on the extent to which the goals of quantum cosmology have yet been achieved. My own impression, also shared by some of my friends, is that the proposals which have been made to realize (i–iii), if they are to be appreciated by those of us who have not been participating in this activity but would like to follow it, deserve and require more extensive explanations than can be given in the introductory part of a status report on the field. Perhaps a carefully structured and prepared question session would be helpful to many of us.

New canonical variables

A more conservative approach towards a quantum theory of gravity which also occupied a sizeable part of this conference, is based on the *new canonical variables*. Its principal characteristic is that *canonical quantization* is attempted *non-perturbatively*, i.e. a quantum structure is given to the full, non-linear field without splitting it into a (kinematical) background and (dynamic) fluctuations thereof. (In this context I should like to remind you of Pauli's remarks in his summary of the Bern conference (my translation): "I should like to recall that in the old paper on field quantization by Heisenberg and myself linearity was not assumed ...It appears to me that the core of the matter is not so much linearity or non-linearity, but rather the circumstance that here (i.e., in GR) a more general group than the Poincaré group (*viz.*, the group of diffeomorphisms) is present ...In quantizing gravity a new situation arises from the uncertainty (Unbestimmtheit) of the lightcone ..."). The hope that such a non-perturbative canonical quantization might now be possible is due to the remarkable discovery that if one first extends and complexifies the standard, unconstrained phase space of gravity, then (i) the constraint submanifold of Einstein's theory turns out to be a proper submanifold of that of Yang-Mills theory; and (ii) the constraint equations become polynomial (of highest degree four). We learned that, contrary to earlier assertions, the polynomial structure is maintained when reality conditions are imposed on the (primarily complex) variables and equations. Instead of modifying Einstein's theory, *complicating* it by increasing the number of dimensions and/or field variables, the order of derivatives or the degree of non-linearity, as is done in alternative theories or some perturbative approaches, the new formulation *simplifies* the mathematics of Einstein's theory without – at the classical level – changing it intrinsically. Because of this, the new formulation, besides allowing some *applications to classical GR* (e.g., self-dual solutions, numerical GR) has made possible the construction of *quantum state functionals* in Bianchi IX and RW models and thus to establish contact with quantum cosmology. Even in the full theory, an infinite-dimensional space of exact solutions to the quantum constraints and a Poisson algebra of (candidates for) observables have been set up. Open questions concern the existence of an inner product on the state space and, most importantly, the identification of *physical observables*. (It would seem to me to be useful to distinguish, not only in this context, between "dynamical variables" and "observables," reserving the latter name to those variables for whom one has an idea how they could, in fact, be observed; in other words, the term observables should be reserved to those variables which establish a link between formalism and "reality." One can associate with a quantum mechanical system infinitely many self-adjoint operators, but only rather few represent observables in the strict sense of the term.)

A major part of the effort, in this as in other quantum gravity approaches, concerns the derivation of a Schrödinger equation involving an *intrinsic time* variable. The theory seems to be capable of providing eventually answers to the question of what constitutes the microstructure of space time, in particular, how a Lorentz metric arises macroscopically. (Already at the classical level the reformulation allows changes of signature.) The field is very actively pursued, so we may expect interesting news at the next GR-meeting.

(Super-)string theory

Still another avenue which might ultimately include a quantum theory of gravity employs *strings* or *superstrings*. (One reason for this expectation appears to be that if one requires conformal invariance to be preserved in quantization, the metric of the multi- dimensional imbedding space has to be Ricci flat in the limit of small curvature. For an interested non-expert like myself the relation of this result to gravity as a macroscopic phenomenon in four- dimensional space time is less than obvious.) As we heard, unfortunately but not surprisingly, physically testable predictions cannot be expected in the near future. Thus here as in other quantization or unification attempts the work will be mainly mathematical for some time, including differential and algebraic topology and conformal invariants. We have been informed that *finiteness* of superstring amplitudes to all orders of perturbation theory, due to the extendedness of the quantized objects, has almost been established, but since the perturbation series (at least for the bosonic string) *diverges*, non-perturbative methods are called for also in this approach. New types of *topological quantum field theories* not involving a metric or a connection, which are also used as test cases in the new variables approach, are thought to indicate how physics might "look" like in a pre-metric era, though I find it hard to recognize the physics in these mathematical models.

A common feature of various quantization schemes is that microscopically there is no metric. A metric, and perhaps even a spacetime manifold, is expected to arise only in some macroscopic limit. How this comes about and which consequences it has for particle physics is, however, by no means clear. The very *fundamental question* remains whether a changed microstructure of some yet unknown kind can cure quantum field theory of its long-standing illness, the infamous divergences.

Alternative theories

Let me mention very briefly only that interest in *alternative theories* of gravity seems now to be focussed mainly on issues related to quantization. Properties of Kaluza-Klein theories (whose energy-functionals might not be bounded below), higher-derivative theories (in particular those with non-quasilinear field equations arising from Gauss-Bonnet terms in the Lagrangian), and others have been investigated. Clearly we have more theories than well-established facts concerning gravity.

Looking back at this conference my impression is that classical general relativity with its experimental, theoretical and astrophysical ramifications is proceeding slowly but firmly, with rather well defined next open problems. Not surprisingly, in view of the conceptual and technical-mathematical difficulties and because of the absence of experimental guidance and control, the same cannot be said for the several imaginative and bold speculative attempts at somehow uniting general relativity with quantum theory. Sometimes I felt that formal theories of this kind are in danger of losing contact not only with real facts, but even with thought experiments; but perhaps this is a transient phenomenon.

I wish to conclude with a light-hearted suggestion. During the general assembly of our society the question was raised whether we should have some visible membership symbol. A tie was rejected. Would it not be progressive to wear a nice little lightcone as an earring, suitable in these days for members of both sexes?

At the very end let us thank the organizers, their helpers, the supporting agencies, the speakers and workshop chairpersons for there efforts. Good luck to all of you until GR-13.